Chemical and Process Industries

Osei-Wusu Achaw • Eric Danso-Boateng

Chemical and Process Industries

With Examples of Industries in Ghana

 Springer

Osei-Wusu Achaw ⓘ
Department of Chemical Engineering
Kumasi Technical University
Kumasi, Ghana

Eric Danso-Boateng ⓘ
School of Chemical and Process Engineering
University of Leeds
Leeds, UK

ISBN 978-3-030-79141-4   ISBN 978-3-030-79139-1   (eBook)
https://doi.org/10.1007/978-3-030-79139-1

This Springer imprint is published by the registered company Springer Nature Switzerland AG
The registered company address is: Gewerbestrasse 11, 6330 Cham, Switzerland

*To my children—Afia, Kwame,
and Kwaku—and to my dear
wife Serwaa Anima.
Also, to the memories of my father and my
mother.
Osei-Wusu*

*To my wonderful children—Godfred,
Christabel, Roselyn, and Eric—and to my
beloved wife Mercy.
To my mother—and to the memory of my
father.
Eric.*

Preface

The chemical processing industries broadly include industries which convert raw materials and intermediary chemical materials into finished chemical products through chemical or physical means. In chemical conversions, the raw materials undergo one or more chemical reactions during the manufacturing process to produce products that have different chemical and, sometimes, varying physical characteristics from the original materials from which they were made. During a physical process, a material transforms into a new one that has different physical characteristics but retains its chemical properties. These industries include the traditional chemical industries, both organic and inorganic; the petrochemical industry; and allied industries in which chemical processing plays a significant part. The modern world economy hinges on the chemical processing industries as these industries convert raw materials, namely oil, natural gas, air, water, metals, minerals and intermediary chemical products into several thousands of different finished and semi-finished products.

The chemical processing and allied industries have traditionally provided the major arena for the employment of chemical engineers where they play leading roles in the design and operation of chemical processes and process units. The training of the chemical engineer, therefore, invariably includes understanding of the design, operation and the technologies underlying the processes of chemical and physical changes that result in the conversion of materials into finished chemical products. The authors draw from established literature and several years' experience in supervising students' placement in industry to introduce the budding chemical engineer to the engineering, technology and chemistry underlying the manufacture of products in the chemical processing industries. The book focuses on the technology and methodology of the processes underlying the production of chemical products rather than on the engineering design of the processes and process units.

The book is written primarily for students studying undergraduate degree in Chemical Engineering even though students in related disciplines and personnel working in the chemical and allied industries would equally find the book very useful. The aim of the book is to present information on the technology and state of industry on the manufacture and offers guidance in the operating of the processing units. Examples of specific chemical products in Ghana are presented. Therefore, each chapter has a section that provides information on the state of manufacture of the particular chemical product which is the subject of that chapter in Ghana.

Each chapter is divided into three sections, namely a discussion of the chemistry and technology of the manufacture of a specific chemical product, a discussion of the state of the industry of the manufacture of the particular chemical product in Ghana, and a bibliography that provides leads for the diligent students to find further information on the subject. Prospective investors looking for opportunities in Ghana will therefore find the book very helpful. The book has 11 chapters, each chapter covering a major chemical processing industry in Ghana.

Chapter 1 covers the manufacturing of different types of soaps and detergents. Raw materials selection and pretreatment methods are covered. The various manufacturing technologies and the mechanisms of the processes are presented in detail. Chapter 2 focuses on the extractive metallurgy industry, with emphasis on the mining industry in Ghana. The chapter covers mining and mineral processing methods. Attention is paid to the unit operations in mineral processing and the environmental and safety issues in the extractive metallurgy industry.

Chapter 3 focuses on the textile and fabric manufacture. Types of textile fibers and the raw materials for manufacturing these fibers are covered in this chapter. Various methods of fiber and yarn formation are presented. The chapter also covers fabric processing methods and the unit operations in textile and fiber manufacture. Environmental concerns in the textile industry are presented in this chapter. Finally, the textile industry in Ghana is covered, with emphasis on the indigenous textile industries. Chapter 4 emphasizes fertilizer technology. It covers the role of specific nutrients in fertilizers on plant growth, and the determination of the nutrient need of plants. The types of fertilizers and their manufacturing techniques and the unit operations involved are presented. The chapter also covers the various fertilizer application methods.

Chapter 5 presents the manufacturing of cement and clay products. Raw material selection, their preparation, and storage are presented. Types of cements, their compositions, manufacturing techniques and the chemistry of cement manufacturing are covered. The chapter also includes clinker formation and the functions of the various raw materials in cement. The chemistry of cement use and safety and environmental effects of cement are presented in this chapter. The chapter also covers quality control indicators of cement, which include chemical and physical specifications and analysis. The chapter also presents refractory materials and the manufacturing techniques. The current cement industry in Ghana is covered in this chapter. Chapter 6 focuses on pulp and paper technology. Types of paper according to ISO 216 paper sizing system are covered. The chapter includes raw materials for making pulp and paper, and the production processes of pulp and paper as well as the problems associated with each pulping process. Environmental concerns and pollution from pulp and paper mills as well as effluent treatment methods are covered in this chapter.

Chapter 7 focuses on crude and refined palm oil manufacture. Socioeconomic and environmental impacts in Ghana are covered in this chapter. The chapter presents the uses of palm oil, and the unit operations and the manufacturing processes of crude and refined palm oil. Mass and energy balances on the unit operations are covered. Chapter 8 presents the production of various industrial gases such as atmospheric gases, Liquefied Petroleum Gas (LPG), natural gas and gases produced by chemical synthesis. LPG characteristics, applications and process description of LPG from refinery by-products are covered in this chapter. The chapter further presents the processing of natural gas.

Chapter 9 presents crude oil refinery and refinery products. It covers the classification and characterization of crude oil. Sulfur recovery processes and catalytic reforming are presented in this chapter. Manufacturing of products such as bitumen and lubrication oil is covered. The chapter also presents refinery waste treatment process as well as oil and gas storage and transportation. Chapter 10 focuses on cocoa processing and chocolate manufacture. The chapter includes harvesting of cocoa, the fermentation process and drying of the cocoa beans. The unit operations in the chocolate manufacturing process are covered in this chapter. Quality control in chocolate manufacture is also presented in this chapter. The chapter also covers the nutritional and health benefits of cocoa and chocolate. The chapter also presents the cocoa and cholate processing companies in Ghana.

Chapter 11, which is the final chapter, focuses on milk and dairy product manufacture. It covers the composition, nutrients and physical properties of milk. The chapter covers collection and storage of milk following milking and includes milk quality and testing methods. The unit operations in milk/dairy processing are covered. The types of dairy products are covered, which include liquid (or fluid), solid, fermented, cultured, concentrated and long-life products. The chapter covers manufacturing of dairy products such as yoghurt, cheese, ice cream and milk powder. The chapter further presents manufacturing of whey products and fractions and covers the membrane filtration technologies utilized in whey product manufacture. Environmental impacts from the dairy industry and wastewater treatment methods are covered.

The authors wish to acknowledge the Teaching and Learning Innovation Fund (TALIF), jointly provided by the World Bank and the Government of Ghana, for the financial support toward the writing of this book. The authors again wish to express their profound gratitude to the management of Kumasi Technical University through whose initiative the authors were able to access the TALIF support.

Kumasi, Ghana

Ing. Prof. Osei-Wusu Achaw, Ph.D., MSc, BSc, MGhIE

Leeds, UK

Dr. Eric Danso-Boateng
Ph.D., MSc, HND, PGCAP, FHEA, AMIChemE

Contents

About the Authors

The authors are experienced chemical engineers with several years' teaching experience in higher education and supervising students' placement in industry.

Ing. Prof. Osei-Wusu Achaw is an Associate Professor of Chemical Engineering at the Department of Chemical Engineering of Kumasi Technical University where he has taught for close to 24 years. He has published several peer-reviewed papers in reputable journals. Professor Achaw is a member of the Ghana Institution of Engineering and served on the Council of the Institution for 4 years.

Dr. Eric Danso-Boateng has taught in Ghana, specially at Kumasi Technical University for about 12 years as a Lecturer/Senior Lecturer, and at the Department of Chemical Engineering of Loughborough University in the UK for a year a half. He is presently a Teaching Fellow at the School of Process and Chemical Engineering of University of Leeds in the UK where he has taught for about 5 years. He has published one edited book and several peer-reviewed papers in high-impact factor journals and has a patent to his credit. Dr. Danso-Boateng is a Fellow of Higher Education Academy, UK, and an Associate Member of Institution of Chemical Engineers.

Soaps and Detergents

Abstract

Soaps and detergents are used in every household for cleansing, and other
purposes such as in cosmetics and pharmaceuticals. This chapter, therefore,
presents the processing technologies for the manufacturing of different types of
soaps and detergents. Ingredients used in soap making are covered in detail
including fats and oils, alkali salts/bases and additives. Before being used in soap
making, fats and oils are pretreated through the processes of degumming,
deodorization, fractionation, hydrogenation and bleaching. The different types of
soaps are covered which are liquid, powdered and bar soaps and the
soap-making methods, namely, neutral fat saponification and fatty neutralization
processes have been presented in detail. The different types of detergents and
their formulation methods have been covered including dishwashing cleaners,
household cleaners, general purpose cleaners, degreasers, bathroom cleaners,
toilet cleaners, glass cleaners, among others. Information on ingredients used in
detergents' manufacture, namely, surfactants and builders (e.g. phosphates,
silicates, zeolites etc.) as well as additives that are added to improve the
performance such as softeners, enzymes, optical brighteners, thickening agents,
has also been discussed.

1.1 Introduction

Soaps and detergents are substances that when dissolved in water can cause dirt
(soils, oils and other stains) to be removed from the surfaces of skins, textiles and
other solid surfaces. Even though soaps and detergents perform the same or similar
functions, they are different compounds. To begin with, they are made from different raw materials. Soaps are made from natural oils and fats, while detergents are
made mostly from synthetic compounds. Also, differences arise in the chemistry

and method of their manufacture and the applications to which they are put. Soaps and detergents belong to a larger group of chemicals called surface-active agents or surfactants. These groups of substances are bipolar in nature, that is, they have hydrophobic tails and hydrophilic head groups. Their bipolar nature renders them soluble in both aqueous and organic phases. They also have the ability to aggregate at the interface of two-phase systems as for instance at the air–water interface. The unique chemical structure of surfactants enables them to associate in solution when at certain concentrations such that they are able to solubilize oil and dirt and also to stabilize dispersions. This latter ability of surface-active agents is the basis of the cleansing ability of soaps and detergents. Furthermore, the ability of surface-active agents to aggregate at the air–water interfaces is the basis of lathering, foaming and sudsing.

1.2 Soaps

Soaps are water-soluble alkali or ammonium salts of fatty acids. Soaps are used primarily for cleansing, but they are also used in cosmetics, pharmaceuticals and other products. They are made from fats and oils, or their fatty acids by treating them chemically with a strong solution of alkali or ammonium salts. Soaps come in various types. Depending on the physical appearance, they are classified as bar, liquid/gel or powered soaps. Based on the intended use of the soap, one can distinguish medicinal soaps from dispenser soaps and general cleansing soaps. Dispenser soaps are used for cleaning specific parts of the body, for instance, the hand and the face.

The specific manufacturing details of soaps vary from manufacturer to manufacturer, but generally procedures are common to all products of a similar kind. Figures 1.1 and 1.2 illustrate the generalized flowcharts for the manufacture of soaps. The main steps in soaps/detergent making include: (i) selection of raw materials; (ii) reaction of the oils/fat with the basic salt; (iii) separating the reaction products where necessary; (iv) drying of the reaction product; (v) homogenizing the dried soap together with other additives; (vi) extruding the mixture to the right form; (vi) cutting and stamping and (vii) packaging the soap.

1.2.1 Raw Material Selection

Raw materials for the manufacture of soap are selected based on chemistry of the materials, human and environmental safety, cost, compatibility with other ingredients and form and performance characteristics of the finished product. The raw materials used in the manufacture of soap include fatty acids and oils, alkali salts and other bases and additives such as fragrances, builders, colorants, antioxidants and antimicrobial agents.

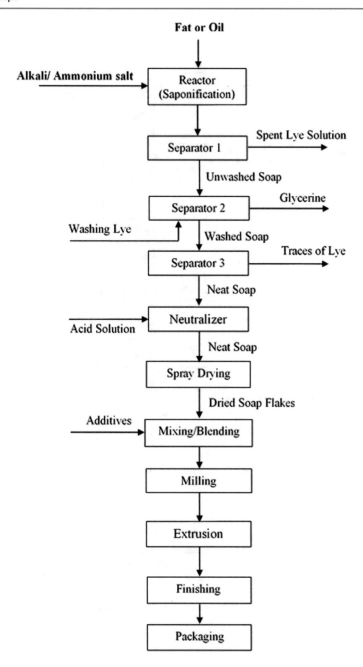

Fig. 1.1 A generalized flowchart for the manufacture of bar soap by the neutral fat saponification method

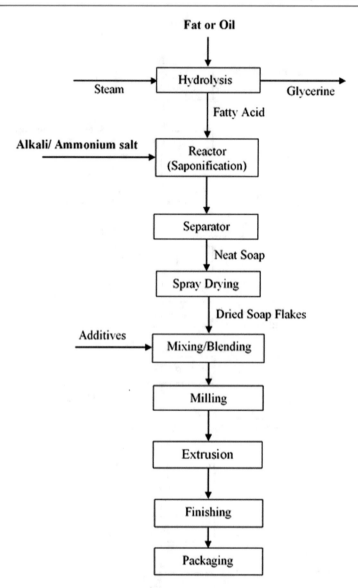

Fig. 1.2 A generalized flowchart for the manufacture of bar soap by the fatty acid neutralization method

1.2.1.1 Fats and Oils

Fats and oils used in making soap are without exception of all from plant and animal origin. These fats and oils contain as their main component a substance called triglyceride, which when reacted with alkali salts form soap. The common fats and oils used in soap manufacture are tallow, coconut oil, palm oil and palm kernel oil. Table 1.1 lists the oils and fats that have been used in the manufacture of soap in the past.

The main components of the fats and oils that react to form soaps are the triglycerides. These are esters formed from glycerol and one to three fatty acids. In a

Table 1.1 A list of oils and fats commonly used in soap manufacture

Oil/Fat	Source	Type fatty acid	Benefit in soap
Almond oil	Nut of sweet almond tree	Oleic acid	Adds extra conditioning to the skin
Apricot Kernel oil	Apricot stone	Oleic acid and linoleic acid	Works well for sensitive skin
Avocado oil	Avocado seed	Palmitic, stearic, oleic and linoleic acid	Emollient in soaps. Has a healing and moisturizing effect on skin. Good for dry skin
Avocado butter	Hydrogenated avocado oil	Palmitic, stearic, oleic and linoleic acids	Gives great skin feel
Calendula oil	Calendula seeds	Calendic acid, linoleic acid, oleic acid and palmitic acid	Useful for health conditions like acne, eczema bruises, spider veins and boils. It has moisturizing properties
Castor oil	Castor beans	Ricinoleic acid	Have emollient properties, lubricating to the skin, acts as humectant
Cocoa butter	Cocoa beans	Oleic acid, palmitic acid, stearic acid	Emollient properties
Jojoba oil	Jojoba seeds	Eicosenoic acid, docosenoic acid and oleic acid	Anti-inflammatory and antioxidant properties
Macadamia nut oil	Nut of macadamia tree	Palmitoleic acid	Excellent moisturizing properties
Sunflower oil	Sunflower seed	Oleic acid	Emollient
Coconut oil	Copra	Lauric acid, myristic acid, palmitic acid, oleic acid, capric acid, caprylic acid and others	Lathering, emollience and tanning properties
Grapeseed oil	Seed of grapes	Linoelic acid	Base oil for aromatherapy

(continued)

Table 1.1 (continued)

Oil/Fat	Source	Type fatty acid	Benefit in soap
Hemp seed oil	Hemp seed	Linoleic acid, omega-6 (LA), alpha linoleic acid, gamma linoleic acid, and stearidinic acid	Soothes and heals dry skin. Protects skin, moisturizer, anti-inflammatory
Kukui Nut oil	Kukui tree nut	Oleic acid, palmitic acid, alpha-linoleic acid and linoleic acid	Beneficial for eczema, acne and other skin conditions. Also, an emollient
Mango Butter	Seed of mango fruit	Palmitic acid, stearic acid and oleic acid	Beneficial for skin rash, dry skin & eczema, insect bites
Meadowfoam seed oil	Seed of meadow foam plant	Brassic acid, erucic acid, gadoleic acid and oleic acid	Gives smooth, soft and silky, feel to soap
Palm oil	Palm nut fruit	Palmitic acid and oleic acid	Base oil for soap
Palm kernel oil	Palm nut seed	lauric acid, mystiric acid and oleic acid	Bubblier leather, provides hardness to soap, moderate moisturizer
Safflower oil	Safflower seed	Linoleic acid	Good for healing wounds, emollient
Shear butter	Shear tree nut	Oleic acid, stearic acid linoleic acid, palmitic acid and arachidic acid	Emollient, soothing to the skin, protects the skin
Tallow	Animals fat (cow/sheep)	Oleic acid, palmitic acid and stearic acid	Base fat for making soap, excellent cleansing properties
Wheatgerm oil	Germ of wheat kernel	Linoleic acid palmitic acid and oleic acid	High in antioxidants and collagen for anti-aging

simple triglyceride, such as palmitin or stearin, all the three fatty acid groups are identical or the same. In mixed triglycerides, two or all three fatty acid groups are different. Most fats and oils contain mixed triglycerides. Generalized chemical structures of a triglyceride, a fatty acid and a glycerol, are shown in Fig. 1.3.

The three fatty acids in the triglyceride structure, namely, $RCOOH$, R^1COOH and R^2COOH can all be different, the same or only two can be the same. The alkyl chain length of naturally occurring triglycerides range from 3 to 22 carbon atoms, but 16 and 22 are the most common. Typically, plants and animals have natural fatty acids that comprise only of even numbers of carbon atoms. However, the action of bacteria can result in the formation of odd and branched chain fatty acids.

Pretreatment of fats and oils

Often it is required, in order to get a commercially appealing soap product of the right color, devoid of bad odor and right quality characteristics, to first pretreat or refine the oil or fat. The pretreatment processes include *degumming*, *deodorization*, *refining*, *fractionation*, *hydrogenation* and *bleaching*.

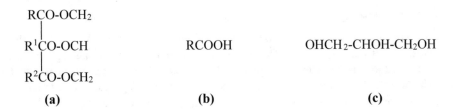

Fig. 1.3 Chemical structures of: **a** a triglyceride, **b** a fatty acid and **c** a glycerol. R, R^1 and R^2 are long alkyl chains

Degumming

This is the process by which complex organophosphorous compounds, generally called phosphatides, contained in fats and oils are removed. The phosphatides are called gums. The process of removing them involves the treatment of the oil or fat with water, phosphoric acid or a polybasic organic acid either individually or in combination to precipitate the gum and subsequently removing the precipitated gums by centrifuging. Alternatively, the precipitate may be absorbed on a bleaching earth or filter out.

Deodorization

The process of removing odors and flavors from oils and fats is called deodorization. This is accomplished by removing substances in oils and fats, which give rise to the odors and flavors by the application of heat, steam or vacuum.

Fractionation

Fractionation is the process of separating oil or a fat into two or more fractions. This is achieved by cooling and crystallizing the oil under controlled conditions. The resulting product, often a mixture of solid (crystals/precipitate) and liquid, is separated by filtration or centrifugation. The fractions obtained usually have different physical and chemical properties from the original oil and have different applications. The basic principle underlying fractionation is the solubility differences existing between the components of triglycerides of oil or fat.

Hydrogenation

This is a process of adding hydrogen to the double bonds of unsaturated fatty acids in oils and fats. This is carried out by reacting oil with gaseous hydrogen at elevated temperatures and pressures in the presence of a catalyst like nickel. The process significantly changes the properties of the fatty acids and therefore the properties and physical behavior of the oil.

1.2.1.2 Fatty Acids

The fatty acids are gotten by splitting the triglyceride(s). The methods commonly employed to split triglycerides into fatty acids and glycerols include enzyme splitting, the Colgate-Emery method, mineral acid catalysis, alkali hydrolysis and interesterification with methanol.

Enzyme splitting

The enzyme process uses naturally immobilized lipase enzyme to catalyze the hydrolysis of the triglyceride. The hydrolysis can be conducted in an organic solvent at a room temperature. The process takes about 24 h during which minimal oxidation and by-product formation occurs. The fatty acid is produced as a free acid and is removed from the lipase using an organic solvent. The reaction process for the enzymatic hydrolysis of a triglyceride proceeds as follows (Eq. 1.1):

$$
\begin{array}{l}
\text{RCO-OCH}_2 \\
\;\;\;\; | \\
\text{RCO-OCH} \quad + \quad \text{H}_2\text{O} \quad \longrightarrow \quad \text{3RCOOH} \quad + \quad \text{OHCH}_2\text{-CHOH-CH}_2\text{OH} \\
\;\;\;\; | \qquad\qquad\quad \textbf{Water} \qquad\qquad\qquad \textbf{Fatty acid} \qquad\qquad \textbf{Glycerol} \\
\text{RCO-OCH}_2 \\
\textbf{Triglyceride}
\end{array}
\qquad (1.1)
$$

The Colgate-Emery method

This method mixes superheated steam with the fat and keeps the mixture at a temperature of 252 °C and a pressure of 48 atmospheres to break up the triglyceride. The process takes place in a nitrogen atmosphere to avoid degradation of the acid produced. The process is very energy intensive and requires, for every pound of oil split, about 340 Btu (359 kJ) of energy. The hydrolysis reaction of triglycerides using steam proceeds as shown in Eq. (1.2):

$$
\begin{array}{l}
\text{RCO-OCH}_2 \\
\;\;\;\; | \\
\text{RCO-OCH} \quad + \quad \text{H}_2\text{O(v)} \quad \longrightarrow \quad \text{3RCOOH} \quad + \quad \text{OHCH}_2\text{-CHOH-CH}_2\text{OH} \\
\;\;\;\; | \qquad\qquad\quad \textbf{Steam} \qquad\qquad\qquad \textbf{Fatty acid} \qquad\qquad \textbf{Glycerol} \\
\text{RCO-OCH}_2 \\
\textbf{Triglyceride}
\end{array}
\qquad (1.2)
$$

Mineral acid catalysis.

This method uses water in the presence of a mineral acid as a catalyst for the hydrolysis. In this process, the triglyceride is heated with water at temperature between 150 and 250 °C and 10 and 25 bar pressure. Splitting is achieved in 6–10 h depending on the nature of the triglyceride feed. The higher the molecular weight of the triglyceride, the slower the splitting process. The reaction process for the acid catalyzed hydrolysis of triglyceride in an aqueous medium is shown in Eq. (1.3).

$$
\begin{array}{l}
\text{RCO-OCH}_2 \\
\;\;\;\; | \\
\text{RCO-OCH} \quad + \quad \text{H}_2\text{O} \quad \longrightarrow \quad \text{3RCOOH} \quad + \quad \text{OHCH}_2\text{-CHOH-CH}_2\text{OH} \\
\;\;\;\; | \qquad\qquad\quad \textbf{Water} \qquad\qquad \textbf{Fatty acid} \qquad\qquad \textbf{Glycerol} \\
\text{RCO-OCH}_2 \quad (\textbf{H}_2\textbf{SO}_4) \\
\textbf{Triglyceride}
\end{array}
\qquad (1.3)
$$

Alkali hydrolysis

In this process, the triglycerides directly react with the alkali followed by neutralization with an acid. During the process, the triglyceride is first converted via the saponification reaction to glycerol and alkali salt of the acid. This is followed by the acidification of the reaction mixture using a mineral acid to produce the fatty acid. The two reactions proceed as follows (Eqs. 1.4 and 1.5):

(a) Reaction of sodium hydroxide and triglyceride:

$$
\begin{array}{c}
\text{RCO-OCH}_2 \\
| \\
\text{RCO-OCH} \quad + \quad \text{NaOH} \quad \longrightarrow \quad \text{3RCOOH} \quad + \quad \text{OHCH}_2\text{-CHOH-CH}_2\text{OH} \\
| \quad\quad\quad\quad\quad \textbf{Sodium} \quad\quad\quad\quad \textbf{Fatty acid} \quad\quad\quad\quad \textbf{Glycerol} \\
\text{RCO-OCH}_2 \quad \textbf{hydroxide} \quad\quad\quad \textbf{salt} \\
\textbf{Triglyceride}
\end{array}
\tag{1.4}
$$

The resultant alkali salt of the fatty acid reacts with a mineral acid to produce a fatty acid according to the reaction in Eq. 1.5.

(b) Reaction of fatty acid salt with sulfuric acid:

$$
\begin{array}{ccccccc}
\text{RCOONa} & + & \text{H}_2\text{SO}_4 & \longrightarrow & \text{RCOOH} & + & \text{Na}_2\text{SO}_4 \\
\textbf{Fatty acid salt} & & \textbf{Sulfuric acid} & & \textbf{Fatty acid} & & \textbf{Sodium sulfate}
\end{array}
\tag{1.5}
$$

Interesterification of triglycerides with methanol

This method is mostly used to produce certain lower molecular weight fatty acids (C_8–C_{10}) by reacting a triglyceride with methanol. The two-step reaction proceeds as follows:

(a) Transesterificaion of a triglyceride using methanol:

$$
\begin{array}{c}
\text{RCO-OCH}_2 \\
| \\
\text{RCO-OCH} \quad + \quad \text{CH}_3\text{OH} \quad \longrightarrow \quad \text{3RCOOCH}_3 \quad + \quad \text{OHCH}_2\text{-CHOH-CH}_2\text{OH} \\
| \quad\quad\quad\quad\quad \textbf{Methanol} \quad\quad\quad\quad \textbf{Fatty methyl} \quad\quad\quad\quad \textbf{Glycerol} \\
\text{RCO-OCH}_2 \quad\quad\quad\quad\quad\quad\quad\quad \textbf{ester} \\
\textbf{Triglyceride}
\end{array}
\tag{1.6}
$$

The ester formed in (a) is hydrolyzed with water in the presence of a mineral acid to produce a fatty acid (Eq. 1.7). Alternatively, the ester may be reacted with an alkali first to form a soap, which is then hydrolyzed with a mineral acid to form the fatty acid as in Eq. (1.8).

(b) Hydrolysis of an ester with water to form a fatty acid:

$$
\begin{array}{ccccccc}
\text{RCOOCH}_3 & + & \text{H}_2\text{O} & \longrightarrow & \text{RCOOH} & + & \text{CH}_3\text{OH} \\
\textbf{Ester} & & \textbf{Water} & \text{H}_2\text{SO}_4 & \textbf{Fatty acid} & & \textbf{Methanol}
\end{array}
\tag{1.7}
$$

(c) Hydrolysis of an ester with sodium hydroxide to form a fatty acid:

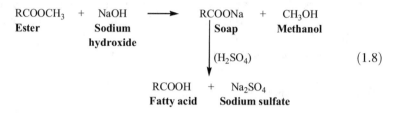

$$RCOOCH_3 \quad + \quad NaOH \quad \longrightarrow \quad RCOONa \quad + \quad CH_3OH$$

Ester   Sodium   Soap   Methanol

hydroxide

(H_2SO_4) (1.8)

$$RCOOH \quad + \quad Na_2SO_4$$

Fatty acid   Sodium sulfate

Step (a) is catalyzed by an alkali, a quaternary ammonium salts or the enzyme lipase.

Fatty acids may also be produced by special methods such as the ozonolysis of unsaturated fatty acids and by chemical oxidation. The fatty acid produced by the splitting of triglycerides is often a mixture of several molecular weight fatty acids and is usually refined into the specific components by distillation before use. Use of distilled fatty acids, which is usually purer specific fatty acids, gives an improved color, odor and quality of the finished product. The nature (chemical structure) of the fatty acid used in the manufacture of soap affects the properties of soap. An increase in the chain length of the fatty acid mostly results in a decrease in the solubility of soap. The presence of unsaturation in the alkyl chain of the fatty acid results in an increase in the solubility of soap. In general, an increase in the water solubility of soap is accompanied by an increase in the softness and mushiness properties of soap. Fatty acids of alkyl chain length of C_{10} to C_{12} give soaps with good lathering properties while those with chain of C_{16} to C_{18} contribute to the cleansing properties of soap. Based on this latter observation, a combination of tallow and coconut oil, which gave fatty acids of chain lengths C_{10} to C_{12} and C_{16} to C_{18}, respectively, has been the preferred choice for the manufacture of soap.

1.2.1.3 Alkali Salts and Other Bases

The most common bases used in the manufacture of soap are sodium hydroxide (caustic soda), potassium hydroxide (caustic potash) and sodium carbonate (soda ash). Other bases also used in the manufacture of soap are quaternary ammonium salts and ethanolamine. The choice of base depends on the specific properties required of the soap. For instance, caustic soda is the preferred base in the manufacture of bar soaps, since it produces hard soaps. For soft and more water-soluble soaps (and, also liquid soaps), caustic potash and ammonium salts are the preferred bases.

1.2.1.4 Other Additives

Several other additives are often added to the mixture during processing to give commercial soaps special properties and appeal. Table 1.2 is a list of the additives used in soap manufacture.

Table 1.2 Additives used in processing of soaps

Additives	Examples	Effect in soap
Builders	Sodium carbonate	Enhances the cleansing power of soap
Colorants	Alkanet root, dyes, pigments	Gives color to the soap
Fragrance	Assorted	Gives good smell to soap
Chelants	Citric acid	Controls metal ions in water
Antioxidants	Citric acid	Retards reaction of soap with atmospheric oxygen
Antimicrobial agents	Sulfoamide	Prevents growth of microbes
Alcohol	Glycerol	Humectant, emollient
Oatmeal powder	Oatmeal powder	Soothing properties in soap
Nettles leaf powder	Nettles leaf powder	Good effect on hair
Kelp powder	Kelp powder	Exfoliating effect
Honey	Honey	Humectant, soothing, emollient properties
Epsom salt	Epsom salt	Therapeutic effect
Citric acid	Citric acid	pH adjuster
Calendula petals	Calendula petals	Soothing to the skin
Fillers	Kaolin, bentonite	Modifies composition and reduces cost

1.2.2 Soap Manufacturing Processes and Technology

Manufacture of soap is premised on two main chemical processes, namely, the neutral fat saponification process and the fatty acid neutralization process. Saponification of fatty methyl esters is another process used in the manufacture of soaps. However, this last method does not yet have the popularity of the first two methods in commercial soap production.

1.2.2.1 Neutral Fat Saponification Process

The process flow diagram for the neutral fat saponification method is shown in Fig. 1.1. In this method, measured quantities of the main raw materials (the raw materials may be pretreated to remove impurities, odor, color and to achieve performance features desired in the finished product), the fat or oil and a base are charged into a reactor. The mixture is intensively mixed because the reactants are immiscible. External heat is applied to maintain the reaction temperature high enough to facilitate a rapid and complete reaction process. The chemical reaction between the fat or oil (a triglyceride) and a base, in this case, sodium hydroxide, is shown in Eq. 1.9.

$$
\begin{array}{l}
\text{RCO-OCH}_2 \\
\quad | \\
\text{RCO-OCH} \quad + \quad 3\,\text{NaOH} \quad \longrightarrow \quad 3\text{RCOONa} \quad + \quad \text{C}_3\text{H}_8\text{O}_3 \\
\quad | \qquad\qquad\quad \textbf{Sodium} \qquad\qquad \textbf{Soap} \qquad\quad \textbf{Glycerin} \\
\text{RCO-OCH}_2 \quad \textbf{hydroxide} \\
\textbf{Triglyceride}
\end{array}
\tag{1.9}
$$

The products of the reaction are neat soap, glycerin and spent lye. The soap product, at the reaction stage, is in liquid form. After the reaction, the mixture is first cooled and subsequently separated in a static separator where the soap is separated from the spent lye solution. The soap at this stage is not clean and is, therefore, washed with washing lye to separate the glycerin from the neat soap. In a commercial practice, this is usually done in a tower in a countercurrent operation. After the washing, the traces of lye in the soap are removed in a centrifugal separator. The resultant soap called a neat soap (which is essentially soap and water) is neutralized in a tank and finally dried in a vacuum spray dryer. Depending on the extent of drying required and on the soap formulation, a single stage or a multi-stage spray dryer may be used. The multi-stage dryers are more versatile allowing for drying of all manners of soap formulations and more efficient. The moisture content of the dried soap is between 8 and 18% (wt) and a total fatty matter of 72–82% (wt). The dried soap, in the form of pellets, is then sent to be processed into a finished product.

The amount of base required to saponify a given amount of triglyceride blend is estimated from the relation below.

$$
m_{\text{NaOH}} = \left[SV * m_{\text{oil/fat}} \left[\frac{MW_{\text{NaOH}}}{MW_{\text{KOH}}} \right] \right]
\tag{1.10}
$$

where m_{NaOH} is the mass of sodium hydroxide needed for the saponification of the triglyceride, $m_{\text{oil/fat}}$ is the mass of oil or fat that must be saponified, SV is the saponification value of the triglyceride, MW_{NaOH} is the molecular weight of sodium hydroxide and MW_{KOH} is the molecular weight of potassium hydroxide.

The saponification value is the amount of potassium hydroxide (KOH) in milligrams required to saponify 1 g of fat or oil. Table 1.3 lists the saponification values of some common oils and fats.

The values in the last two columns of the table indicate the amount in grams of NaOH and KOH, respectively, required to saponify 1 g of oil or fat. To apply Eq. (1.10), use SV values for sodium hydroxide in the table. Notice that the equation can also be used to calculate the mass of potassium hydroxide required to saponify a given amount of oil or fat. It is noted that, in practice, excess oil or fat, in the region of 5–8% (wt), is usually used for the saponification reaction. In this case, there will be unreacted oil or fat in the soap product. This is deliberately done in order to produce a soap that is mild on the skin. If more oil is used than needed, the resulting

Table 1.3 Saponification values of some common oils and fats

Oil/Fat	Saponification value (mg KOH/g oil or fat)	NaOH (g)	KOH (g)
Aloe vera oil	185–200	0.135	0.191
Beef tallow	–	0.140	0.196
Canola oil	–	0.123	0.173
Cocoa butter	188–200	0.136	0.192
Cocoa oil	173–188	0.127	0.179
Coconut oil (organic)	250–264	0.178	0.252
Cod liver oil	–	0.132	0.185
Corn oil	–	0.135	0.190
Fractionated coconut oil	325–340	0.234	0.329
Goat tallow	–	0.139	0.195
Jojoba (organic)	91–93	0.065	0.091
Margarine	–	0.136	0.190
Neem oil	175–205	0.134	0.188
Palm kernel oil	220	0.155	0.218
Palm oil	190–205	0.139	0.195
Shea butter	170–190	0.126	0.178
Shea oil	170–195	0.128	0.181
Soybean oil	190	0.134	0.188

soap goes rancid quickly. If on the other hand, stoichiometric amount of lye is used, the soap produced is often harsh on the skin. Allowing a certain percentage of fats and oils within your recipe to remain unsaponified is called *superfatting*.

1.2.2.2 The Fatty Acid Neutralization Process

This is essentially an acid–base reaction in which the fatty acid is reacted with a measured quantity of the base. This reaction is faster than the reaction of triglycerides with alkalis. The generalized chemical equation for the reaction of an organic acid with sodium hydroxide as the base to form soap is as follows:

$$RCOOH \quad + \quad NaOH \quad \rightarrow \quad RCOONa \quad + \quad H_2O \quad\quad (1.11)$$

The amount of base (e.g. sodium hydroxide) required to neutralize a given fatty acid blend is estimated from the relation:

$$m_{NaOH} = [\text{weight of fatty acid} \times 40]/MW_{fattyacid} \quad\quad (1.12)$$

where

$$MW_{fattyacid} = 56.1 \times 1000/AV \quad\quad (1.13)$$

AV is the acid value of the fatty acid blend. It is equal to the mass in milligram of potassium hydroxide required to neutralize 1 g of fatty acid.

To carry out the process, measured quantities of the fatty acid and the base are fed into a reactor, where heat and intensive mixing are applied until the reaction is completed. The progress of the reaction is monitored by tracking the conversion of the base in the reaction mixture. This can be done by taking samples from the reaction mixture and titrating against an acid with phenolphthalein as an indicator. Alternatively, the progress of the reaction can be tracked by electric potential measurement for alkalinity of the reaction mixture. At the end of the reaction, the neat soap is dried in a vacuum spray drier. The product from the dryer comes in the form of pellets, which is subsequently processed into a finished product. It is noteworthy that the direct product of the reaction in this process is a neat soap, which is quite unlike the previous process where the reaction product must go through several purification processes before a neat soap is obtained.

1.2.2.3 Preparation of bar Soap

One of the final forms into which the base soap may be processed into and sold is the bar soap. Bar soaps are made from the pellets (base soap) obtained from the vacuum dryer. The equipment needed for this operation are mixers, millers, extruders, chain cutters and a press. The soap pellets from the dryer are first thoroughly mixed with the relevant additives for instance, color, fragrance, builders, fillers and synthetic detergents in a mixer (also called amalgamator). The mixed soap is then milled into a uniform and homogeneous product. Finally, the milled product is extruded in an extruder into a continuous slug. A chain cutter is used to cut the slug into pieces that are stamped in a press into the desired size and shape of soap bars.

1.2.2.4 Preparation of Soap Powders

Powdered soaps are produced by spray drying. The neat soap slurry is heated and pumped to the top of a tower where it is sprayed through a very fine sprayer nozzle under high pressure to produce small droplets. The droplets fall through a current of hot air in the tower, forming hollow granules. The dried and fine particles are collected from the bottom of the spray tower where they are screened to achieve a relatively uniform size.

1.2.2.5 Preparation of Liquid Soaps

Liquid soaps are formulated by reacting potassium and ammonium salts with fatty acids or triglycerides to yield soaps that are highly soluble. Additives such as thickeners, fragrance, color, preservatives, stabilizers and pearlizing agents are thereafter added in the right amounts. Liquid soaps are normally packaged in bottles for sale.

1.2.2.6 Manufacturing Methods for Soap

The manufacturing process may be batch or continuous. There are two types of the batch process, the full boiled method and the semi-boiled method. The full boiled method is the most common method of soap manufacture. In this method, an excess alkali (also called lye) is combined with oil. Heat is applied and intensive mixing is applied to the reaction mixture until the oil is fully saponified. Subsequently, saltwater solution is added to the soap mixture. This causes the mixture to separate with the soap on top and the impurities, excess lye and glycerin at the bottom. The bottom liquid is drained off and the glycerin, a valuable substance, is recovered from it. The saltwater solution is applied to the soap and separated about three or four times before the soap is fully saponified and ready for further processing. In the semi-boiled method, also referred to as 'hot process' method, stoichiometric quantities of the oil and lye are mixed in a reactor and heated until full saponification occurs. The glycerin produced as a result of the saponification reaction is not separated but actually made part of the soap mass. This method is used when soap is going to be made into a transparent soap. The soap mixture is finally dried and formulated into a finished product. Handmade soaps are sometimes prepared by this method. This is the method commonly used by indigenous soap makers. Instead of an oil or fat, a fatty acid could be used as the starting raw material. In this case, the product of the reaction is neat soap without any glycerin. The continuous method requires that the raw materials are fed and processed continuously, and the products drawn off continuously at the end of the process line.

1.3 Detergents

Detergents are mixtures of surface-active agents of synthetic origin and other additives which are formulated for the purpose of cleansing. In a broader sense, a detergent is a substance or a mixture of substances intended to assist cleaning. Detergents are chemical compounds that can effectively remove oil stains and dirt from the surface of solids and cloth materials by breaking them into smaller particles and thereby making them soluble. Defined in this manner, the term covers all substances including soap that are used for the purposes of cleansing dirt and grease from surfaces. In everyday usage and in commercial circles, however, the term is used for a class of substances of synthetic origin used for cleaning purposes. Therefore, for clarity and for avoidance of doubt, the subsequent discussions will refer to synthetic detergents. In particular, as soaps were discussed in the previous section of this chapter.

Detergents are more resistant to calcium salt formation in hard water and generally provide better wetting, cleansing and surfactancy than soaps. Although the main use of detergents is for cleaning, they are also used in several other products including pharmaceuticals, shampoos, cosmetics, antiseptics, petroleum recovery, foods, cleaning products, toothpaste, vitamin and tablets. The form and composition of a detergent formulation is determined by the application to which it is to be put. Detergents come in several forms such as powdered, liquid, gel, sticks, sprays, sheets, pastes and bars. The composition of a detergent is determined mainly by the material to be cleaned, the nature of dirt to be cleaned and by the apparatus to be used (manual or machinery). The various formulations are laundry detergents, dish washing detergents and household detergents.

1.3.1 Ingredients Used in the Formulation of Synthetic Detergents

Irrespective of its form or use, detergent products are complex mixtures of several ingredients. The backbone of any detergent formulation, however, is surfactants and builders. Other additives designed to maximize the performance of the main ingredients are also included. These include bleaches, bleach activators, softeners, enzymes, scouring agents, fabric brighteners, anticaking agents, thickening agents, pearlescent agents and opacifying agents, among others. The role of these ingredients and specific examples of the ingredients are discussed below.

1.3.1.1 Surfactants

The word surfactant is a shortened form of the expression 'surface active agents'. These are organic chemicals that are capable of changing the properties of water. They lower the surface tension of water and enable a cleaning solution to wet the surface of clothes and other solid surfaces, so dirt can be readily loosened and removed. Surfactants also emulsify oily soils and keep them dispersed and suspended, so that they do not settle back on the surface from which they were removed. Surfactants derive their unique activity from their molecular structure (Fig. 1.4). The characteristic features of a molecule of a surfactant are a hydrophilic (water attracting) end and a hydrophobic (oil attracting) tail. Due to this unique structure, a surfactant acts as an emulsifier by bridging the water and oil phases, thereby breaking oil into tiny droplets suspended in water. The disruption of the oil film allows the dirt particles to be solubilized. Surfactants can be tailored for

Fig. 1.4 Molecular structure of surfactant

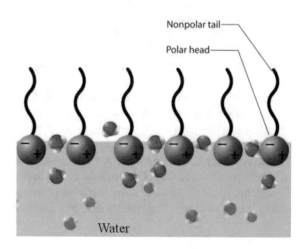

Fig. 1.5 The different types of surfactants

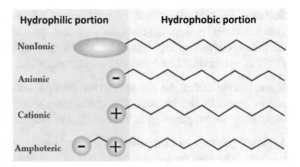

effective use in either aqueous or nonaqueous systems, depending upon their solubility characteristics. An optimum hydrophile–hydrophobe balance generally exists for a specific detergent application and for the compound class used. This optimum composition is arrived at through evaluation of the required hydrophile–hydrophobe characteristics of the application. An example is alkylbenzene, which is slightly surface active in nonaqueous media but insoluble and ineffective in water. However, upon sulfonation of alkylbenzene into alkylbenzesulfone, it acquires characteristics that make it highly water soluble with excellent surfactant characteristics but insoluble in petroleum solvents. If alkylbenzene is di- or tri-sulfonated, the compound becomes more water soluble but loses much of its surface activity.

The two major surfactants used in detergents are linear alkylbenzene sulfonates (LAS) and the alkyl phenol ethoxylates (APE). As shown in Fig. 1.5, surfactants are classified into four main groupings, namely, as; anionic, cationic, nonionic or zwitterionic (also known as amphoteric surfactants).

i. **Anionic surfactants**.

Anionic surfactants are negatively charged molecules that determine the effectiveness of the detergent. These anions react with basic components, thereby resulting in better cleansing action. In detergents, anionic surfactants are used for cleaning automobiles, washing metal articles, washing of clothes and as hair cleansers. The main anionic surfactants are:

Sulfonates: The sulfonate group—SO_3M, attached to an alkyl, aryl or akylaryl hydrophobe, is a highly effective solubilizing group. Examples are sodium alkylbenzene sulfonates (**ABS**), which are the most widely used non-soap surfactants, short-chain alkylarene sulfonates, lignosulfonates, naphthalene sulfonates, o-olefin sulfonates, petroleum sulfonates, sulfonates with ester, ether and amide linkages and fatty acid ester sulfonates. Sulfonates interact moderately with the hardness of Ca^{2+} and Mg^{2+} ions. They are used in the manufacture of light and heavy-duty liquid detergents and as spray-dried detergents for institutional and industrial cleaning applications and as car washing compounds. They can be tailored for specific applications by the introduction of double bonds, ester or amide groups either into the hydrocarbon chain or as substituents. Sulfonates are among the high-volume surfactants used in the preparation of detergents.

Sulfates: The sulfates represent the sulfuric acid half ester of an alcohol. The sulfate group is more hydrophilic than the sulfonate group because of the presence of an additional oxygen group. Examples are alcohol sulfates, ethoxylated and sulfated alcohols, ethoxylated and sulfated alkylphenols and sulfated natural oils and fats.

N-Acylsarcosinates: An example in this group is sodium N-lauroylsarcosinate. The amido group in the hydrophobic chain lessens interactions with water hardness ions. They are prepared from the reaction of fatty acids with sarcosine.

Acylated protein hydrolysates: These are mild surfactants used in personal care products. They are prepared by acylation of protein hydrosylates with fatty acids or acid chlorides.

Phosphate esters: Examples are the mono- and di-esters of orthophosphoric acid and their salts. Unlike the sulfonates and the sulfates, the resistance of the alkyl phosphate esters to acids and hard water is poor. That is, their calcium and magnesium salts are insoluble. In the acid form, the esters show limited water solubility, although their alkali metal salts are more soluble. The surface activity of phosphate esters is good but somewhat lower than their phosphate-free precursors. At higher temperatures, however, the phosphate surfactant is significantly more effective. Phosphate surfactants find applications in specialty situations due to their high costs and limitations. Their applications include dry-cleaning compositions where solubility in hydrocarbon solvents is a particular advantage, textile mill processing where the stability and emulsifying power for oil and wax under highly alkaline conditions are necessary, and in industrial cleaning, compositions where tolerant for high concentrations of electrolyte and alkalinity are required. They are also used as corrosion inhibitors.

The anionic surfactants mostly used in detergents are the alkyl sulfates, alkyl ethoxylate sulfates and soaps.

ii. Cationic Surfactants

Cationic surfactants have positively charged heads when in solution. Three types of cationic surfactants can be distinguished based on their applications. Those used in fabric softeners where they provide soft feel to fabric. An example of cationic used for this purpose is esterquat. In laundry detergents, they are used in conjunction with anionics to boost the performance of anionic surfactants and leading to improved dirt removal systems, and especially very efficient stain removal. An example of a cationic surfactant used in this category is mono-alkyl quaternary system. In household and bathroom cleaners, cationic surfactants play a role as disinfecting and sanitizing properties of detergents. Examples of cationic surfactants are primary, secondary or tertiary amines, benzalkonium chloride (BAC), cetylpyridinium chloride (CPC), polyethoxylated tallow amine (POEA) and alkyl trimethylammonium salts, namely, cetyl trimethylammonium bromide (CTAB), hexadecyl trimethylammonium bromide, cetyl trimethylammonium chloride (CTAC).

iii. **Nonionic surfactants**

The surfactants in this category do not carry any electrical charge. This makes them resistant to water hardness deactivation. Among the most commonly used nonionic surfactants are ethers of fatty alcohols. Nonionic surfactants are excellent grease removers. Their major area of application is in laundry detergents, hand dish-washing liquids and in household cleaners. Nonionic surfactants are often used alongside anionic surfactants as they complement each other's cleaning action. When present, they help in making surfactant systems less hardness sensitive. Examples include cetyl alcohol, stearyl alcohol, cetostearyl alcohol, oleyl alcohol, octaethylene glycol monododecyl ether, pentaethylene glycol monododecyl ether, glucoside alkyl ethers and glycerol alkyl esters among others.

iv. **Amphoteric (Zwitterion) surfactants**

These are special class of surfactants. The molecules contain both positive and negative charges. This bestows them with properties of both cationic and anionic surfactants. Amphoteric surfactants anionic at pH > 7, cationic at pH < 7 and have no charge at pH = 7. They are compatible with both cationic and anionic surfactants. Amphoteric surfactants offer high performance only at neutral pH. They are, thus, preferred for use in personal care formulations such as hair and body shampoos, bubble baths and other toiletries and cosmetics. Amphoteric surfactants are relatively expensive because of the raw materials involved and the production costs. Examples of surfactants in this group are imidazolinium derivatives and amino alcohols and their derivatives, e.g. diethylcarboxymethyl coconut fatty acids, propylamide ammonium betaine, alkylamide betaine and cocomidopropylamine oxide.

Evaluation of Surfactants

Surfactants are evaluated based on their performance in the following areas; foamability, wetting, emulsifying, detergency, skin and eye irritation and compatibility with /other surfactants. For the four surfactants discussed, their performance in five of these indicators is listed in Table 1.4.

1.3.1.2 **Builders**

In addition to surfactants, other backbone substances of detergents are builders. They are added to detergents to extend their cleaning performance across a wide range of use conditions. Combining surfactants and builders lead to a synergistic effect to boost total detergency and cleaning efficacy as compared with an equal amount of either compound alone. The most important function of detergent builders is as sequestering agents. Other characteristics required of good builders are:

i. It must be able to control water hardness and metal ions in water.
ii. It must contribute to the final product alkalinity.
iii. It should be able to provide buffer capacity in the proper pH range.

Table 1.4 Performance indicators of the different types of surfactants

Indicator	Ionic surfactant	Cationic surfactant	Nonionic surfactant	Amphoteric surfactant
Foaming	Very good	Good	Low to average	Very good
Emulsifying	Good	Poor	Very good	Very good
Eye and skin irritation	Mild	Very mild	–	–
Detergency	Very good	Poor	Very good	Very good
Compatibility	Incompatible with cationics	Incompatible with anionics	Very good	Very good

iv. It must have deflocculating capability.
v.   It must be compatible with other formulation ingredients and additives.
vi. It must be safe to use by consumers.
vii. It must not hurt the environment.
viii. It should not pose any problem to the processing of the detergent and must itself be easy to process.
ix. Its performance and cost must be adequate.

Builders are used singly or in combination with other builders to contribute unique properties for final product performance enhancement. Some of the most commonly used builders are listed below.

i. **Phosphates**

These include sodium tripolyphosphate-STP ($Na_5P_3O_{10}$), which is the major builder ingredient used in heavy duty laundry detergents, automatic dishwashing formulations and industrial and institutional cleaners. It has superior detergent processing, solubility and hardness ion sequestering characteristics. Another phosphate builder is tetrapotassium pyrophosphate, which is used in industrial and institutional liquid detergents. It has good solubility and sequestering characteristics. Others in this category are tetrasodium pyrophosphate-TSPP ($Na_4P_2O_7$), sodium trimetaphosphate, trisodium phosphate and tripotassium phosphate.

ii. **Silicates**

Principal among these are sodium silicate and potassium silicate. They are used either as solids or in solution. They find use as emulsification, buffering, deflocculating and anti-redeposition agents. They also provide corrosion protection to metal paints in washing appliances and enhance the sequestering ability of other builder systems. Their performance depends on the ratio, silicate/alkali (SiO_2/Na_2O), selected for the formulation. For example, for dry blending applications, the desired ratio is 1:1. For laundry and automatic dishwashing applications, the desired ratio is 2:1.

iii. **Sodium citrate ($Na_3C_6H_{25}O_7$).**

It is the preferred builder in liquid consumer laundry detergents. It sequesters water hardness ions and deflocculates soils. It has limited use in powdered products.

iv **Sodium carbonates**

They are used mostly in granular laundry detergents, automatic dishwashing formulations, hard surface cleaners and in presoak formulations. It precipitates hardness minerals, provides alkalinity for detergents and facilitates soil dispersion. It is sometimes used in combination with zeolite as a substitution for sodium tripolyphosphate.

v. **Zeolites (e.g. $Na_2.Al_2O_3.2SiO_3.4.5H_2O$)**

Zeolites are the most widely used builders for laundry detergents. They are not soluble in water but reduce water hardness by ion exchange. The calcium (Ca) ions pass through the zeolite pore opening and exchange with the sodium (Na) ions. Magnesium (Mg) ions, because of their larger size are not removed. They are used in conjunction with other builders such as sodium carbonate to control magnesium and other ions present in water.

vi. **Sodium nitrilotriacetate ($N(CH_2COONa)_3$**

It has excellent sequestrant and chelating ability. It is, however, considered to be carcinogenic. Therefore, in some countries, it is not allowed in personal care, hand dish-washing detergents and foods. It is nonetheless widely used in industrial and institutional cleaning products. It is normally used in detergents and as a supplement to phosphates in spray-dried detergents.

1.3.1.3 Detergent Additives
Other additives used in detergents and the functions they play are listed in Table 1.5.

1.3.2 Laundry Detergents

A laundry detergent is any of the various detergents designed for use in washing clothes. In general, these help in the breakup and removal of dirt and stains in the laundry and are available as liquids, powders, gels, sticks, sprays, pumps, sheets or bars. All laundry detergents are soluble in water. They function in a wide range of temperature and water conditions. They are formulated with the objective that they can remove soil and stain from fabrics to bleach fabrics and as fabric softening agents. They are also used for conditioning fabrics and as disinfectants. Liquid laundry detergents are particularly good at removing greasy and starchy food stains. Powdered laundry detergent is especially effective on heavily soiled clothing. Enzymes are often added to laundry detergents to improve their effectiveness in breaking down complex food stains.

Table 1.5 Additives employed in commercial detergents

Substance	Functions in detergents
Sodium sulfate (Na_2SO_4)	Improves finished product flow characteristics in dry mix and agglomerated products. In spray-dried products, it acts as an inert 'filler' and aids in density control and flow characteristics
Sodium chloride (NaCl)	Improves finished product flow characteristics in dry mix and agglomerated products. In spray-dried products, it acts as an inert 'filler' and aids in density control and flow characteristics
Sodium carboximethyl cellulose ($C_8H_{15}NaO_8$)	It reduces soil redeposition during detergent application (at formulation levels of 0.1 to 0.5%)
Optical brightners	They improve the whiteness or brightness of fabrics. They are used in both liquid and powdered products. All detergents contain one or more optical brighteners
Hydrophobes	Used in liquid detergents to improve the solubility of less soluble components and also improve storage stability. They are also used occasionally as viscosity modifiers during spray-dried processing. Examples are ammonium, potassium or sodium salts of toluene, xylene or cumene sulfates
Pearlescent agents	Gives aesthetic feel to fabrics. Examples are bismuth and titanium dioxide-coated mica
Opacifying agents	They reduce translucence, modify the viscosity characteristics of liquids and provide a creamlike texture or pearlescent effect on products. Examples are water-soluble salts of styrene and maleic anhydride
Thickening agents	They provide rheological modification to automatic dishwashing liquids. An example is betonite clay. Use level in automatic dishwashing liquids is 0.5 to 1.0%
Anticaking agents	Prevent caking in detergent systems, especially in humid climates. Examples include re sodium benzoate, tricalcium oxide, calcium stearate, and microcrystalline cellulose
Enzymes	They are used to remove stains. They break down carbohydrates, proteins and other soil materials to forms readily removed by detergent. They are used in presoak detergent systems and some liquid and powdered laundry detergents
Fillers	Used to adjust the active matter in the detergent to the doses used. Examples include sodium sulfate in powders, water and other solvents in liquids

Like all detergents, the key ingredient in laundry detergents is the surfactant. Most laundry detergents use anionic or nonionic surfactants or a mixture of the two, although cationic surfactants are sometimes used in laundry detergents. Cationic and anionic surfactants are not used together in detergents as they are incompatible. The usual content of surfactants in a typical detergent is about 8–18%. Other additives are added to the surfactants to improve its efficiency. For instance, laundry detergents may have ingredients to help control the pH of the wash water. Sodium carbonate or sodium bicarbonate is often added to solid detergents to neutralize acidic materials that may enter the wash water and, thus, help maintain its pH. Builders such as trisodium orthophosphate, monosodium orthophosphate or a form of tripolyphosphate (TPP) may be used to enhance ('build') the surfactant effect. They lower the water hardness by scavenging the calcium and magnesium ions and adsorbing them or chelating them. In place of the phosphates, other chelating agents may be used. Sodium carbonate precipitates insoluble calcium carbonate and magnesium carbonate. Organic chemicals similar to ethylenediaminetetraacetic acid (EDTA) can be employed, for example, nitriloacetic acid (NTA). Borates and ion exchange materials may also be used as builders in detergents. The usual content of builders in a typical detergent is about 20–45%.

Many detergents contain bleaches. Sodium hypochlorite-based bleach additives are more commonly used in certain conditions. Peroxide-based bleaches are also used. These contain compounds that release hydrogen peroxide when in use. Examples of such compounds are sodium percarbonate and sodium perborate. Peroxide bleaches either need higher temperature (60 °C or more) to become effective or a suitable catalyst or activator (e.g. manganese or iron complexes or Tetraacetylethylenediamine—TAED), which lowers the required temperature down to 40 °C or even to room temperature. The usual content of bleaches in a typical detergent is about 15–30%. Optical brighteners, usually at a concentration of about 0.1%, may also be added to laundry detergents to improve the whiteness of the cloth. These compounds act by converting some ultraviolet radiation to blue light and, thus, optically offsetting the yellowing of the material. Fillers are added to detergents to modify the physical properties of the material. In solid detergents, sodium sulfate or borax can be used to make the powder-free flowing. Other additives used in liquid detergents include alcohols, corrosion inhibitors and anti-foaming agents. Alcohols increase the solubility of the compounds and decrease the mixture's freezing point. Corrosion inhibitors when present in detergents protect the washing machines and prolong their lifetime while anti-foaming agents reduce foam production. The percentage composition of fillers in a typical detergent is about 5–45% (wt).

Enzymes are used in some laundry detergents to help in the removal of biological stains. Often, the enzymes produced by the bacteria *Bacillus subtilis* and *Bacillus licheniformis* are used. The content of enzymes can be as high as 0.75%. Some laundry detergents have fabric softeners and a perfume or color ingredients to give better smell or to give a detergent some color.

1.3.3 Dishwashing Products

These include detergents for hand and machine washing as well as some specialty products. They are available in liquids, gels, powders and solids. Hand dishwashing detergents remove food soils, hold soil in suspension and provide long-lasting suds that indicate how much cleaning power is left in the wastewater. Automatic

Table 1.6 Ingredients used in dishwashing detergents

Detergent ingredient	Functions
Phosphates	Dissolve calcium and magnesium ions to prevent 'hardwater'-type limescale deposits
Oxygen-based bleach	Breaks up and bleaches organic deposits
Nonionic surfactant	Reduces the surface tension of water. Emulsifies oil, lipid and fatty foods. Ensures that droplets are not formed on drying
Enzymes	Split-up and dissolve protein-based food, oil, lipid and fats
Anti-corrosion agents	Prevent corrosion of dishwasher components
Anti-foaming agents	Prevent foam formation. Foam interferes with the washing
Perfume	Gives good flavor to the detergent
Anti-caking agents	Prevent cake formation of the detergent

dishwashing detergents on the other hand, in addition to removing food soils and holding them in suspension, tie up hardness minerals, emulsify grease and break up and bleach organic deposits. Different kinds of dishwashing detergents contain different combinations of chemicals and additives. Besides chemical detergents for dishwashing, biodegradable detergents also do exist. The latter are more environmentally friendly than other detergents. In Table 1.6, there is a list of specific ingredients of dishwashing detergents.

1.3.4 Household Cleaners

Household cleaners are products that are used to clean soft or hard surfaces in the home. Laundry and dish items are not included in the category of items for which household cleaners are used. Household cleaning products play an important role in personal and public health. With their ability to loosen and remove soil from surfaces, household cleaning products help in achieving a good personal hygiene, in reducing the presence of germs and, hence, potential diseases and also extend the useful life of household products. These products, thus, make our homes and workplaces more healthy and pleasant.

In formulating household cleaners, two principal factors must be taken into consideration, namely, the nature of soil or dirt that must be removed from the surface and the nature of the surface. The range of soils or dirt varies considerably to include generic dust to clay, oil, grease, soap scum, watermarks, lime scale, mold and mildew. Equally the surfaces that one faces can vary considerably from wood to ceramic, porcelain, enamel, glass, marble, stainless steel, metal, carpets and plastics. This wide variety of soil and surface means that a whole range of household cleaners may be formulated to deal with specific combination of soil and surface. General purpose cleaners also exist that are formulated to deal with all manner of soil and surfaces. Other factors that must be considered in formulating household cleaners are the effect on the environment of the detergent and the safety of use of the product or its toxicity to the user. Household cleansing products are formulated in different forms, such as liquids, gels, powders, solids, sheets, pads or sprays.

1.3.4.1 Type of Household Cleaners

Household cleaners are some of the most widely used consumer products. Apart from plastic and synthetic fiber materials, there is probably no other classes of chemical products that people come into contact with more frequently. Household cleaners are extremely diverse, ranging from general purpose cleaners, some of which are represented as being capable of virtually any cleaning job, to specialized cleaners, such as glass cleaners or tub and tile cleaners. Similarly, the ingredients found in this class of products are also diverse, ranging from simple soap to proprietary formulations of petrochemical surfactants, solvents and complexing agents. Accordingly, these classes of products are grouped based on their applications or composition—ingredients from which the products are made. Table 1.7 lists the different uses to which household detergents are put.

Table 1.7 Classification of household products according to use

Product use category	Areas of application
General purpose	Multipurpose surface cleaners labeled as multipurpose, intended for use in a variety of applications in the home. Examples include wall cleaners, floor cleaners, disinfecting cleaners, cleaner-degreasers and concentrated cleaners.
Bathroom cleaners	Cleaners intended primarily for use on bathroom surfaces. These include tub and tile cleaners, mildew stain removers, shower cleaners and disinfecting bathroom cleaners.
Disinfectants	Products that play the dual role of cleaning and disinfecting surfaces, for example, antimicrobial liquid sprays or concentrated germicides.
Scouring cleansers	Surface cleaners also act as abrasives. Examples are scouring powders, scouring pastes or liquids.
Glass cleaners	Cleaners specifically meant for glass surfaces. These are often marketed as pump spray, aerosol or liquid glass cleaners.
Carpet/upholstery cleaners	Cleaners specifically designed for use on fabrics that cannot be removed for laundering or dry cleaning. These are formulated as liquids, foams or dry powders.
Stain removers	Products designed to remove spots. Bleaches are not included in this category. Examples are cleaning fluids, stain sticks, enzyme spot removers.
Toilet bowl cleaners	Products designed for the sole purpose of cleaning the toilet bowls. These include liquid or crystal acid-based cleaners, detergent cleaners. For automatic toilet cleaners, these are formulated as blocks, tablets or controlled release bottles.

1.3.4.2 Ingredients in Household Detergents

Ingredients found in household cleaners are grouped into five. These are surfactants, builders, solvents, antimicrobials and miscellaneous ingredients.

i. Surfactants

Surfactants or surface-active ingredients are the wetting and foaming agents, which form the basis for most aqueous cleaners. While anionic, nonionic and amphoteric surfactants are used in household detergents formulated for cleaning purposes, cationic surfactants serve as antimicrobial agents in household detergents. For household cleaners, numerous examples abound in this category including:

Anionic surfactants: Linear alkylbenzene sulfonate (sodium dodecylbenzene sulfonate, dodecylbenzene sulfonate, sodium laurylbenzene sulfonate) alpha sulfo methyl ester (alpha sulfo acid ester) alkyl polyglucoside (alkyl polyglycoside) alcohol sulfates (lauryl sulfates) alcohol ether sulfates (lauryl ether sulfates, laureth sulfates) lauryl sarcosinate, and soaps.

Nonionic surfactants: Alcohol ethoxylates (linear alcohol ethoxylates, primary alcohol ethoxylates, ethoxylated alcohols, alcohol polyethylene glycol ethers), coconut-based surfactant, unspecified (probably nonionic), lauryl amine oxide, nonylphenol ethoxylates, octylphenol ethoxylates, coconut diethanolamide (cocoamide DEA).

Cationic surfactants: This group includes dialkyl dimethyl ammonium chlorides, alkyl dimethyl benzyl ammonium chlorides, alkyl dimethyl ethylbenzyl ammonium chlorides, hexadecyl trimethyl ammonium bromide and quaternary ammonium chlorides.

Amphoteric Surfactants: Betaines (hydroxy-imidazoline and N-alkyl betaine, lauryl betaine, amido propyl betaine), amphoacetates (disodium cocoamphodiacetate, sodium cocoamphoacetate), glycine (quaternized fatty acid amides glycine), alkyl group glycine (diamine-ethyl-group) and di (alkyl amino-ethyl group), Lauryllactam imidazolium salts, disodium salt.

ii. Builders

Builders regulate the pH of the washing solution, soften water by removing metal ions and regulate foam height. They may be organic or inorganic compounds whose presence improves the performance of the surfactants. Examples used in household cleaners are acetic acid, calcium carbonate, calcium chlorate, calcium chloride, calcium hydroxide, citric acid, diethanolamine, monoethanolamine, potassium hydroxide, potassium silicate, sodium metasilicate, potassium hydroxide, sodium bicarbonate, sodium bisulfate, sodium carbonate sodium chloride, sodium citrate, sodium EDTA (tetrasodium EDTA), sodium hydroxide, sodium sesquicarbonate, sodium silicate, sodium sulfate, sodium tripolyphosphate, tetrapotassium pyrophosphate, triethanolamine and trisodium phosphate.

iii. **Solvents**

Solvents are added to help dissolve oil and grease. Antimicrobials are pesticides that kill bacteria, fungus or mildew on surfaces. Some common solvents found in household detergents are acetone, almond oil, ammonia (ammonium hydroxide), apricot kernel oil, t-butyl alcohol 1,2-butylene oxide, citronella oil, citrus oil (d-limonene, orange oil, lime oil), diethylene glycol monobutyl ether, (2–2-butoxyethoxy) ethanol, butyl diglycol and dimethoxymethane, dipropylene glycol methyl ether, ethanol, ethylene glycol ether, ethylene glycol ethyl ether, ethylene glycol monobutyl ether (2-butoxyethanol), eucalyptus oil, glycerine (1,2,3-propanetriol), glycol ethers, hexylene glycol, isopropanol, lavender oil, mineral oil, naphtha (petroleum distillates), peppermint oil, pine oil (pinene), propylene glycol, propylene glycol ethers, propylene glycol methyl ether (1-methoxy-2-propanol), rosemary oil, toluene, 1,1,1-trichloroethane and xylene.

iv. **Antimicrobial agents**

Some antimicrobial agents found in household detergents include calcium hypochlorite, dialkyl dimethyl ammonium chlorides (alkyl can include octyl, decyl, didecyl), alkyl dimethyl benzyl ammonium chlorides, alkyl dimethyl ethylbenzyl ammonium chlorides, calcium hypochlorite, glutaraldehyde, phenol, o-benzyl-p-chlorophenol, o-phenyl, sodium dichloro-s-triazinetrione, sodium hypochlorite and sodium trichloro-s-triazinetrione.

v. **Miscellaneous agents**

All other ingredients used in household products have been placed in the category called miscellaneous. Examples are abrasives, fragrances, dyes, thickeners, hydrotopes, and preservatives. Among these are aloe vera, carbon dioxide (propellant), chalk, 1-(3-chloroallyl)-3,5,7-triaza-1-azoniaadamantane chloride, (Dowicil 75, Quaternium 15), clay, denatonium benzoate (Bitrex), enzyme, amylas enzyme, proteinase, extract of berberis, extract of marigold, feldspar, fluoraliphatic acid salt, hydrochloric acid, hydroxyacetic acid, isobutene, magnesium oxide, methylparaben, methyl salicylate, oxalic acid, phenol, o-benzyl-p-chloro phenylmethanol (phenylcarbinol), phosphoric acid, propane, propylparaben, silica, amorphous silica, crystalline sodium cumene sulfonate, sodium naphthalene sulfonate, sodium octane sulfonate, sodium perborate (borax), sodium xylene sulfonate, styrene maleic anhydride resin, sulfamic acid, urea, witch hazel and xanthan gum.

1.3.4.3 Typical Formulations for Various Products

i. **General purpose cleaners**

The variety of dirt encountered by general purpose cleaners can be characterized as oils, fats, waxes, food residues, dyestuffs and tannins, silicates, carbonates (limestone), oxides (sand, rust), soot and humus. The ingredients include surfactants, builders, organic polymers, solvents, viscosity regulators, pH buffers,

Table 1.8 Typical formulations of general-purpose household cleaners

Type I: Powdered cleaners
The key ingredients are anionic or nonionic surfactants, sodium carbonate, sodium silicates or metasilicates, phosphates or aluminosilicates

Type II: Weakly alkaline liquids
Anionic or nonionic surfactants, alcohols, glycols, glycol ethers, citrates, sodium EDTA, citrus oil, pine oil or other essential oils, sodium hydroxide, amines, dyes, fragrances, preservatives

Type III: Disinfecting cleaners
Ingredients in Type II plus quaternary ammonium compounds, sodium hypochlorite, pine oil or phenolics

Type IV: Multi-purpose spray cleaners
Ingredients in Type II above plus glycol ethers and alcohols

Type V: Cleaner/degreasers
Nonionic surfactants, citrus oil or d-limonene

anti-microbials, hydrotropes, dyes and fragrances. General purpose cleaners are generally formulated as powders, alkaline liquid cleaners, disinfecting cleaners, spray cleaners and cleaner/degreasers. Table 1.8 lists typical ingredients for each group of general-purpose cleaners.

ii. **Bathroom Cleaners**

Bathroom cleaners are formulated to clean dust, sand, street dirt, oil and fat found on bathroom floors and walls. They also are designed to clean wash room contaminants such as calcium and rust deposits from the water, metal corrosion products, soaps and lime soaps, hair and fibers. For cleaning bathroom floors and walls, a weakly alkaline all-purpose cleaner similar to those described above for general purpose cleaners is typical, though, for bathroom cleaners, the presence of disinfectant chemicals is perhaps more common (see Table 1.9).

iii. **Scouring cleansers**

Scouring cleansers use the mechanical action of abrasives to assist in the cleaning process. Scouring cleansers may be solids (powders) or liquids. Ingredients used in the different types are listed in Table 1.10.

Table 1.9 Typical ingredients found in bathroom cleaners

Type I: Acidic cleaners
Key ingredients of acidic cleaners include inorganic or organic acids (e.g. phosphoric acid, citric acid and hydroxyacetic acid), anionic or nonionic surfactants, glycol ethers, alcohols, citrates and sodium EDTA

Type II: Alkaline cleaners
The ingredients of alkaline cleaners are the same as in acidic cleaners except that in place of the acids the alkaline compounds sodium carbonate, sodium hydroxide and sodium hypochlorite are used

Table 1.10 Typical ingredients found in scouring cleansers	**Type I: Powder cleansers** Anionic surfactants, nonionic surfactants, organic polymers, sequestering agents, alkaline salts/bases, bleaching agents, dyestuff, fragrances and abrasives
	Type II: Liquid cleansers Anionic surfactants, nonionic surfactants, organic polymers, sequestering agents, alkaline salts/bases, dyestuff, fragrances, solvents, preservatives, skin protection additives, viscosity regulators, pH regulators, hydrotropes, water and abrasives

iv. Glass cleaners

The main ingredients of these products are water, alcohols and glycol ethers, with surfactants being a very small part of the mixture. Additives such as dyes and silicone are sometimes added to the formulation.

v. Carpet and upholstery cleaners

The products under this group fall under two general categories: liquid shampoos or powders. Even though both types can be used on upholstered furniture, shampoos are preferred. Shampoos work in a three-way action. First, when applied, it generates copious amounts of foam. The foam then lifts and holds dirt from the carpet or upholstery. Finally, the mixture of foam and dirt is vacuumed away. In both carpet and upholstery cleaning, the material being cleaned cannot be rinsed. The liquid foams containing surfactant mixtures are designed for high foaming, foam stabilizers and usually resins to harden the residues for easy vacuuming.

Surfactants used in these products are sodium or lithium salts of dodecyl sulfate, alpha-olefin sulfonates, alkali salts of fatty acid monoethanolamide sulfo succinic acid half-esters and fatty alcohol polyethyleneglycol ether carboxylic acids. The most effective surfactants are half esters of sodium sulfosuccinates used alone or with fatty alcohol sulfates. Foam stabilizers can be fatty acid ethanolamides or long-chain fatty alcohols. The hardening resins are usually styrene maleic resins. These products may also contain alcohols such as ethanol and isopropanol and glycol ethers such as ethylene glycol monobutyl ether.

vi. Stain removers

Stain removers are used to remove spots and stains from fabrics that can be laundered. The active compounds used in these formulations are surfactants, solvents or enzymes. Some formulations are only a mixture of surfactant and enzyme or surfactant and solvent. Some types of stains can be removed by applying concentrated liquid detergents or by use of concentrated citrus solvents.

Common surfactants found in these products are sodium dodecylbenzene sulfonate (LAS), ethoxylated C12-C15 alcohols, alpha sulfo methyl ester, linear secondary alcohol ethoxylates and nonylphenoxy polypropyleneoxy polyethyleneoxy ethanol (commonly known as an EO-PO polymer). The builders include both

Table 1.11 Typical ingredients found in toilet bowl cleansers

Type I: Liquid toilet bowl cleansers
Hydrochloric acid, nonylphenol polyethyleneglycol ethers, cetyl dimethylbenzylammonium chloride, sodium silicate and water.

Type II: Powdered toilet bowl cleansers
Sodium hydrogen sulfate (bisulfate), linear alkylbenzene sulfonate, sodium chloride and sodium carbonate/bicarbonate.

diethanolamine and triethanolamine. Proteinase and amylase may also be present as well as sodium hypochlorite bleach.

vii. **Toilet bowl cleaners**

Toilet bowl cleaners are usually acidic and may be liquid or powder. Many of these products are considered corrosive. Table 1.11 lists the ingredients of the different types of these products.

1.4 Manufacture of Detergents

The choice of process route and the nature of equipment used during the manufacture of detergents depend on the detergent product. Commercial detergent products are formulated as liquids or solids (powered form).

1.4.1 Powdered Detergents

They are produced by spray drying, agglomeration, dry mixing or a combination of both.

1.4.1.1 Spray Drying

A significant portion of powdered detergents is manufactured by spray drying. The process is made up essentially of the following steps: (i) selection of formulations, (ii) slurry preparation, (iii) slurry atomization, (iv) drying, (v) conditioning of the product, (vi) post additions and (vii) packaging. The flowchart of the process is shown in Fig. 1.6.

Materials to be selected for the formulation depend on the nature of product required and on the processing route and equipment to be adopted. Once the formulation is determined, the next and the most important step in the spray drying process is the slurry preparation. The slurries are prepared in crutching units. The process may be batch, semi-batch or continuous. To deliver a product of consistent quality, the order of addition of the raw material, the concentration of solids in the slurry, slurry temperature, viscosity and aeration must all be carefully controlled. Solid levels of the slurry are kept between 64 and 72% to reduce the heat requirement during the subsequent treatment in the spray tower, and also to

Fig. 1.6 Flowchart for making powdered detergents by spray drying

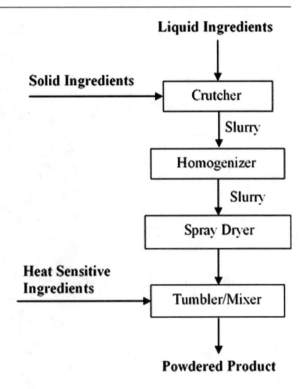

maximize the tower output. After the slurry is prepared, it is charged to a homogenizer where any large lump of materials or gritty particles in the slurry are desized and screened to prevent plugging of the spray tower nozzles during atomization.

Subsequently, the slurry is deaerated in a booster pump and thereafter pumped through nozzles placed at the top of the spray tower to be atomized. Pump pressures between 31 and 124 atmospheres are employed during atomization. A spray tower is essentially a contained heat source that removes moisture at a uniform rate dependent upon the quality and quantity of feed to the dryer. If feed levels are not controlled, variability in the density, moisture level, aging and/or packaging characteristics increase.

As shown in Fig. 1.7, the atomized droplets of the slurry enter at the top of the spray tower where either in a countercurrent or concurrent mode, it is contacted with a hot gas which dries the droplets by evaporating water from them. The slurry feed rate is determined by the slurry feed facilities, solid level and drying capacity. In general, a countercurrent tower will produce a higher density detergent particle than a concurrent tower. In most towers, the air flow rate and the temperature are adjusted at the entry ports to maintain balanced air and temperature patterns. There are numerous advantages to be accrued from the use of the spray towers to produce powdered detergents. For instance, the product density can be varied, the detergent granule solubility is considerably improved, increased production rates are

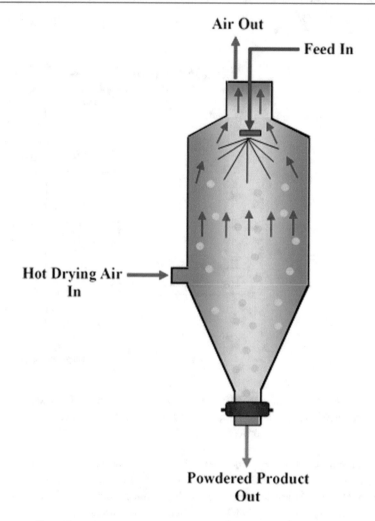

Fig. 1.7 An illustration of the spray drying chamber. The hot air and the feed are fed in a countercurrent mode

achievable and multiple formulations can be obtained using the same equipment. The only major disadvantages of the spray tower are its initial high capital cost, the energy-intensive nature of the process and the relative inflexibility of some spray tower units. The product from the spray tower is usually hot in the temperature range of 75–120 °C and has free surface moisture levels of between 1 and 3%. If the powder is left in this state caking or product flow problems will develop. To avoid this, the hot powder is first conditioned or cooled to an appropriate temperature. As a final step, other additives like perfume and heat-sensitive ingredients, for instance, enzymes, surfactant and bleaches are added to the conditioned product. The post addition is accomplished by metering out the additives and the product from the spray dryer into a rotary drum or some other mixing device where they are thoroughly mixed to a uniform product before packaging. The post addition ingredients usually comprise about 2–15% of the total detergent product.

1.4.1.2 Agglomeration Processing

Agglomeration consists of blending dry raw materials with liquid ingredients in the presence of a binder to form attractive granular product with enhanced aesthetic and flow characteristics. The essential steps in this process are shown in the flowchart of Fig. 1.8.

The dry raw materials are first premixed and thereafter metered into an agglomeration unit where it is sprayed with surfactants or silicates to begin building the agglomerated particles. The wet product from the agglomerator is subsequently conditioned to remove excess moisture and thereafter screened to remove oversized particles. Finally, it is blended with additives such as fragrance, bleaches, enzymes and dyes under agglomeration conditions.

Equipment used for agglomeration include horizontal ribbon and vertical mixers, rotary drums, zig-zag mixers and pan agglomerators. Horizontal ribbon mixers are used primarily for 'dry mix' formulation, where only limited agglomeration occurs. The operation could be batch or continuous. Vertical mixers use high-speed agitation to intimately mix and uniformly blend liquids and dry raw materials prior to

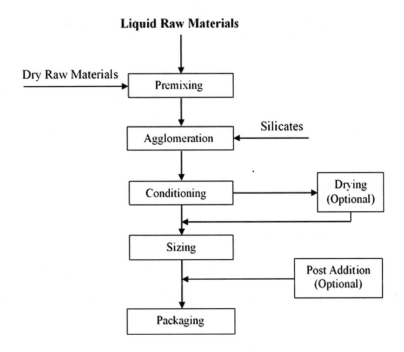

Fig. 1.8 Flowchart for making powdered detergents by agglomeration processing

discharge. The contact time is less than 5 s. Vertical mixers are preferred due to low space requirements and high efficiency of operation. The rotary drum units use baffles that 'roll' or 'lift' the product for uniform dispersion of liquid feeds. Pan and zig-zag agglomerators are used for small-scale specialty product manufacture.

The vertical and rotary drum agglomerators are preferred in the detergent industry because of improved agglomeration, increased liquid feed dispersion levels and optimized product uniformity. These units can also be used for post addition of raw materials.

1.4.1.3 Dry Mixing (Dry Blending) Processing

This process is used to prepare detergents from dry raw materials by mixing them in a suitable unit. The raw materials are first metered and fed into a blender where they are thoroughly mixed to a uniform product. A little amount of water is sometimes added to facilitate the blending process.

1.4.2 Preparation of Liquid Detergents

Liquid products contain approximately 50–60% water, with the remainder being a combination of surfactants, builders, foam regulators, hydrotropes, anti-redeposition agents, whitening agents, corrosion inhibitors, colorants and perfume. They may sometimes also contain antistatic and fabric-softening ingredients. To achieve a stable product, careful selection of ingredients and blending of the raw materials are required. For liquid detergents, the following variables must be carefully controlled:

Viscosity: The product must be pourable and must have consistent viscosity characteristics from batch to batch. This can be achieved by selecting the proper solvent of hydrotrope.

Clear-cloud point: Hazing of product and separation of the product upon storage in cool temperature should be avoided. This can be achieved if the composition has sufficient solubility.

Freeze–thaw stability: The formulation should be such that it does not freeze, nor phase separate at freezing temperatures.

Liquid detergents are produced in relatively simple equipment batchwise or by continuous methods. The raw materials, solid and liquids are carefully metered into a blender where they are thoroughly blended. In order to ensure uniformity and stability of the finished product, stabilizers are added during the manufacturing process. In-line or static mixers are the main equipment used. High concentrated liquid detergents are produced by using high-energy mixing processes in combination with stabilizing agents.

1.5   Soaps and Detergent Manufacture in Ghana

Long before Europeans introduced modern soaps and soap-making processes to Ghana, indigenous peoples made soap from crude palm oil and wood ash—later cocoa pod and plantain peel. The indigenous method essentially consisted of first filtering out solid particulates from an aqueous solution of wood ash and boiling the 'filtrate' with a measured amount of oil until a solid blackish-brown mass, also called potash soap, is formed. This was later cooled and molded for storage and use. Soap making at that time was practiced on a small scale and not as an industry in the sense it is known today. This practice of making soap exists even to today. In particular, potash soap continues to be made and used in some village communities using literally the same kind of raw materials and procedures as was used several hundred years ago. However, a number of factors have combined to make potash soap to lose its appeal to the generality of today's Ghanaian society. In the first place, it is very tedious to prepare the main raw material, caustic potash, from wood ash (mostly from cocoa pod or plantain peel). This has made the availability of this raw material scanty on the market. Again, the shelf life of the potash soap is quite short. Finally, indigenous potash soap has inconsistent quality characteristics. Thus, with the introduction of laundry soap, made from purified caustic soda or caustic potash and vegetable oil or animal fat, most local producers switched over to the production of this brand of soap. Indeed, in Ghana today, laundry soaps of all kinds are made from caustic soda or caustic potash and a variety of animal fat and vegetable oils. Table 1.1 lists the variety of vegetable and animal fats from which laundry soaps can be made. Common vegetable oils used in Ghana for making laundry soap are palm oil, palm kernel oil, coconut oil and shear butter, which are for the most part locally produced. Sodium hydroxide, sodium carbonate, potassium hydroxide and other ingredients for making laundry soap are mostly imported into the country.

In modern Ghana, soaps and detergents are produced commercially by both small-scale enterprises and large-scale companies. The small enterprises mostly produce and sell in small quantities for local and community markets. These produce and market mostly laundry soaps, laundry detergents and household detergents using imported raw materials and rudimentary technology. Most of these companies are not registered, as such data on the extent of these enterprises are difficult to come by yet one finds all manner of products by these companies. Four companies produce soaps and detergents in commercial quantities in Ghana. Unilever Ghana Ltd, located in Tema, was the first to establish soap and detergent manufacture in Ghana as far back as 1948. Today, Unilever produces and markets several brands of soaps and detergents in Ghana and beyond. In Ghana, its flagship products include the Key soap—a bar soap, and the Omo laundry detergents. It also produces and markets the Lux, Fa, Rexona, Sunlight and Geisha brands of toilet soaps. For a long time, Unilever was the only major producer of soaps and detergents in Ghana. In 1976, PZ Cussons, also based in Tema, set up and began production of the Duck brand of bar soap, which is a laundry detergent. PZ Cussons also markets the Imperial leather brand of toilet soap. Another company, the Appiah Menka Complex located in Kumasi, produces and markets the Apeno brand of laundry soap. Finally, Ameem Shangari of Cape Coast also operates a soap plant that produces and sells the Ameem laundry soap on the Ghanaian market. Other commercial soaps and detergent producers, albeit on a small scale, in Ghana include A. Biney and Company Ltd who are producers of the A. Biney brand of liquid soap, disinfecting detergents and stain removers. Also, in this last category is Ray Company limited of Kumasi, which produces and markets the Ray brand of bar laundry soap. What is significant of note among all these producers is that almost exclusively the detergent manufacturers import their raw materials into the country for use. The toilet soap and laundry soap producers on the other hand derive the fats and oils from local market even though the purified caustic soda and other additives are imported into the country.

Further Readings

Ali, M. F., El Ali, B. M., & Speight, J. G. (2005). Handbook of industrial chemistry. McGraw-Hill Education.

Appleton, H. A., & Simmons, W. H. (2007). The handbook of soap manufacture. Fairford: Echo Library.

Bhatia, S. C. (2009). Perfumes, soaps, detergents and cosmetics, vol 1: Soaps & detergents (1st ed.). New Delhi: CBS Publisher and Distributors.

Chattopadhyay, P. K. (2003). *Modern technology of soaps, detergents and toiletries: with formulae and project profiles* (2nd ed.). National Institutive of Industrial Research.

EIRI. (2017). Complete technology book on soaps detergents cleaners and fragrance with formulae. Boi-Green Books.

Flick, E. W. (1989). Advanced cleaning product formulations: Household, industrial, automotive (Vol. 1; 1st ed.). Norwich, New York: William Andrew Publishing

Maine, S. (1995). *Soap book: Simple herbal recipes*. Interweave Press.

NCPS Board of Consultants & Engineers. (2019). Detergents and disinfectants technology handbook (2nd ed.). Niir Projects Consultancy Services.

NIIR Board of Consultant Engineers. (2013). *The complete technology book on detergents* (2nd ed.). NIIR Project Consultancy Services.

NIIR Board. (2013). Handbook on soaps, detergents and acid slurry (3rd ed.). Asia Pacific Business Press Inc.

Showell, M. S. (Ed.). (2005). *Handbook on detergents: Formulation, Part 4* (Vol. 128). CRC Press.

Simmons, W. H., & Appleton, H. A. (2007). The handbook of soap manufacture. Echo Library.

Smulders, E. (2002). *Laundry detergents* (1st ed.). Wiley-VCH Verlag GmbH.

Spitz, L. (2004). *Sodeopec: Soaps, detergents, oleochemicals and personal care products*. Taylor and Francis.

Spitz, L. (Ed.). (2016). *Soap manufacturing technology* (2nd ed.). Academic Press and AOCS Press.

Thomssen EG (2019) Soap making manual: A practical handbook on the raw materials, their manipulation, analysis and control in a modern soap plant. Good Press.

Umbach, W. (Ed.). (1991). Cosmetics and toiletries: Development, production and use (1st ed.). Chichester: Ellis Horwood

Zoller, U. (Ed.). (2008). *Handbook of detergents, Part E: Applications*. CRC Press.

The Extractive Metallurgy Industry

2

Abstract

Several minerals are extracted from the earth, namely, base metals, precious metals and materials such as coal, limestone, salt etc. In this chapter, the mining and processing methods of gold, a precious mineral and bauxite, which is a valuable mineral, are covered in detail. Several processes are involved in mining and extraction of these minerals from the earth's crust. There are similarities in the processing methods, but there are also key differences concerning the processing of the ores. Two methods used to mine gold, which are surface and subsurface mining, are discussed. The later involves digging tunnels or shafts into the earth buried deposits of ore, while surface mining consists of removing surface foliage, dirt and if required layers of bedrock from the land to reach buried deposits. Processing for gold involves a number of unit operations including crushing/grinding of the ore, gravity concentration, amalgamation, retorting, floatation and agglomeration, roasting, leaching/extraction of gold from the ore; these have comprehensively been covered. Gold mining could pose serious risks to personal and the environment, if not handled properly. The impact of mining on the environment and methods of reclaiming the degraded environment are presented in this chapter. The extraction of alumina from the earth crust and the methods and technologies employed in the processing bauxite into alumina have also been covered. Finally, the nature of the mining industry in Ghana, especially, the mining of ore-bearing rock, the extraction of gold from the ore-bearing rocks and the economics of the industry in the country have been discussed.

© The Author(s), under exclusive license to Springer Nature Switzerland AG 2021
O.-W. Achaw and E. Danso-Boateng, *Chemical and Process Industries*,
https://doi.org/10.1007/978-3-030-79139-1_2

2.1 Introduction

The extractive metallurgy industry refers to those industries that are engaged in the mining and processing of valuable minerals from the earth, usually from an ore body, vein or seam. Materials recovered by mining include base metals (e.g. copper, zinc, tin, aluminum, etc.), precious metals (e.g. gold, diamond, silver, etc.), iron, uranium, coal, limestone, oil shale, rock and salt. In Ghana, the valuable minerals mined are gold, bauxite, diamond, manganese and to a much lesser extent silver. Production levels of these minerals in Ghana in 2005 and 2013 are shown in Table 2.1. Since 2008 or so, crude oil and limestone mining have also been taking place in Ghana.

2.2 Economic Impact of the Mining Industry in Ghana

Until the discovery of oil, the mining industry dealt mostly with mineral extraction and, therefore, the industry was synonymous with the extractive metallurgy industry. The statistics on the mining industry therefore right away reflects the actual situation of the extractive metallurgy industry in Ghana for the affected years. Recently, however, mining of limestone and crude oil has assumed significance in the industry. The discussions in this section and indeed the whole chapter, however, focus exclusively on the mining and processing of minerals.

Table 2.1 Production levels of minerals mined in Ghana between 2005 and 2013. (adapted with permission from U.S. Geological Survey by Omayra Bermúdez-Lugo. @2016 USGS)

Mineral	Production levels in 2005	Production level in 2013	Major mine owners	Mine location
Gold	66,852 kg	88,376 kg	AngloGold Ashanti, Newmont Gold Fields, Gold Star, etc.	Obuasi, Takwa, Bibiani, Kenyasi, Abirem, Damang, Bogoso, etc.
Manganese	1,719,589 kg	1,912 million tons	Ghana Manganese Co. Ltd.	Nsuta, W/R
Bauxite	606,700 tons	912,000 tons	Ghana Bauxite Co.	Awaso, W/R
Diamond	1,065,923 carats	169,000 carats	Ghana Consolidated Diamonds (GCD)	Akwatia, E/R
Silver	3,519.04 kg	3,300 kg	NR	NR

ER = Eastern Region. W/R = Western Region. NR = Not Reported

Table 2.2 Income from the mining industry January–December 2005 (adapted from Ghana Minerals Commission, 2007)

Mineral/Producer	US$ million January–December 2005	%
Gold large scale	804.9	81.3
Gold small scale	96.3	
Total (gold)	**901.2**	**91.0**
Diamond large scale	2.9	
Diamond small scale (PMMC)	31.8	
Total (diamond)	**34.7**	**3.5**
Bauxite	14.4	1.5
Manganese	39.2	3.9
Silver	0.8	0.081
Grand total	**990.3**	

The mining industry in Ghana makes a substantial contribution to the economy. Specifically,

(i)   The contribution of Ghana's mining sector to the country's gross domestic product (GDP) increased from 1.3% in 1991 to an average of about 14.0% in recent years.

(ii)   Export earnings from minerals averaged 35%, and the sector is one of the largest contributors to Government revenues through the payment of mineral royalties, employee income taxes and corporate taxes.

(iii)   The minerals sector accounts for about 11% of fiscal receipts by the Internal Revenue Service (IRS).

(iv)   Gold mining remains the highest contributor in the sector, with large-scale gold mining accounting for over 80% by value of the total income from the sector. The other important minerals are diamond, bauxite, manganese and silver (see Table 2.1).

(v)   In 2005, gold production accounted for about 95% of total mining export proceeds. Table 2.2 lists the direct income from the sector between January and December 2005.

(vi)   The sector contributed GH¢17. 1 billion to the country's GDP in 2017 as compared with GH¢15.8 billion in 2016 and over 40% to its total foreign exchange earnings.

2.2.1   Benefit Received from Mining Companies

The main benefits received from mining companies operating in these communities are the following.

(i)   **Mineral royalties**

This is the biggest of all the benefits accruing to the mining communities. It emanates from payments made by mining entities to the government. Mining companies pay 3% of the gross revenue of minerals won during their mining operations. A portion of this amount is recycled into the communities in which mining activities take place through their respective District Assemblies, Traditional Authorities and Stools. Table 2.3 lists the direct beneficiaries of the royalties and the relative amounts due to each beneficiary.

(ii) **Property tax**

Property rates as the name indicates are levies that are imposed on buildings and plants that are fixed to the ground.

(iii) **Operating permit fees**

This is paid by all businesses in a community. It is monies paid by companies, so that they would be allowed to set up to operate businesses.

(iv) **Ground rent**

This is paid to the Office of the Administrator of Stool lands and considered as stool land revenue. It is paid annually. The current rate is one cedi (Gh¢ 1.00) per acre. Table 2.4 lists the recipients and the extent of their benefits from the Ground Rent.

(v)   **Voluntary contributions from mining companies**

These are non-regulatory projects and payments made to the communities where these mines operate.

Table 2.3 Disbursement of mineral royalties (adapted from Administrative fiat of 1999 (letter no. AB.85/156/01), Government of Ghana)

Beneficiary		Share (%) of total amount
Government in Consolidated Fund		80%
Minerals Development fund		10%
Office of the Administrator of Stool Lands		10% of total amount
The Administrator of stool Lands takes 10% of the amount received to cover administrative expenses The remaining 90% is distributed as follows		1%
District assemblies	55%	4.95%
Stools	25%	2.25%
Traditional councils	20%	1.80%
Total		100%

Table 2.4 Disbursement of Ground rent (Stool Lands Revenue) (adapted from 1992 Constitution of Ghana)

Beneficiary		% Share of total amount collected
Office of the Administrator of Stool Lands *The Administrator of Stool Lands takes 10% to cover administrative costs* Remaining 90% is shared as follows		10%
Beneficiary	Share %	
District assemblies	55	49.5
Stools	25	22.5
Traditional authority	20	18.0
Total		100

2.3   The Legislative Regime of the Mining Industry

The overall legislative framework for the mining sector in Ghana is provided by the Minerals and Mining Law of 1986 (PNDCL 153), as amended in 1994 and 2005. In accordance with the Minerals and Mining Law, mining companies must pay royalties and may also pay corporate taxes at standard rates. It also sets up bodies and assigns responsibilities to these bodies as follows:

  i. Under the Minerals and Mining Act, 2006, Act 703, every mineral in its natural state in, under or upon any land in Ghana, rivers, streams, water courses throughout the country, the exclusive economic zone and any area covered by territorial sea or continental shelf is the property of the Republic of Ghana and is vested in the President in trust for the people of Ghana.
 ii. **The Ministry responsible for Mines**
     The ministry now responsible for the mining sector is the Ministry of Forestry and Mines.
     The ministry oversees all aspects of Ghana's mineral sector and is responsible for granting mining and exploration licenses. Within the Ministry;

     • The Minerals Commission has responsibility for administering the Mining Act, recommending mineral policy, promoting mineral development, advising the Government on mineral matters and serving as a liaison between industry and the Government.
     • The Ghana Geological Survey Department conducts geological studies.
     • Ghana National Petroleum Corporation (GNPC) is the Government entity responsible for petroleum exploration and production.
     • The Precious Minerals Marketing Corporation (PMMC) is the Government entity responsible for promoting the development of small-scale gold and

diamond mining in Ghana and for purchasing the output of such mining, either directly or through licensed buyers.

- The Mines Department has authority in mine safety matters.

iii. **The Ghana Chamber of Mines**

This is a private association of operating mining companies. The Chamber provides information on Ghana's mining laws to the public and negotiates with the mine labor unions on behalf of its member companies. All mine accidents and other safety problems also must be reported to the Ghana Chamber of Mines.

2.4 Mining and Mineral Processing Methods

A number of processes are involved leading to the mining and eventual extraction of a mineral from the earth's crust. There are general similarities irrespective of the mineral under consideration, but there are also major differences among these minerals, especially where the processing of the ores is concerned. The processes generally include any number of the following steps, namely:

  i.   Prospecting/exploration to discover the ore.
 ii.   Figuring out the size and grade of the deposit.
iii.   Feasibility study.
 iv.   Construction of mine infrastructure.
  v.   Operation of the mine.
 vi.   Processing of ores where necessary.
vii.   Refining of the mineral.
viii. Reclamation of mined land.
 ix.   Treatment of and/or disposal of waste.

The specific processes for the extraction of specific minerals are discussed in the following sections.

2.4.1 Gold Mining in Ghana

2019 data indicate that Ghana ranks as the world's 8th gold producing nation. In Africa, Ghana has surpassed South Africa to become the leading gold-producing nation. Gold production in Ghana is undertaken by both large-scale and small-scale mining companies. These companies jointly produced 2.81 million ounces of gold in 2017. According to some estimates, the small-scale and artisanal mining account for about 34% of the gold output in Ghana and about 6% of Ghana's GDP. The top ten leading gold-producing companies in Ghana in 2018 are listed in Table 2.5. Without exception, all these companies also process the gold-bearing ore into refined gold. Besides those presented in Table 2.5, there are other companies

Table 2.5 The top ten (10) gold producing companies in Ghana in 2018 (adapted from Ghana Chamber of Mines, 2019)

Mine company	2018 rank	Ounces of Au	2017 rank	Ounces of Au	2016 rank	Ounces of Au
Gold Fields, Tarkwa	1	524,869	1	566,400	1	586,036
Newmont Ahafo	2	436,106	3	349,000	3	348,861
Newmont Akyem	3	414,427	2	473,000	2	470,313
AngloGold Ashanti Iduapirem	4	253,487	5	228,000	4	214,196
Kinross Chirano	5	226,370	4	246,027	5	211,954
Asanko Gold	6	223,152	7	205,047	6	162,802
Perseus Edikan	7	217,219	6	208,226	7	153,208
Abosso Gold fields Damang	8	180,844	8	143,585	8	147,720
Golden Star Wassa	9	149,698	9	137,234	9	114,380
Golden Star Prestea	10	75,087	10	130,331	10	89,673
Endeavor Nzema	–	–	11	115,621	11	87,710
Adamus Resources	11	103,731	–	–	–	–
Total		2,806,597		2,802,247		2.558,853

currently exploring for gold in Ghana and who, it is hoped would mine and process gold ore in the future. Missing from Table 2.5 is AngloGold Ashanti Obuasi mine, which until 2019 was closed down but has been resuscitated in 2019.

Gold is mined in Ghana through one of two methods, namely, surface mining and subsurface mining. *Surface mining* involves removing (stripping) surface vegetation, dirt and, if necessary, layers of bedrock from the land in order to reach buried ore deposits. Surface mining techniques are open-pit mining, strip mining and mountain top removal. During subsurface mining tunnels or shafts are dug into the earth to reach ore deposits buried in the ground. The deposits consisting of ore-bearing rock and waste are excavated and brought to the surface through the tunnels or shafts. At the surface, the ores are separated and taken to processing plants, while the rock debris and other waste materials are disposed of. Subsurface mining is classified in three ways, namely, the access shafts used, the extraction method used and the technique used to reach the ore deposit. Based on access shafts, subsurface mining can be grouped into, *drift mining* (horizontal tunnels), *shaft mining* (vertical tunneling) and *slope mining* (uses diagonally sloping access shafts).

2.4.2   Unit Operations in Gold Processing

In Ghana, with the exception of the deposits in Bonte in the Ashanti Region, all the other deposits occur in an ore form and require comprehensive physical and chemical treatment processes before the commercial gold product is produced. The gold deposits occur mostly as oxide ores and/or refractory ores. *Oxide ores* are those in which the ore material has been well oxidized or weathered such that in its free milling state about 95% of the gold in the ore can be extracted by cyanidation when the ore is ground such that about 80% of the milled particles have a size of less than 75 µm. *Refractory ores* apply to ores where direct treatment by cyanidation gives unacceptably low gold recoveries. The low recoveries associated with refractory ores may be due to any number of the following reasons:

i. The gold may be locked up in unoxidized minerals, often as solid solution within sulfides, and cannot be adequately liberated, even by fine grinding.
ii. The gold occurs in the presence of other minerals, e.g. pyrrhotite, arsenopyrites and others that consume unacceptably high quantities of reagents.
iii. The gold occurs in the midst of carbonaceous materials, which adsorb gold during leaching.

The extraction of gold involves several process steps depending on the nature of the ore in which it is found. The method adopted generally depends on:

i. the mineralogical characteristics of the ore, i.e. whether the ore is an oxide ore or a refractory ore;
ii. the concentration of gold in the ore;
iii. whether it occurs in the native form or not.

The major and often applied unit operations in the processing of gold ores are comminution (size reduction which is carried out by crushing and milling), concentration (floatation, gravity settling, etc.), extraction and purification. For refractory ores, additional processing steps like acidic pretreatment, roasting, and hydrometallurgical oxidation may be required. Generalized flow sheets for the extraction of gold from oxide and refractive ores are shown in Figs. 2.1 and 2.2, respectively. The significance of the various unit operations and the means of attaining them are explained after the flow sheets.

2.4.2.1 Crushing/Grinding

Ores often occur in nature as big lumps that are not directly suitable for most extraction processes. However, most concentration processes, e.g. leaching and floatation, are feasible and effective only when small-sized particles are used. Therefore, the unit operations of crushing and grinding are carried out to get the ore into sizes that will render them useful for subsequent operations. The process of reducing the size of ore particles is termed *comminution*. The two common unit operations employed are crushing and grinding. Comminution to particle size of about 6 mm (maximum) is termed *crushing*. The process of further reducing the particle size to less than 6 mm is referred to as *grinding*. The two operations are done using crushers and grinders, respectively.

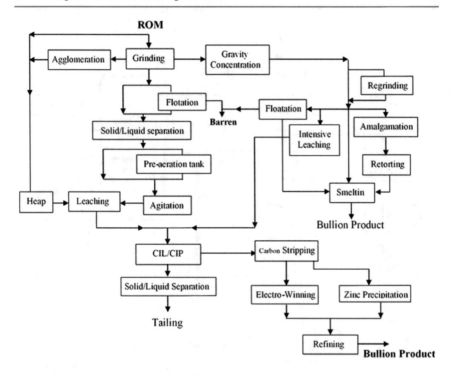

Fig. 2.1 Typical flow diagram of process routes for the extraction of gold from oxidized ores

Crushers may be classified into four groups, namely, (i) those for coarse or primary breaking, e.g. jaw and gyratory crushers; (ii) those for intermediate or secondary crushing, e.g. cone crushers, disk crushers, reduction gyratory crushers and spring roll crushers; (iii) those for fine crushing, e.g. gravity stamp mills and (iv) those for special uses. *Grinding* may be done in one of two ways, namely, closed circuit or open circuit operation. In open circuit, the material to be ground is fed once through the grinder. In closed circuit, a series of cyclones is used to separate the particles such that larger particles are returned to the grinder. In this latter operation, only particles that meet a required grade emerge out of the system. Trash including wood, fiber and plastic are directed to the overflow of the cyclone.

2.4.2.2   Gravity Concentration

Gravity concentration makes use of the difference in density of materials to separate them. In the gold mining industry, it is widely used for the recovery of free gold, whose relative density is much higher than that of the barren material accompanying it. Separation by gravity concentration is achieved by the use of sluice boxes. The sluice box is an open channel with a riffled bottom. The ore to be processed is mixed with water at the head of the sluice. The slurry runs down the sluice and the gold, being heavier than the other minerals present, settles in between the riffles

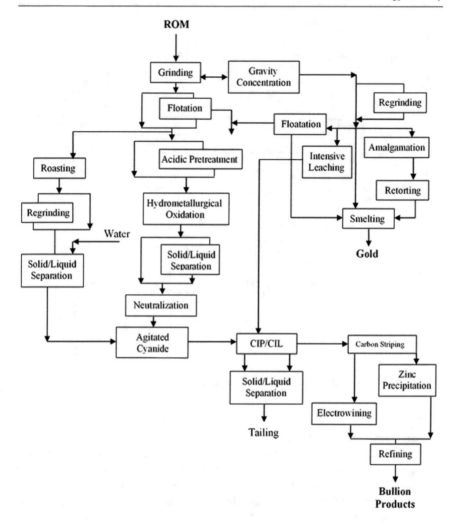

Fig. 2.2 Generalized process routes for gold extraction from refractory ores

while the lighter components of the ore are carried away by the flowing water. The difference between the specific gravity of gold (r.d.19.3) and that of the gangue (2.6–2.75) results in a satisfactorily high degree of concentration and of extraction, where gold is found in particle sizes between 600 and 30 μ, gravity concentration remains effective. The method is not viable for treating ore grades below 10 g Au/t or where a high proportion of gold is locked in sulfides. The sluice box is a batch-type machine and must, therefore, be periodically shut down, so that the gold concentrate that settles down is collected for further processing. The efficiency of this method depends on the rate of removal of the gold concentrate from the sluice box. The recovery efficiency is high if the concentrate is removed more often.

2.4.2.3 Amalgamation

In spite of environmental and human health concerns about the use of mercury, the method is still used in the recovery of gold from gravity concentrates since there are no suitable alternatives. However, its application in industry has considerably reduced. The amalgamation process takes advantage of the *selective solubility of gold in mercury* to separate it from other components of an ore. In one variation of its application, ore is crushed and ground in stamp mills, and later in tube mills. The crushed ore, as fine as talcum powder, is hydrated to form a viscous pulp that is subsequently passed over copper plates (4.57 m long by 1.52 m wide) at an incline of about 18%, which is coated on their upper surface with mercury. By virtue of its high relative density, the gold sinks through the pulp to contact with the mercury and is caught and converted into amalgam. This amalgam (a solution of gold in mercury that is in the form of a thick paste) is scraped off the plates and retorted to yield gold sponge. Despite regular scraping, hard layers of gold amalgam accumulate on the plates. This can be removed by softening the amalgam with steam, which allows it to be scraped off without exposing the copper. In Ghana, this method is widely used by indigenous small-scale gold prospector operators also called 'galamsey' operators.

2.4.2.4 Retorting

The gold recovered after the amalgamation process must be rid of the mercury before smelting. Also, the mercury when recovered can be recycled to extract more gold. Again, if not removed from the gold, the vapor from the mercury, which is considerably toxic, represents hazard during smelting. Mercury has a higher vapor pressure than many other metals. This observation is exploited in separating mercury from other precious metals by a simple distillation procedure. The distillation is carried out in units called retorts specially designed for mercury distillation. A retort is a closed box with a long narrow bent spout. The boiling point of mercury is 357 °C, and, typically, temperatures of 600–700 °C are applied to vaporize the mercury. The retort temperature is ramped up gradually to enable the amalgam to dry completely before the mercury is vaporized and to allow time for the diffusion of mercury to the solid surface. The retort temperature is held at 600–700 °C for 2–3 h to ensure complete evaporation. Mercury removal efficiency in excess of 99% can be achieved in this manner. Retorts are normally operated under a slightly negative pressure and the mercury vapor is exhausted into water condensation system where it is cooled below the boiling point of liquid mercury and is collected to avoid re-evaporation. The mercury produced has little quantities of gold, silver and other metals but is usually of sufficient purity to be reused for amalgamation. Retorting of mercury amalgams yields a mercury-free product in the form of sponge gold, which can be treated directly by smelting for the removal of residual base metal impurities.

2.4.2.5 Floatation and Agglomeration

These are ore concentration processes that make use of the differences in the ability of mineral particles to adhere to the surface of air bubbles to separate them. The two

operations differ from each other not in the fundamental physicochemical phenomenon of selective adhesion, but in the physical means that are employed to put
selective adhesion to practical use. During flotation (froth flotation), adhesion is
affected between gas bubbles and small particles in such a way that the specific
gravity of the mineral–air association (mixture) is less than that of the pulp, so that
they rise to the surface of the pulp. The floating mineralized froth is, subsequently,
mechanically skimmed off from the pulp. In another variation of floatation termed
skin floatation, the ore particles adhere to the free water surface. The particles in
skin flotation are larger than those involved in froth flotation. In agglomeration, the
association formed between the particles and bubbles is heavier than the water but
lighter and coarser than the particles adhering to the water. What this means is that
the agglomerates settle on top of the non-agglomerated particles. The former is
separated from the non-agglomerated particles by a flowing film gravity concentration method. An illustration of the flotation process is shown in Fig. 2.3.

In practice, flotation/agglomeration is brought about by agitating an ore pulp in a
vessel and blowing air through it. In the process, froth is produced. Some ore
particles attach themselves to the air bubbles and are concentrated in the froth,
which is skimmed off at the surface (in the case of flotation) or settles on top of the
gangue in the case of agglomeration and is subsequently separated by flowing film
gravity concentration. To stabilize the froth formed, that is, to slow down or prevent
the froth from breaking up and thereby disrupting the process, substances called
frothers are sometimes added to the pulp. Substances that have the alcohol, ketone,
aldehyde and phenol functional groupings are most suitable as frothers. Examples

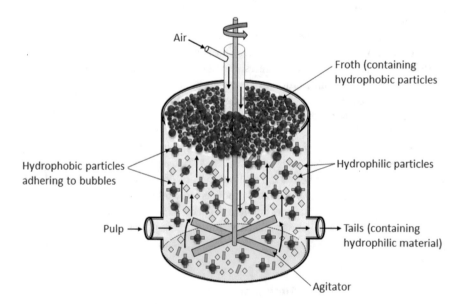

Fig. 2.3 An illustration of the flotation process

of frothers used in the gold extraction industry are pine oil, cresylic oil, 'fused oils', higher alcohols, eucalyptus oil and camphor. Froth flotation is one of the most outstanding mineral beneficiation techniques in the mining industry.

2.4.2.6 Roasting

This is the process by which a refractory ore is chemically converted to an ore form that is amenable to leaching by application of heat in the presence of oxygen or some other element or reagent. Refractory ores are difficult to convert directly into the metal. Also, recoveries during leaching from refractory ores are very low. To avert these, a refractory ore is first rendered into a form such that subsequent treatment for the extraction of the metal, for example, leaching, reduction and other processes would be facilitated. This can be achieved by roasting the ore. Roasting may be done with or without prior concentration by flotation of the ore. The products of the roasting operation are oxides, chlorides, sulfates of the metal and volatile oxides of sulfur, arsenic and tellurium. Roasting is different from calcinations, whereas the latter seeks to decompose the ore, the former only converts the ore to a different chemical form. Oxidizing roasting is the most important roasting process in the extractive metallurgy industry. It proceeds as follows:

$$2MS(c) \; + \; 3O_2(g) \; \underset{\Delta}{\rightarrow} 2MO(c) \; + \; 2SO_2(g) \qquad (2.1)$$

MS = Mineral sulfide; MO = Mineral oxide.

2.4.2.7 Acidic Pretreatment

An alternative process to rendering a refractory ore amenable to leaching is by a special chemical treatment of the ore. One such step is the acidic pretreatment using oxidizing reagents such as nitric acid.

2.4.2.8 Hydrometallurgical Oxidation

An alternative to roasting as a means to rendering refractory ores amenable to subsequent gold extraction steps is hydrometallurgical oxidation. Two such processes are *pressure oxidation* and *bio-oxidation*. In pressure oxidation, a slurry of the refractory ore is subjected to high pressures and temperatures in an autoclave. In the process, the gold in the ore is leached into solution and after neutralization the gold can be subsequently recovered by cyanidation or Carbon-In-Leach (CIL). Bio-oxidation is a pre-cyanidation treatment of refractory gold ores or concentrates. In bio-oxidation, a refractory sulfide run-of-mine (ROM) ore or concentrate is contacted with a strain of a bacterial culture for a suitable treatment period under optimum operating environment. The bacteria oxidize the sulfide minerals, thus liberating the occluded gold for subsequent recovery via cyanidation. The bacterial culture used is a mixed culture of *Thiobacillus ferrooxidans, Thiobacillus thiooxidans* and *Leptospirillum ferroxidans*. For optimum activity, the pH of the medium must be between 1.0 and 2.0, at a temperature of between 30 and 45 °C, and a steady supply of oxygen and carbon dioxide.

2.4.2.9 Leaching/Extraction of Gold from Ore

Leaching is a gold extraction process that aims at producing a gold solution as an intermediate product by dissolving out the gold from a gold-bearing ore using an appropriate reagent (see Fig. 2.4). In current industrial practice, leaching is effected using an alkaline cyanide solution. Gold dissolution can also be effected by

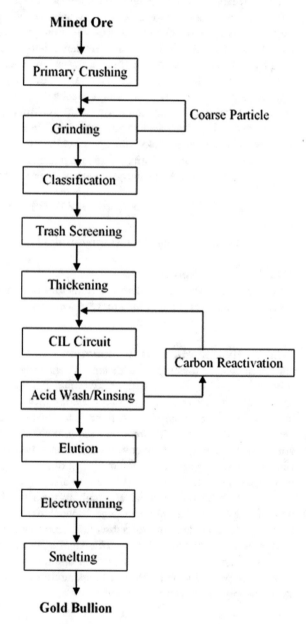

Fig. 2.4 A schematic of gold extracting processes at a gold mine in Ghana

chlorinating the ore at high temperatures and pressures. However, the chlorination method has not caught with industry because of the problems associated with the handling of chlorine, namely, corrosion of vessels and equipment parts and environmental concerns. Cyanidation is the most widely employed method used in the extraction of gold. In the cyanidation process, dilute alkaline cyanide (NaCN) solutions are used to dissolve (leach) gold from gold-bearing ores. The manner of the leaching of the ore is applied in one of the following forms, also depending on the nature of the ore:

i. Agitated leaching

This is used for ground ore slurries and for retreating old mine tailings, which contain residual minerals. The resulting product is either subjected to solid/liquid separation to recover gold or may be treated 'in-pulp' with carbon (CIP) or resin (RIP) for gold recovery. The in-pulp process can be incorporated into the leach circuit. This latter process is referred to as Carbon-In-Leach (CIL) or Resin-In-Leach (RIL), respectively.

ii. Heap leaching

The heap leaching process involves placing the ore in a stationary heap and making a solvent (cyanide) to percolate through the solid. The primary benefit of heap leaching is that it can be carried out with little size reduction and, hence, significant cost savings. In general, ore of selected rock size is placed on a *leach pad*, which is saturated with a weak cyanide solution. The pads are constructed by laying impermeable plastic sheeting over a gently inclined, compacted graded soil. A thin layer of sand is first placed onto the plastic sheeting before the ore is stacked onto the leach pads. The layer of sand not only serves as an insulator but also protects the leach pad. The gold (and other metals) is then leached from the stack where they are gathered in a recovery pond for subsequent processing. For single-stage leaching, two steps are involved, namely,

- contact of solid and solvent during which the solute is transferred to the solvent from the solid;
- separation of the resulting solution from the residual solid.

The solvent phase is called the extract. The residual wet solid material is termed the raffinate. The solute can be found in both the solid and liquid phases in the raffinate.

iii. Vat and intensive leaching

Vat leaching is essentially a flooded heap leach in a suitable vessel. Intensive leaching involves leaching at high temperatures and pressures. These are rarely used leaching procedures due to the high costs involved.

2.4.2.10 Recovery of Leached Gold

Leaching gold ores and concentrates with cyanide result in a solution-containing anionic metal cyanide complexes from which gold must be recovered. Traditional gold recovery was achieved by precipitation from solution. Now, and increasing so, the carbon-in-pulp technology is used.

i. Cementation (precipitation) method

The most important method of gold recovery from dilute solutions arising from liquid–solid separation is zinc precipitation. In this process, the pregnant gold solution is contacted with a dust of zinc and pumping the resultant emulsion through a filter. The gold is precipitated out of the solution by the zinc and is filtered using frame filters. The gold precipitate and excess dust remain in the filter frames while the barren solution passed through the canvas filter cloth into a storage tank for reuse in the cyanide plant. By connecting the canvas filter to a vacuum, the process can be improved as this allows the removal of oxygen from the pregnant solution before adding the zinc dust. Vacuum is achieved by connecting the leaves of the filter to a pipe leading to the suction of a centrifugal pump. The pregnant solution is first gravitated to a Crowe tank, where *zinc dust* and *lead nitrate* are added continuously and then into a circular tank, in which the canvas frames are completely immersed.

ii. Adsorption with activated carbon (carbon-in-pulp)

The CIP method makes use of the tremendous physical affinity 'activated' carbon has for gold (it can attract 7% of its weight in gold), which readily attracts to its surface in a cyanide solution. The finely ground ore (typically about 75 μm particle size) and the slurry of fine ore and water (the 'pulp') are treated with cyanide in large tanks that are stirred mechanically or by air agitation. Activated carbon is used to adsorb the gold directly from the cyanided pulp, which flows continually through a number of vessels in series, and the carbon is transferred intermittently by pumping in the opposite (countercurrent) direction. The quantity of gold in the pulp decreases downstream of the process, while the gold loading on the carbon increases upstream. Therefore, the gold loading in the first tank has the highest value. The CIP process consists of three essential stages: *adsorption* in which the dissolved gold in the pulp is loaded onto aerated carbon; *elution*, in which the gold is removed from the carbon into an alkaline cyanide solution and *electro-winning*, in which the gold is removed by an electrical process from the alkaline cyanide solution and deposited on steel wool electrodes. The carbon is then treated with acid to remove contaminants, after which the acid itself is treated. Both are then recirculated into the adsorption–elution circuit. When the leach and adsorption circuits are combined, the process is described as carbon-in leach (CIL).

iii. **Carbon stripping**

The gold that is recovered from the leach solution by adsorption onto activated carbons must be desorbed from the carbon for subsequent processing to win the gold. The process of desorbing the gold from the carbon is called striping. The process is based on identifying and using a medium in which the affinity between the carbon and gold cyanide complex is sufficiently weakened to detach the metal ion complex into solution. The commonly used stripping methods are listed in Table 2.6.

Table 2.6 Methods of stripping gold from carbon

Stripping method	Description of method
Atmospheric Zadra	The method was named after J. B. Zadra who first developed the process in the USA. The process consists of circulating 1% sodium hydroxide (aq) and 0.1% cyanide solution (aq) upflow through a stationary bed of loaded carbon at a flow rate of about 2 bed volumes per hour at about 93 °C. Gold that was previously adsorbed is desorbed from the carbon into the solution by a reversal of the adsorption kinetics. The process takes about 48–72 h before the gold content of the carbon can be reduced to about 3 oz. Au/ton of carbon. The pregnant solution is sent to an electrowinning cell, where the gold is recovered from the solution. The solution leaving the cell, now depleted of gold, is heated and sent back to the carbon bed to be reused. Similarly, the carbon is reactivated by heating and recycled. A drawback of this method is its low rate of desorption. Indeed, it is much slower than the alternatives. The slow rate of adsorption necessitates larger carbon inventories and larger equipment than other faster processes. Another disadvantage of the process is the buildup of ions in the eluant after continuous recycling. This results in the reduction in stripping efficiency of the eluant. To alleviate this problem, most operations routinely bleed a fraction of the strip liquor inventory and replenish with fresh solution.
Pressurized Zadra	This method is a variant of the atmospheric Zadra method and is based on the observation that at higher temperatures than that applied in the Atmospheric Zadra process, it is possible to increase the rate of gold desorption. However, at higher temperatures, the process must be operated under pressure, hence, the name Pressurized Zadra Stripping. In practice, a solution containing about 1% sodium hydroxide and 0.1% sodium cyanide at about 138 °C and 4.4 atm is circulated through a pressure vessel filled with loaded carbon at a flow rate of 2.0 bed volumes per hour. The time required for pressure stripping is between 10 and 14 h. It is observed from this method that it is possible to increase stripping efficiency by increasing the process temperature up to 180 °C beyond which the cyanide begins to decompose. As such, most plants applying this method often fix the upper temperature for the process at 149 °C.

(continued)

Table 2.6 (continued)

Stripping method	Description of method
Anglo American Research Laboratory (AARL)	This method involves first, washing the gold-adsorbed carbon with an acid solution followed by a freshwater wash to remove residual acid. The carbon is then soaked in a hot solution (at a temperature of 90 °C) of 3% sodium cyanate and 1% sodium hydroxide for about 30 min. High-quality fresh water at about 110 °C is then pumped through the pressurized stripping vessel to produce a pregnant eluant. The need for a pressurized vessel is because of the high temperatures required of the eluant water. Gold is recovered from the pregnant eluant by electrowinning, and the barren eluant is discarded.
Alcohol Stripping	When alcohol is added to the conventional Zadra solution, better desorption rates are gotten. Alcohols found to be effective are ethanol, methanol and isopropyl alcohol. The main disadvantage of the alcohol stripping process is the potential for fires. In particular, the electrowinning process has the potential to develop sparks and is, therefore, very susceptible to catching fire. Ethylene or propylene glycol is frequently used instead of alcohol to increase the speed of atmospheric pressure Zadra stripping. Also, the glycols are inflammable. A typical protocol involves passing a solution of 20–25% ethylene or propylene glycol and 2% (wt.) sodium hydroxide over a bed of carbon after heating to about 88 °C. The flow rate of the solution is 2 bed volumes per hour. Glycol consumption is about 20–40 gallons per ton of carbon stripped.
Micron Elution	This method consists of pretreating the loaded carbon with a caustic cyanide solution followed by elution with an alcohol mixture. The elution unit functions like a distillation tower. The length of the column is loaded with carbon. Base of the tower is a heater, and at the top, an overhead condenser and a reflux pump. Loaded carbon is first presoaked with sodium cyanide/sodium hydroxide solution at ambient temperature. After the presoak solution is drained from the carbon bed, alcohol is added at the base of the column and heated. First, the vapor of the alcohol passes through the carbon bed and subsequently a condensate of the vapor is collected and returned through the bed to elute the gold.

iv. **Carbon regeneration**

An important component of this technology is the recycle of the activated carbon. Often after the leach circuit, it would have lost most of its activity. However, the improved economy of the process requires that it be regenerated and reused. This is achieved by reactivating the stripped carbon by the passage of high-temperature steam through a bed of the carbon and subsequently heating it in a kiln. After this, it can now be sent to the beginning of the leach cycle to be reused.

v. **Electrowinning**

This process uses the principles underlying electrolysis and the electromotive series to extract metals from electrolytic solutions by the passage of direct electrical current through the solutions. Gold winning takes place in electrowinning cells. The cells contain the electrolyte solutions of metal ion complexes of gold and cyanide and carry electrodes (a cathode and an anode). Usually, the cathodes are of steel wool and the anodes are made of stainless steel. Current is passed through the electrolyte through the electrodes. The eluate produced during the elution process is pumped through the *electrowinning cells* where the gold plates onto the steel wool cathodes as gold particles. Since silver is often found together with gold in the ore, the gold-plated cathodes often contain silver which alongside other impurities must be removed from the gold. High-pressure sprays are used to remove the deposited gold particles together with other metal impurities from the cathodes. The slurry of gold and impurities are subsequently dried and then smelted in an induction furnace at 1400 °C. Fluxes are usually added to the gold particles before smelting to enhance the separation of impurities from the gold. The fluxes form a slag of impurities that are removed. The molten gold is poured into molds and allowed to cool down. The gold bars are finally removed from the molds, cleaned, stamped and sold. The rate at which gold can be extracted (i.e. plated onto the cathode) from solutions depends on several factors, including the concentration of gold in the electrolyte, the size of the unit in terms of current and cathode area and the presence of other metal species in the electrolyte solution.

2.5 Environmental and Safety Issues in the Extractive Metallurgy Industry

2.5.1 Environmental Effects of the Extractive Metallurgy Industry

Beginning from the point where the ore is mined from the earth down to the stage when the gold is finally purified by smelting, the processes involved impacting the environment in ways that are not always favorable. The effect of the activities leading to extraction of gold may be grouped into two, namely, the impact of mechanical activities like removal of earth during the extraction of ore-bearing rocks and the impact of chemicals and tailings during the extraction of gold from the ore. Specifically, the effects of these activities include:

i. Erosion of exposed hillsides, mine dumps, tailings dams and resultant siltation of drainages, creeks and rivers significantly impact surrounding areas and communities.
ii. Destruction of wilderness and attendant disturbance of ecosystems and loss of biodiversity.

iii. The dumping of the runoff in surface waters or in forests poses an equal danger to the environment.

iv. In farming areas, activities of mining may disturb or destroy productive grazing and croplands.

v. Mining may produce noise pollution, dust pollution and visual pollution, which are nuisance and harmful to nearby settlements and workers.

vi. Contamination of the area surrounding mines due to the potentially damaging compounds and metals removed from the ground with the ore is always a threat.

vii. Contamination of soil, groundwater and surface water by chemicals employed or exposed during the extraction of gold. The contamination arising from the leakage of chemicals affects the environment and health of the local population. The result can be unnaturally high concentrations of some chemicals such as arsenic, sulfuric acid and mercury over a significant area of surface or subsurface.

In view of the foregoing, after the ore has been mined and the gold extracted, a number of outstanding issues remain that must be dealt with before closure of the mine. Among these are how to treat the degraded land, how to treat the cyanide-bearing wastewater before discharge into the environment or recycled (where necessary), how to treat the ore tailings (spent ore) before discharge and in the case of heap leach processes what to do with the residual heap. Necessarily, these issues must be addressed because: (i) a degraded land if not reclaimed could mean a lost land or a long period of fallow before land can be reused. Also, unclaimed land can pose danger to those using the land as huge manholes could become traps to those trespassing the land; (ii) cyanide contaminated soils and water post danger to flora, fauna and residents of affected communities because cyanide is highly poisonous if taken in; (iii) Government regulation requires that standards be met in land state and cyanide remnants in soils and water before mines can be closed.

2.5.1.1   Reclaiming Degraded Land

Land rehabilitation attempts to return land affected by mining to some degree of its former state. Efforts employed to restore land degraded by mining activities include:

i. Waste dumps are flattened and stabilized against erosion.

ii. If the ore contains sulfides, it is usually covered with a layer of clay to prevent access of rain and oxygen from the air, which can oxidize the sulfides to produce sulfuric acid.

iii. Top soil is used to cover landfills and later vegetation is planted on it.

iv. Vegetation cover of dumps is protected from denudation by fencing them against intrusion by livestock.

v. Dams and tailings are first left to dry up. They are then filled with waste rock, followed by clay if necessary, and then soil. Finally, it is planted to stabilize it with vegetation.

2.5.1.2   Treating Cyanide Contamination

Cyanide contamination of soil and water is another key environmental problem that the gold extraction industry must deal with. Treating cyanide contamination requires a consideration of many issues, like what to use to perform the treatment. This depends on the form of cyanide, the nature of the medium in which the cyanide is, availability of the requisite materials to treat the cyanide, the economy of the process to be adopted to do the treatment and finally government regulation. Other considerations are the method of treatment to be adopted, how long the treatment will last, the precaution to be adopted during treatment and the means of assessment of the efficiency of treatment. An overview of methods for treating cyanide is presented in Table 2.7.

Measurement of cyanide concentration in a medium depends on the form of the cyanide. Cyanide is usually measured in three forms. These are free cyanide, which is the cyanide anion (CN^-) present in solutions and usually bonded to cations such as Na, K, Ca or Mg or HCN. In general, free cyanide is difficult to measure. The second is in the form of weak acid cyanide (WAD), which are cyanide complexes that upon treatment with a weak acid (pH = 4.5) gives off HCN. It includes free cyanide, simple cyanide and weak cyanide complexes of zinc, cadmium, silver, copper and nickel. The third form is total cyanide, which includes all forms of cyanide (free and complex) including iron, gold, cobalt and platinum complexes. Not all methods can measure each of these forms of the cyanide. Indeed, most measurement techniques are unable to accurately determine the cyanide concentration in a medium.

2.5.2   Safety

Subsurface mining has always had issues to contend with. Prominent safety concerns in the mining industry include:

i.   Exposure to harmful gases, heat and dust inside subsurface mines, which are the result of poor ventilation, can lead to harmful physiological effects, including death.

ii. Gases in mines can also poison the workers or displace the oxygen in the mine, causing asphyxiation. For this reason, some countries require that mine workers use gas detection equipment in groups of miners. The equipment must be able to detect common gases such as CO, O_2, H_2S and the percentage Lower Explosive Limit of these gases.

iii. High temperatures and humidity may result in heat-related illnesses, including heat stroke, which can be fatal.

iv. Dusts in mines can cause lung problems such as silicosis, asbestosis and pneumoconiosis.

Table 2.7 Methods of treatment of cyanide

Method	Description of process
Rinsing	Rinsing is applied to cyanide found in heap leaches. Freshwater or recycled (treated) water is sprinkled over the heap over a time to dissolve or extract cyanide in the heap. The rinsate is subsequently treated. Chemicals may be dissolved in the water to oxidize and facilitate the extraction of cyanide from the heap. The water may contain dissolved chemicals that could oxidize the cyanide and facilitate its extraction from the heap. In this case, further treatment of the rinsate may not be necessary.
Sulfur process	Cyanide (CN^-) in solution is oxidized into cyanate (CNO^-) using sulfur dioxide (INCO Method) or ferrous sulfate (Noranda Method) in the presence of copper ions. This later method is suited for ore with a significant concentration of arsenic and antimony. The former method is slow at low temperatures and does not remove cyanate, ammonia or thiocyanate and does not sufficiently remove toxic metals.
Alkaline chlorination	Chlorine or a hypochlorite solution is used to oxidize the cyanide to cyanate. The method is applicable to clear wastewaters and slurries. The pH is maintained in the alkaline range using lime. Precipitated metals are removed in a clarifier before discharging the wastewater. The method does not remove iron cyanides. Also, it leaves chloramines and free chlorine in the solution that is toxic to fish.
Hydrogen Peroxide process	Hydrogen peroxide is used to oxidize cyanide to cyanate in the presence of copper ions. The method is applicable to wastewater. When applied to slurries, reagent requirements increase. Ammonia, which is a by-product of the process, is toxic to fish. The method requires specially designed equipment as hydrogen peroxide is a strong oxidizer and gives rise to explosion and fire in the presence of combustible materials. Also, it is expensive and, therefore, has cost implications.
AVR Cyanide recovery process	The method is applicable to barren solutions. Sulfuric acid is added to the cyanide-containing solution to generate HCN, which is then absorbed in NaOH to generate NaCN. The NaCN can then be recycled.
Natural degradation	When left unattended natural processes combine to degrade cyanide. The processes include microbial generation of ammonia and cyanate in soil; volatilization of cyanide in solution after absorbing CO_2 and SO_2 from atmosphere and consequent formation of acid; cyanide hydrolysis in soils; anaerobic degradation of cyanide; complexation with stable metal ions. Efficiency of cyanide degradation may be slower at the bottom of ponds and in the interior of heaps.
Biological treatment	Microbial agents, e.g. *pseudomonas pseudoalcaligenes,* are applied to treat the cyanide. The microbes transform cyanide, first, to ammonia and then to harmless nitrates. The method can be used to treat barren solutions, cyanide wastewater and heap leach pad solutions.

v.   Since mining entails removing dirt and rock from its natural location creating large empty pits, rooms and tunnels, cave-ins are a major concern within mines as they result in accidents.

vi.   The presence of heavy equipment in confined spaces also poses a risk to miners, in spite of modern improvements to safety practices.

vii.   Without sufficient knowledge and proper training anyone who attempts to explore abandoned mines face the threat of danger. Stagnant water in abandoned mines may hide deep pits and trap gases. These pose a significant hazard to intruders.

viii.   Entrance to an old mine can be very dangerous as surrounding rock and earth may have been weakened by the weather.

ix.   Old mine workings and caves often lack oxygen in the air making them hazardous. This condition in mines is called *blackdamp*. Again, old mines can contain deadly gases making them very dangerous.

x.   The noise made by the heavy equipment used in the mines combined with the confined workspace of underground mines can cause hearing loss in miners.

In Ghana, the Mineral Commission through its Mining Inspectorate Division has the responsibility to seeing that players in the industry comply with safety regulations. Among others, the Commission is responsible for:

i.   Enforcing compliance with occupational health and safety regulations.

ii.   Inspection of areas of mining operations to ascertain whether any nuisance is created in the area by the mineral operation.

iii.   Control of mines, mills and other mineral treatment plants, ensuring that wasteful mining or ore treatment is not used in these operations.

iv.   Examination and enquiries into mining-related accidents and incidents.

v.   Compilation of mine accidents statistics and dissemination of information thereon.

vi.   Technical inspection, control and enforcement of technical regulations for mining.

vii.   Inspection of distribution, storage and handling of explosives, enforcement of Explosives Regulations both within and outside the mining sector.

viii.   Training of personnel handling explosives.

ix.   Certification of mine management, certification for blasting and operations of winding engines.

x.   Technical and Health and Safety Supervision of small-scale diamond mining.

2.6   Bauxite Processing

Alumina is the base material from which aluminum is made. Alumina itself is extracted from bauxite, which is a naturally occurring mineral ore. Rocks that contain high proportions of aluminum hydroxide are called bauxite. Aluminum occurs in three different mineral forms in bauxite, namely,

Gibbsiteα-Al$_2$O$_3$.3H$_2$O
Diasporeβ-Al$_2$O$_3$.H$_2$O
Boehmiteα-Al$_2$O$_3$.3H$_2$O

Other mineral forms that occur alongside aluminum in bauxite are oxides of iron, titanium and silica. Table 2.8 lists the minerals commonly found in bauxite.

The extent of the presence of these other minerals and the crystal form of alumina in the bauxite determines the process route and the technology applied to extract alumina. The most economical method for the manufacture of alumina from bauxite is the Bayer process. In this process, bauxite is first ground and then combined with caustic soda, lime and steam to produce sodium aluminate liquor. Impurities are subsequently filtered out of the liquor, and alumina hydrate is precipitated out of the mixture. The alumina hydrate is finally calcined to remove moisture and bonded water.

The resulting alumina is ready for smelting into aluminum. A generalized flowchart for the Bayer process is shown in Fig. 2.5.

As a first step in the Bayer process, the bauxite ore is crushed to particles sizes of about 20 mm and subsequently milled in ball and rod mills to allow for better liquid–solid contact during extraction. Following the size reduction, recycled caustic soda is not only added to the ore to produce slurry for easy transport of the ore but also to initiate the digestion of the ore. Lime is also added at this stage to

Table 2.8 Minerals commonly found in bauxite

Mineral forms in bauxite	Chemical formula
Gibbsite (hydragillite)	α-Al$_2$O$_3$·3H$_2$O
Boehmite	α-Al$_2$O$_3$·H$_2$O
Diaspore	β-Al$_2$O$_3$·H$_2$O
Hematite	α-Fe$_2$O$_3$
Goethite	α-FeOOH
Magnetite	Fe$_3$O$_4$
Siderite	FeCO$_3$
Ilmenite	FeTiO$_3$
Anatase	TiO$_2$
Rutile	TiO$_2$
Brookite	TiO$_2$
Halloysite	Al$_2$O$_3$·2SiO$_2$·3H$_2$O
Kaolinite	Al$_2$O$_3$·2SiO$_2$·2H$_2$O
Quartz	SiO$_2$

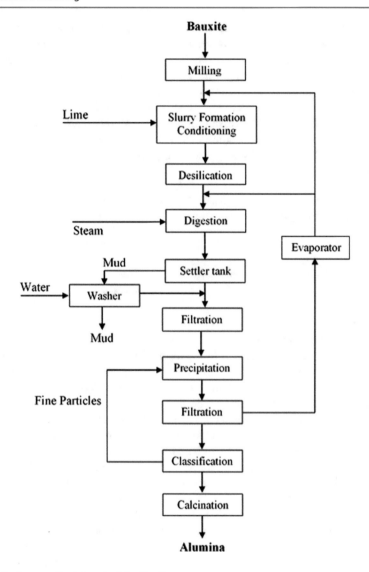

Fig. 2.5 A generalized flowchart for the Bayer process

condition the ore and to control phosphate present in the ore. The slurry is pumped
into pretreatment tanks where it is heated and held at atmospheric pressure to
facilitate desilication of silica present in the ore by the caustic. The products of the
desilication process pass out into the mud of the ore. The desilication also helps to
reduce the buildup of scales in the pipes and tanks.

After desilication, the slurry is pumped into the digester to extract the aluminates. The digestion is achieved by addition of steam and caustic soda to the slurry. Conditions of digestion for the monohydrate bauxite (Diasporic) are different from conditions required for the digestion of the trihydrate bauxite (Gibbsitic). The digestion of the monohydrate bauxite occurs at higher temperatures and pressures of 250 °C and 3500 kPa, respectively. The trihydrate bauxite is usually digested at a temperature of 180 °C and a pressure of 500 kPa. Alumina in the bauxite forms a concentrated sodium aluminate solution leaving undissolved impurities, principally inert iron and titanium oxides, and silica compounds. The digester reactions occur as follows:

$$\underset{\textbf{Caustic soda}}{2NaOH} \; + \; \underset{\textbf{Gibbsite}}{Al_2O_3.3H_2O} \; \longrightarrow \; \underset{\substack{\textbf{Sodium} \\ \textbf{aluminate}}}{2NaAlO_2} \; + \; \underset{\textbf{Water}}{4H_2O} \qquad (2.2)$$

$$\underset{\textbf{Caustic soda}}{2NaOH} \; + \; \underset{\textbf{Diaspore}}{Al_2O_3.H_2O} \; \longrightarrow \; \underset{\substack{\textbf{Sodium} \\ \textbf{aluminate}}}{2NaAlO_2} \; + \; \underset{\textbf{Water}}{2H_2O} \qquad (2.3)$$

The digester vessels are sized to optimum holding time to extract a very high percentage of the alumina in the bauxite. About 30% of the bauxite mass remains in suspension after digestion. This is in the form of a thin red mud slurry of silicates and oxides of iron and titanium suspension. To separate the sodium aluminate formed from the rest of the slurry, the mixture is sent to a clarification unit. The unit is made up of three distinct steps. The first is a settler tank in which the liquor containing the aluminate is separated. Flocculants, for example, starch, are usually added to the settler feed stream to improve the rate of mud settling. The aluminate liquor overflows the tank and is first filtered before sending to the precipitator. The residual mud is sent to washers where freshwater is added to recover the soda and aluminate content of the mud. To remove carbonate (Na_2CO_3), slaked lime is added to the dilute caustic liquor during the washing process. The carbonate forms by the reaction of caustic with compounds in bauxite and also from the atmosphere. The reaction reduces the ability of the liquor to dissolve alumina. The caustic soda is regenerated when lime reacts with the carbonate. The reaction produces calcium chloride, which, been insoluble, precipitates out and is removed with the waste mud. The reaction of lime and sodium carbonate proceeds as follows:

$$\underset{\substack{\textbf{Sodium} \\ \textbf{carbonate}}}{Na_2CO_3} \; + \; \underset{\textbf{Line}}{Ca(OH)_2} \; \longrightarrow \; \underset{\substack{\textbf{Calcium} \\ \textbf{carbonate}}}{CaCO_3} \; + \; \underset{\textbf{Caustic soda}}{2NaOH} \qquad (2.4)$$

The liquor from the settler tank contains traces of fine mud and is filtered in Kelly-type constant pressure filters using polypropylene filter cloth. Slaked lime slurry is used to produce a filter cake. Mud particles in the liquor are held onto the filter leaves for removal and treatment in the mud washer. With all solids removed, the pregnant liquor leaving the filter area contains alumina in clear supersaturated solution. The alumina dissolved in the liquor is recovered by precipitation of crystals. The alumina precipitates as a trihydrate according to the following reaction:

$$\begin{array}{ccccccc}
2NaAlO_2 & + & 4H_2O & \longrightarrow & Al_2O_3.3H_2O & + & 2NaOH \\
\textbf{Sodium} & & \textbf{Water} & & \textbf{Alumina} & & \textbf{Caustic soda} \\
\textbf{aluminate} & & & & \textbf{trihydrate} & &
\end{array} \qquad (2.5)$$

The precipitation is achieved by charging and cooling the pregnant liquor in precipitation tanks. The process is facilitated by seeding the liquor with fine crystalline trihydrate. When the tanks are agitated, the alumina trihydrate crystallizes out of the solution in varied crystal sizes. The entry temperature of the liquor into the tank and the temperature gradient across the tanks, the seed rate and caustic concentration are manipulated to control the size of crystals formed. The formed crystals are filtered from the liquor. The liquor is regenerated in evaporators and returned to the digester. The crystals are classified, and the finer crystals are returned to the precipitation tanks to serve as seed for the precipitation of incoming liquor. The coarser crystals are finally calcined to drive out unbound and bound moisture to form the alumina product. Alumina forms from alumina hydroxide according to the following reaction.

$$\begin{array}{ccccc}
Al_2O_3.3H_2O & \longrightarrow & Al_2O_3 & + & 3H_2O \\
\textbf{Alumina trihydrate} & & \textbf{Alumina} & & \textbf{Water}
\end{array} \qquad (2.6)$$

Further Readings

Adams, M. D. (Ed.). (2005). Advances in gold ore processing (Vol. 15; 1st ed.). Elsevier Science.

Adams, M. D. (Ed.). (2016). Gold ore processing: Project development and operations (Vol. 1; 2nd ed.). Elsevier Science.

Administrative fiat of 1999 (letter no.AB.85/156/01), Government of Ghana.

Bermúdez-Lugo, O. (2016). The mineral industry of Ghana. U.S. Geological survey minerals yearbook—2013. USGS, 22.3 p.

Dunne, R. C., Kawatra, S. K., & Young, C. A. (Eds.). (2019). SME mineral processing & extractive metallurgy handbook, vol. 1 & 2. Society for Mining, Metallurgy and Exploration (SME), USA.

Fuerstenau, M. C., & Han, K. N. (Eds.). (2003). Principles of mineral processing. Society for Mining, Metallurgy and Exploration, Inc. (SME), USA.

Gupta, A., & Yan, D. (2016). Mineral processing design and operations: An introduction (2nd ed.). Elsevier.

Jain, R. (2015). *Environmental impact of mining and mineral processing: Management, monitoring, and auditing strategies* (1st ed.). Butterworth-Heinemann.

Marker, B. R., Petterson, M. G., McEvoy, F., & Stephenson, M. H. (Eds.). (2005). Sustainable mineral operations in the developing world. Geological Society of London (GSL).

Minerals Commission. (2007). Statistical overview of the mineral industry (2006 report).

Ramana Murty, V. V. (2014). *Operational handbook of mineral processing* (2nd ed.). Denette & Co.

Ramkrishna Rao, G. S. (2014). Mineral processing techniques basics and related issues. Zorba
Rao, D. V. S. (2017). *Textbook on mineral processing*. Scientific Publisher.
Republic of Ghana—The 1992 Fourth Republican Constitution of Ghana. Section 267(6).
Wills, B. A., & Napier Munn, T. J. (2006). *Mineral processing technology: An introduction to the
practical aspects of ore treatment and mineral recovery (In SI/Metric Units)* (7th ed.). Elsevier
Science & Technology Books.

Textile and Fabric Manufacture

3

Abstract

Textiles have many uses in the household, industry and transportation industry. They are used for making clothes, bags, carpeting, towels, upholstered furnishings, beds, filtration, conveyors, vehicle tyres, car seat covers, sails and many more. Fibers are classified into two groups, namely, natural fibers that are harvested from plants (e.g. cotton, linen) or animals (e.g. wool, silk) or minerals (asbestos), and man-made fibers synthesized by chemical processes (e.g. chitin, viscose, glass fibers, ceramic fibers). Different methods are required to process natural and synthetic fibers into a yarn that could further be used to produce textiles. Unlike natural fibers, man-made fibers are made from polymers of natural origin produced from processing of natural materials (e.g. wood pulp), and those of synthetic origin obtained by organic polymerization reactions. Extrusion or spinning process is used to convert the polymers to fibers. The two major routes of producing fabrics are knitting and weaving. This chapter covers in detail the raw materials for the production of fiber, the different types of fiber and the methods and technology used for the manufacture of textiles from fiber. The textile industry poses major risks to the environment. The treatment protocols required to mitigate the release of chemicals, dust particles, metals, etc. into the environment have been considered in this chapter. An overview of the textile industry in Ghana is also presented.

3.1 Introduction

Any flexible material made by interlacing fibers is called a textile. Textiles are formed by weaving, knitting, crocheting or knotting yarns in a well-designed pattern or by pressing fibers together. Textiles must be distinguished from fabrics and clothes. While a fabric is any material made by weaving, knitting, crocheting or bonding, a cloth is a finished piece of fabric that can be used for a purpose such as a bedsheet. Defined in this manner, a fabric is actually a special kind of a textile. The range of uses to which textiles are put is quite wide. They are used for making clothes, bags and baskets. In the household, they are used in carpeting, upholstered furnishings, window shades, towels, table covers, beds and a host of other uses. In industry, they are used in such processes as filtration, and as conveyor belts among others. In the transportation industry, they are used in vehicle tyres, as seat covers, as kites, balloons, sails and parachutes, as apparels, beddings, furnishings and for various other applications.

Textiles chosen for industrial purposes, and chosen for characteristics other than appearance, are called *technical textiles*. These include textiles for automotive applications, medical textiles (e.g. textiles used as medical implants), geotextiles (i.e. textiles for reinforcement of embankments), agrotextiles (i.e. textiles for crop protection) and protective clothing (e.g. textiles used against heat and radiation for firefighting clothes, against molten metals for welders, stab protection and bullet proof vests).

Textiles are made from materials called fibers. **A fiber is a flexible, fine material characterized by a high length to width ratio**. Fibers are obtained from several sources and may be classified into two broad groups as natural and man-made fibers. These can further be grouped according to Figs. 3.1 and 3.2.

The natural fibers are harvested as agricultural products of plants or animals or mined as minerals. On the other hand, man-made fibers, as the name suggests, are either purely synthetic materials often made from organic polymers which themselves are made from elementary chemical units called monomers; or are

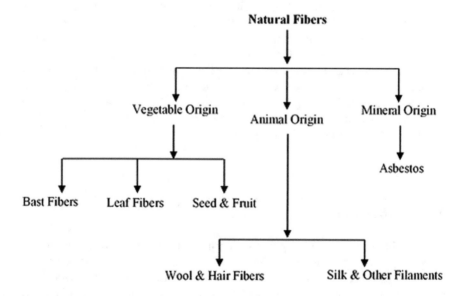

Fig. 3.1 Classification of natural fibers

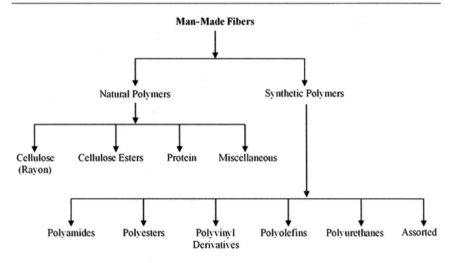

Fig. 3.2 Classification of man-made fibers

synthesized from inorganic minerals; or are made by processing natural raw materials (e.g. rayon from wood pulp).

The first step in the manufacture of textiles is to get a fiber, and in a form, which can be turned into a yarn. Once the fiber has been turned into yarn, the process of making a fabric is generally the same irrespective of the source of the fiber. While natural fibers are harvested, synthetic fibers must be synthesized by chemical processes. Processing natural and synthetic fibers into a form (a yarn) that could further be used to produce textiles therefore is significantly different for the two materials. A generalized flowchart for the manufacture of a textile or fabric is shown in Fig. 3.3.

3.2 Raw Materials of Textile Fibers

The raw materials for making textile fibers are either of natural origin or man-made. The natural fibers are harvested from plants, animals and minerals. The man-made fibers are of two kinds: (i) natural polymer fibers in which the fiber-forming material is of natural origin and (ii) synthetic fibers in which the fiber-forming material is a synthetic polymer.

3.2.1 Natural Fibers

Natural fibers from animal origin are almost invariable all derived from animal hair. These include wool, which is derived from sheepskin and cashmere fibers from

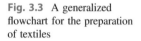

Fig. 3.3 A generalized flowchart for the preparation of textiles

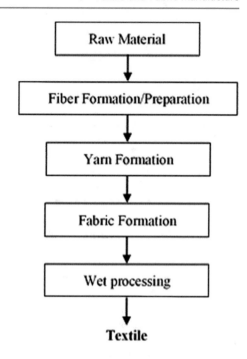

goat. Other animals whose hair is used in making textile fibers include camels, yaks, mohair, angora, llama, alpaca, vicuna and huarizo. The only natural fiber from animal origin not derived from animal hair is silk, which is obtained from the silkworm. Silk fibers are longer than any other natural fiber. The length of a silk filament may vary anywhere from 900 to 1200 m.

Natural fibers from plant origin include cotton, which is probably the best known in this category. Cotton fibers are relatively short and have a light color varying from crème to yellowish. Others in this group include: (i) linen that is obtained from the stem of the flax plant; (ii) bast fibers namely, jute, hemp, kenaf, urena, ramie and nettle; and (iii) leaf fibers, namely, sisal, henequen, abaca (Manila) and others.

The only known natural fiber of mineral origin is asbestos.

3.2.2 Man-Made Fibers

In contrast to natural fibers, man-made (artificial and synthetic) fibers are produced chemically. Artificial fibers are made from materials that exist already of animal, plant, vegetable or mineral origins by different chemical processes. Those of animal origin include chitin, which is produced from shellfish shells. Those of plant origin are viscose or rayon, which are wood fibers, cellulose acetate or triacetate and polylactide (PLA), which is produced from corn.

Artificial fibers of mineral origin include;

Glass fibers—These display varying properties depending on their composition. The main constituents of these fibers are SiO, CaO, MgO, Al_2O_3 and B_2O_3.

Ceramic fibers—The main components of these fibers are Al_2O_3, SiO_2 and B_2O_3 which relative amounts determine the properties of the fiber.

Silicon fibers—These have 99.9% SiO_2.

Boron fibers—These are obtained by the deposition of boron on the tungsten or carbon stem.

Carbon fibers—Obtained from pyrolysis of either acrylic or petroleum pitch.

Metallic fibers—Obtained from metallic compounds. The properties depend on the specific compound from which it is formed.

Synthetic fibers are produced from synthetic macromolecules, which are themselves obtained by chemical synthesis. The wide variation of available polymers, chemical and macromolecular structures means that fibers of different properties can be made from these compounds. The more common polymers in this category are polyamides, polyesters, polypropylenes, polyethylene, aramids, polyurethanes and polyvinyl chlorides, among others.

3.3   Fiber Formation

Fiber formation is the process whereby fibers are formed either through natural processes or from raw materials in the case of synthetic fibers. Natural fibers are formed from natural processes occurring in plants and animals and are harvested as agricultural products. Man-made fibers, however, are either formed from polymers of synthetic origin or from those of natural origin. The synthetic polymers are products of organic polymerization reactions. Natural polymers, on the other hand, are products of processing of natural materials, e.g. wood pulp. To form the fibers, the polymers must be taken through a process called 'extrusion' or 'spinning'.

3.3.1   Formation of Man-Made Fibers

Man-made fibers are synthesized or are processed from natural and synthetic raw materials. Consequently, producers of man-made fibers have greater control of the characteristics of the fibers that are produced. They can also be modified both physically and chemically and tailor made to achieve desirable properties. This versatility of man-made fibers has broadened their area of applications and made them very popular with manufacturers and consumers alike.

   The starting materials for all man-made fibers are polymers. The polymers are turned into fiber by the process of extrusion or spinning. The term extrusion is preferred to spinning as the latter term is a borrowed one from the process of turning staple fibers into yarn. A key equipment in the making of man-made fibers is the *spinneret*. The spinneret is a plate with many tiny holes through which the polymer in a liquid form is passed at high speeds. Often, the polymers when synthesized are in a solid form and to be able to extrude them into fibers they must first be converted into a liquid form either by dissolution in a solvent or by melting. There are three common methods of making man-made fibers. These are *wet extrusion*, *melt extrusion* and *solvent-dry extrusion*. The method used for any particular fiber depends on the ease of conversion of the polymer from the solid state into liquid and also on the behavior of the polymer when subjected to heat treatment. If the polymer can be converted into liquid form easily by heating without damage, then the melt extrusion method may be used. If, on the other hand, it is liable to decomposition or damage when heated but can be dissolved in a low boiling point solvent which can later be easily evaporated then the solvent-dry method is used. If, however, the polymer will suffer damage upon heating and cannot be dissolved in a low boiling point solvent then the wet extrusion method is used.

3.3.1.1 Melt Extrusion Method

Melt extrusion is the simplest of the man-made fiber production processes. The method is mostly used for thermoplastic polymers (i.e. polymers that melt when heated). Among the polymers produced by this method are polyamides (e.g. Nylon 6 and Nylon 66), polyester, polypropylene and polyethylene. These polymers have the characteristics that they can melt to at least 30 °C above their melting points without decomposition. This ensures that the molten polymers have sufficient fluidity to be extruded.

The method involves first heating the polymer to melt it, then filtering the melt to remove foreign particles, extruding the melt through a spinneret and finally cooling the emerging streams of molten polymer in a stream of air current. The heating is done in a Melter in an atmosphere free of oxygen. Oxygen must be prevented from the Melter as it can attack the molten polymer. This is achieved by carrying out the heating in a nitrogen atmosphere. The polymer to be melted is fed from a hopper in the form of chips into the Melter, which is heated to the correct temperature depending on the melting point of the polymer. The molten polymer from the Melter collects into a pool from where metering pumps at high pressures are used to pump the melt through pipes into the spinning head assembly. Before the melt reaches the spinneret, it is filtered to remove foreign particles that could potentially block the holes of the spinneret and hence interrupt the flow of the molten polymer, which could lead to the formation of broken filament fibers. The pool of molten polymer, pump, filter, spinneret assembly and spinneret are kept at 20–30 °C above the melting point of the polymer. The spinneret is a metal plate, roughly 50 mm in diameter containing fine holes. It may contain anywhere from 1 to 50 holes if extruding continuous filament yarns, or over 100 holes if extruding tows for staple yarn production. The molten filaments from the spinneret emerge into a stream of cooling air current, which rapidly solidifies them. The solid filament travels down a conditioning tube to cool further, and a lubricant is applied to lubricate and create non-electrostatic properties for the fiber.

3.3.1.2 The Solvent-Dry Extrusion Method

This is the least used of the major fiber production techniques. The method is used for polymers that for one reason or the other are not amenable to the melt extrusion processes. In this method, the solid polymer is converted to a liquid state by dissolving in a volatile solvent to form a solution. The solution is subsequently extruded into a stream of heated gas where the volatile solvent is rapidly evaporated away leaving the filaments. Examples of the fibers produced by this method are cellulose diacetate, cellulose triacetate, polyacrylonitrile, modacrylics and polyvinyl chloride (PVC).

It is important in this method to get the concentration of the solution right in order to get a filament of the right strength and flexibility. After dissolution, the solution is first filtered and pumped into storage tanks where bubbles are removed from the solution by application of vacuum. When necessary, pigments may be added to the solution at this stage to provide color for the fiber. Finally, the solution is pumped into heated extrusion heads in which one or more spinneret are located. The polymer concentration in the solution is about 25% (wt). At this concentration, however, the viscosity of the solution is quite high, and it is therefore heated so that it will flow sufficiently to facilitate its extrusion through the tiny holes of the spinneret. Before the solution is extruded it passes through an additional filter in the spinneret assembly unit over the top of the spinneret itself. This serves to safeguard any unwanted particles passing to the spinneret and disrupting the extrusion process. The extrusion heads are often sited within the spinning cell so that the evaporating gas passes around them. Metering gear pumps, one for each spinneret, are used to direct the solution to the spinneret. The extrusion occurs at a very high speed, in the order of 500–1000 m per minute. This causes the pressure on the spinneret plate to be great, consequently thicker spinneret plates, often made of stainless steel, are used. The solution emerges from the spinneret into a heating cell where a current of hot gas (often air or nitrogen), either in the direction the filaments are moving or countercurrent to it, heats the filament or evaporates the solvent away. The heating cell is long enough to allow sufficient solvent to evaporate away such that the filaments do not fuse together nor stick to any surface they come into contact with. The filament emerging from the heating cell is contacted with an applicator that lubricates on the filament to reduce both friction and static formation in subsequent operations. Finally, the filaments must be washed to remove residual solvent from the extrusion stage and then dried.

An important component of this method is the solvent recovery and recycle processes that must be done for economic and environmental considerations. Approximately, 3–6 kg of solvent is extruded per kilogram of fiber. It is possible to achieve solvent loss of less than 10% with the use of very efficient solvent recovery methods. The solvent dry extrusion method is most economical for the production of continuous filament yarn. This is because the dynamics of the method allows for only a small number of filaments to be produced from a spinning head. The holes in the spinneret are spaced and this makes for wide spacing of the filament to allow for evaporation of the solvent without any fibers sticking together. The temperatures used in this method are lower than that for melt extrusion, so this process can be used where the polymer would be degraded by high temperatures. The limitation of this method is that it depends on the existence of a volatile solvent for the dissolution of the fiber. Among the solvents commonly used are acetone for cellulose diacetate, and dichloromethane for cellulose triacetate fibers. Dimethyl formamide is another solvent used for some acrylic fibers, but its use is limited as it is not particularly volatile. It boils at 153 °C, so the air into which the fibers are extruded must be heated to between 130 and 200 °C to be able to adequately evaporate the solvent. Another disadvantage of the method is that the solvents used are expensive, toxic, flammable and potentially explosive when mixed with air.

3.3.1.3 The Wet Extrusion Method

Among the polymers for which this method is applicable are acrylic polymers and cellulose. Examples of fibers made by this approach are courtelle, acrilan and viscose rayon. The method is used for the preparation of fibers from those polymers, which do not melt without decomposition and which will neither dissolve in volatile solvents. Consequently, they are dissolved in high boiling solvents and extruded into a chemical bath of the same solvent where the filaments solidify. Examples of solvent used in this process are dimethyl formamide, dimethyl acetamide or aqueous solutions of inorganic salts such as zinc chloride or sodium thiocyanate. There are two variations of the wet extrusion method. These are: (i) the purely physical process method and (ii) the chemical regeneration and physical method.

i. **Physical Process Method**

   In this method, the polymer solution or dope is first deaerated and then filtered before being extruded through the spinnerets. The pressure necessary for the extrusion is provided by a gear pump, which also serves as a metering device. The spinneret itself is immersed in a coagulating bath that contains a dilute solution of the solvent in water. The bath is of such composition that the tiny streams of polymer undergo both physical and chemical changes during which the polymer precipitates upon entering it, forming first a gel and subsequently a fiber. To avoid corrosion of the spinneret, it is made of corrosion-resistant precious metal alloys. The size of a hole in the spinneret is smaller than what pertains to spinnerets used for the melt extrusion and the wet-dry extrusion processes. Large numbers of holes per spinneret can be used due to the liquid coagulation in the extrusion bath. As many as 20,000 holes per spinneret of size of 7.5 cm diameter face or 167,000 holes on a rectangular faced spinneret. The method is therefore economical for the production of staple fiber and tow. Due to low extrusion rates of the method, continuous cleaning, drawing and after treatment of the fiber are possible. An important aspect of this method is the recovery and recycling of the solvents used, which is always desirable as the solvents tend to be expensive.

ii. **Chemical Regeneration and Physical Process Method**
This process involves, first, a chemical step followed by a physical step hence the name. The method is used for the production of viscose rayon fiber. The polymer from which the fiber is made, cellulose, cannot be dissolved in its original form as such it is first chemically modified by a reaction with a suitable reagent before it is dissolved. The resultant solution is extruded in the same manner as the physical processes described earlier, except that the coagulation bath contains an acid, which neutralizes the sodium hydroxide in the modified cellulose. Full regeneration of the filaments to cellulose does not occur in the bath. This allows them to be drawn and so orientate the cellulose molecules while they are still plastic. The need for chemical modification of the cellulose arises from the fact that it is obtained from wood pulp and therefore has very high molecular weight. At that high molecular weight, they cannot be taken through the extrusion process. Therefore, it is first chemically changed to soda cellulose by reaction with sodium hydroxide. In that form, it is oxidized using oxygen to reduce the molecular chain length to suitable levels. The degraded soda cellulose is further treated with carbon disulfide to form cellulose xanthate, which when dissolved in alkali solution forms a brown solution called viscose. The viscose is filtered, then allowed to stand for a few days during which time the xanthate degrades back to cellulose. It is this solution that is filtered and subsequently extruded in the manner described earlier.

3.3.2 Preparation of Natural Fibers

The natural fibers cannot be used in the form in which they are harvested. They must first be cleaned and rendered in a form such that they can be used to make a yarn. The nature of treatment that these fibers must be taken through before making yarns out of them depends on the source of the particular fiber. The specific methods used to process the different natural fibers are described below.

3.3.2.1 Cotton Fiber
The raw harvested cotton is taken through several physical processes before it is ready to be spun into yarn. These processes are ginning, picking, carding and sliver combing.

Ginning: The cotton from the farm contains seeds and these are removed in a cotton gin. The process of removing the seeds is called *ginning*. The cotton gin is made up of a number of different rollers with teeth on them. As the cotton passes through the gin, the seeds are removed. The seedless cotton is then put into bales and sent for further processing.

Picking: Following the removal of the seeds, the vegetable matter in the cotton is removed in a process called *picking*. The machine used for this process is called a *picker'*. In the picker, the cotton from the cotton gin is first beaten with a beater bar to fluff it up and piled onto a screen, from where it is fed through rollers to remove the vegetable matter.

Carding: The carding machine consists principally of one big roller, which is surrounded by smaller ones. All the rollers are covered with small teeth which get progressively finer along the line of passage of the cotton. Large bats of cotton from the picker are fed into the carding machine where they are lined up nicely to make them easier to spin. The cotton leaves the carding machine in the form of large ropes called *sliver*.

Combing: The slivers, several at a time, are first combined in a combing machine to get a consistent size. This, however, produces a very thick rope of cotton fibers. Therefore, directly after being combined the slivers are separated into rovings. It is from these rovings that cotton yarns are spurn. When the roving must be processed by machine, they are separated such that they are about the width of a pencil.

3.3.2.2   Wool Fiber

The processes described here apply generally for all animal hair fiber with slight modifications; therefore, the preparation of individual animal hair fibers would not be described separately in this section. After shearing, the wool is taken through the physical processes of skirting, cleaning, carding and combing before it is ready to be turned into yarn.

   Skirting: During this process, the fleece (sheared wool) that is unsuitable for spinning or requires extra effort to spin into yarn are disposed of. Short fleece and fleece smeared with sheep dung fall into this category.

   Cleaning: This is done to rid the fleece of vegetable matter (sticks, twigs, burs and straw) and lanolin. Actually, vegetable matter could be prevented from the fleece by having the sheep wear a coat all year round. When spinning 'in the grease' is the goal then only the vegetable matter is cleaned (picked) from the fleece. In this case, the lanolin remains in the fleece and is only washed away after spinning or allowed to remain and not removed at all when a water repellent garment is the target. When cleaning is intended to remove both vegetable matter and lanolin, then the fleece is washed in soap either by swishing it around in a tub and rinsing until the fleece is clean of soap and dirt or soaking the fleece in hot water in a washing machine made for the purpose and melting the lanolin away. It is important to note that the fleece must not be rubbed against itself when washing. Otherwise, it may become felt and would be impossible to spin into yarn.

   Carding: Carding is the process of getting the fleece into a form that will facilitate spinning. Carding when done by hand yields a loose woolen roll of fibers called a *rolag*. Carding can also be done using a drum. In the latter case, a mat of flat fibers of rectangular shape called a *bat* is the product. Carding with a mill returns a fleece in a roving, which is a stretched bat. The roving is very long and often the thickness of a wrist. Carding by hand or by drum is a very lengthy process, therefore carding is often done using a carding mill.

Combing: Combing is yet another method used, albeit less often, to align the fibers parallel to the yarn prior to spinning. This method is good for spinning a worsted yarn. Rolags from hand-carding are useful for producing woolen yarns.

3.3.2.3 Bast Fibers

Bast fibers are obtained from the stalks of dicotyledonous plants. Examples of such plants are jute, flax, hemp, sunn or sann, kenaf, nettle, urena and ramie. The fiber is extracted from the stalk of the plants mostly by microbial means in a process called *retting*. Other processes, for instance *dressing*, are sometimes used to extract the fiber from the stalk.

i. **Processing of Jute and Flax Fibers**

   The methods for the extraction of the fibers from jute and flax (commonly called *linen*) are similar. In both cases, retting is the method used to extract the fiber. During retting, the harvested stalk of the plant is exposed to microbial action in clean, slow-moving water, which causes the fiber bundle to separate from the woody stem. They are then washed and dried. Acid is produced from the retting process, which would attack any metal container. As such retting of flax stalks is made to take place in a plastic, concrete or wooden container.

   Retting could also be carried out by submerging the stalks in a standing pool of water in a closed container at about 26.7 °C. After 4 h, a complete change of water is effected, and 8 h after that the scum is washed off the top by the addition of more water. Subsequently, the scum is washed off every 12 h until the retting process is over. At 26.7 °C, the retting process takes about 4 or 5 days. If the temperature is lower than 26.7 °C, the process takes longer. The end of the retting process is marked by a soft and slimy feel of the bundles and a popping out of a few of the fibers out from the stalks. It is better not to let the bundles sit in the water long enough than to let them sit there too long, as they always can be submerged again if found to be wanting later, but the reverse problem cannot be solved. In this case, the fibers are rotted away with the stalks. In the case of ramie fiber, after retting, the ribbons of fiber are first stripped from the stem. The resultant product called *china grass* is then boiled to separate the ribbons into fibers for spinning.

ii. **Dressing the flax**

   In this process, the flax is broken, scotched and hackled. The fibers are thereafter removed from the straw and cleaned well enough to be spun. As a first step, the straws are broken into short segments using a breaking machine. It is then hanged vertically, and the edge of a knife is scrapped along the fibers to pull away the pieces of stalk (this latter process is called scotching). Finally, the fiber is pulled through various sized hackles (a bed of nails-sharp, long, tempered, polished steel pins driven into wooden blocks at rectangular spacing). At the end of this process, the fiber emerges split and polished and ready to be carded or spun.

3.3.2.4   Leaf Fibers

The leaves of some monocotyledonous plants are held in shape and strengthened by fibers that run through the length of the leaf. The fibers can be extracted from the leaves by mechanical or microbial processes or both. The fibers so extracted are used in making ropes, cordage and for the production of textile fabrics. Leaves of commercial importance in making fibers include sisal, henequen and abaca (Manila).

i. **Processing of the Sisal Fiber**

Sisal is a leaf fiber that comes from the plant Agave sisalana. The plant is indigenous to Central America but is now cultivated in East Africa, Mexico, Haiti, Brazil and other regions in South America. The leaves of the plant are harvested when it is about two-and-half years old. Each leaf may contain about 1000 strands of fibers. The sisal plant has a life span of about 7–10 years during which it produces between 200 and 250 leaves. Each leaf has a composition of 4% fiber, 0.75% cuticle, 8% other dry matter and 87.25% moisture. Sisal fiber is mostly made by a process called *decortication*. In this process, the harvested sisal leaves are fed to a rotating wheel set with blunt knives. In the wheel, the leaves are crushed and beaten until only the fiber remains. The other parts of the leaves are then washed away by water. Finally, the fiber is washed and dried either in the sun or by artificial means (e.g. oven drying). Proper drying is important as the quality of the fiber depends to a large extent on its moisture content. Artificial drying has been found to result in generally better grades of fiber than drying with the sun.

The fiber may also be extracted from the leaf by retting (earlier described under bast fibers) or by mechanically scrapping the fibers from the leaves or by retting followed by scrapping. In these instances too, the extracted fiber must be first washed and subsequently dried before further processing.

3.3.2.5   Silk Fiber

Unlike vegetable and hair fibers which are produced in relatively short lengths, silk fibers are produced in long filaments, sometimes as long as 3 km. Consequently, silk fibers are processed differently from other natural fibers. The silk fiber is obtained from silkworms. The silkworm is actually a larva of a domesticated moth called **Bombyx mori**. There are other silkworms from which silk fiber can be obtained. Silk obtained from silkworms other than the Bombyx mori are called *wild silks*. Examples are the Tussah silk, Antheraea yama-mai silk produced in Japan and the Attacus ricini white silk found in the American and Asian continents.

The Bombyx mori silkworms breed on the mulberry leaves. When the silkworm is ready to transform itself into moth, it extrudes liquid silk in a continuous strand which hardens on coming into contact with air before forming cocoon around the silkworm. The cocoon is really a mass of twin filament of silk fiber. The silk fibers are retrieved by unwinding the filaments from the cocoons in a process called **reeling**. During reeling, the cocoons are first soaked in hot water, which kills the moth and softens the sericin gum, a cementitious material that holds the filaments together in the cocoons allowing the smooth unwinding of the cocoon. To unwind the cocoon, a revolving brush is used to search for the tip of the filament in the cocoon. When the end of the filament has been picked, it is drawn through a guide along with the filaments from other cocoons and reeled steadily off the cocoons which are left floating in the hot water to keep them softened.

Silk is wound up by the reeler in the form of skeins. These are made up into bundles of about 2.7 km, called 'books', which are packed into bales for shipment. At this stage, the individual filaments are still cemented together by the sericin. This silk may be used for some purposes without further treatment. Often, however, two or three of these multi-filament strands are twisted (spun) together into heavier threads. The process of twisting silk filaments together is called ***throwing***. During reeling, and subsequent spinning and weaving, the natural gum, sericin, is left on the silk filament. It acts as a cover to protect the fibers from mechanical injury. It may be removed from the finished yarns or fabrics by boiling with soap and water. After degumming, the silk acquires a beautiful luster. As much as one-third of the weight of the fabric may be lost when the gum is removed in this way. Raw silk with the gum on is called '*hard silk*', when the gum is removed it is called '*soft silk*'. Foulard fabric, georgette, chiffon and crepe de chine are examples of fabrics woven from hard silk and afterward degummed.

3.3.2.6 Processing of Asbestos Fiber

An asbestos is a generalized name given to several natural minerals, which occur in a fibrous crystalline form. It is really a rock, which has been subjected to an unusual treatment. As a result, instead of crystallizing in the normal way, it has done so in the form of fibers. The fibers in asbestos are all packed tightly together alongside each other, giving a grainy structure to the rock, resembling wood. Table 3.4 lists the important minerals of asbestos.

Table 3.4 Types of Asbestos

Type	Variety	Chemical name	Appearance
Anthophyllite	–	Magnesium iron silicate	Thin plates
Amphibole	Tremolite	Calcium magnesium silicate	Greyish, brittle fibrous crystals
	Actinolite	Calcium magnesium silicate	Greenish fibrous crystals
	Crocidolite	Iron sodium silicate	Bluish, long flexible fibers
	Mountain leather	–	Leathery sheets or matted fibers
	Amphibole asbestos	–	Greenish fine fibrous crystals
Serpentine	Chrysolite	Hydrated silicate of magnesium	Greenish-brown fine silky fibers
	Picrolyte	Hydrated silicate of magnesium	Inflexible fibrous crystals

Fig. 3.4 A generalized
flowchart for the preparation
of yarns from asbestos

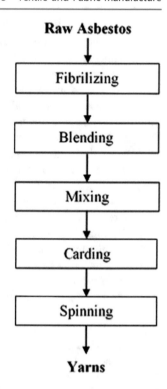

A flowchart for processing of asbestos fibers is shown in Fig. 3.4. As a first step, the mined raw asbestos is crushed in a pan crusher or in a rotating wheel to open up the fibers. Care is usually taken during this process so that the fibers are not broken unduly. Further opening is done in a toothed roller machine after which dirt and powdered rock are removed by means of grids. At this stage, the asbestos may be blended with other fibers before carding. Carding is done in a carding machine containing rotating brushes covered with steel bristles. In the carding machine, short fibers and impurities are removed, and the long fibers are delivered as a loose web or sheet of fiber. As the sheet leaves the machine, it is split into narrow ribbons or rovings, which are wound onto spools and sent for further processing.

3.4   Yarn Formation

A textile yarn is an assembly of fibers of substantial length and relative cross-section and/or filaments with or without twist. All textile yarns are made from either man-made or natural fibers. The method by which a yarn is made depends on the fiber type from which it is made. The process of making a yarn from a fiber is called *spinning*. Based on the fibers from which they are made yarns can be divided into two broad groups, namely, staple spurn yarns and continuous filament yarns. Several examples in each group abound are shown in Fig. 3.5.

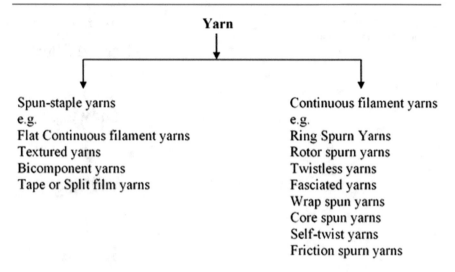

Fig. 3.5 Examples in each of the two broad groups of yarns

3.4.1 Spurn-Staple Yarn

Spurn-staple yarns are spun from staple fibers, i.e. fibers of definite length (usually 10–500 mm). All natural fibers are staple fibers. The only exception is silk. Some man-made fibers also belong to this stock. The basic principles underlying the manufacture of all staple spurn yarns are the same. It is in the detailed application of the principles that vary based on the yarn type. The process consists essentially of drawing (aligning the individual fibers of the sliver-the product from the carding process) and then subjecting them to drafting and finally spinning the drafted material. A simple flowchart of the process is shown in Fig. 3.6.

3.4.1.1 Drawing
Drawing is the process of straightening and aligning the individual fibers in the sliver from the carding process. Usually, the fibers in a carded sliver are arranged in a disorganized and random way. They give mostly weak yarn if not pretreated before spinning. If strong and useful yarns are to be obtained from them, it is important that the fibers in the sliver are properly oriented prior to spinning. The drawing process helps achieve this. This process is carried out in machines called *drawframes*. During drawing the slivers are made to travel in-between a series of pairs of rollers of increasingly greater surface speed. In so doing, the fibers in the sliver slide past one another and hence become straighter and more parallel to each other. Another outcome of the drawing process is that it reduces the irregularity present in individual slivers; it also improves the uniformity of slivers made from blended fibers.

Fig. 3.6 A simplified
flowchart for yarn formation

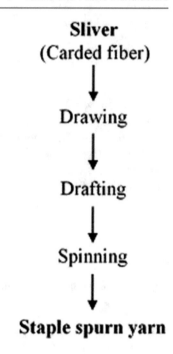

Sliver
(Carded fiber)

↓

Drawing

↓

Drafting

↓

Spinning

↓

Staple spurn yarn

3.4.1.2   Drafting

This is a process during which the cross-section of the sliver is reduced, and its length increased. In the process, the linear density of the sliver is significantly reduced. This is achieved by causing the fibers in a cross-section of a sliver to slide along the axis of the sliver relative to each other. The main objective of the drafting process is to draw out the material to the required fineness before spinning. The most common method of drafting is *roller drafting*. In this method, a roving or sliver is passed between a set of two pairs of rotating rollers one after the other. The second set of rollers (the output rollers) is made to move faster than the first set of rollers (input rollers). The relative speeds of the two sets of rollers determine the extent of reduction of the thickness and the increase in length of the product material relative to the feed sliver/roving. For example, if the second set of rollers moves six times faster than the first set then the thickness of the product material would be one-sixth of the feed material and its length would be six times that of the feed material. Drafting is often carried out not as an individual stand-alone operation but as a component of carding, combing, drawing or reduction.

3.4.1.3   Spinning

The process of turning the loosely held fibers in the drafted material into a yarn with the strength and characteristics required for the intended application is called *spinning*. This is achieved by the insertion of a twist into a strand of fibers. The twisting process generates the forces necessary to prevent the fibers from slipping past each other. It is this resistance to slipping that gives the yarn its axial strength. It forces the generally parallel arrangement of fibers into a spiral of helical formation. Twist is inserted by rotating one end of a strand of fibers relative to the other end. One complete rotation of 360° inserts one turn of twist. Spinning is the final stage in the process of yarn formation from staple fibers. During staple spinning, three objectives are achieved namely;

(i)   the input material is drafted (stretched) to the required linear density;
(ii)   the required amount of twist is inserted in the yarn; and
(iii) to wind the yarn onto a package, which is suitable for handling, storage and transport and is capable of being unwound at high speed during subsequent processing.

3.4.2 Continuous Filament Yarn

The product from the man-made fiber processing is, in the first stage, a continuous filament yarn or tow. During the man-made fiber processing, the fiber product comes off as a fiber of indefinite length called a *filament*. This product, by its nature and characteristics, is for all intents and purposes, a yarn. It is appropriately, therefore, called a *continuous filament yarn* or *flat filament* because it does not have the feel and look associated with staple spurn yarns of natural origin. A tow is a collection of thousands of parallel-lying continuous filaments of man-made fibers. To be used in fabrics and other applications, the flat filaments are therefore first processed to confer on them such characteristics as bulkiness, better cover when used in fabrics, greater comfort, softness of handle, opacity, improved thermal insulation, better moisture absorption, and improve surface characteristics, which are associated with staple spurn yarns. To achieve this, two approaches are used, namely, to introduce crimp and deformation (texture) to the yarn by a process called *texturing* or to cut up the filaments into short lengths (staple fiber) and subsequently spin them on conventional spinning machines. The latter process is called *tow-to-top conversion*.

3.4.2.1 Texturing
The main objective of the texturing process is to confer on the yarn properties similar to those exhibited by yarns made from natural fibers. This is attained by using mechanical means to deform continuous filament yarns and at the same time create crimps in the yarns. The deformation in the yarn is made irreversible by taking the deformed yarn through a heating and cooling cycle. The four main methods used for making textured yarns are described below.

i. False Twist Texturing

In this process, which is a continuous one, the yarn is passed through a twisting mechanism, which imparts temporary twist to the yarn traveling towards it and removes it as it passes beyond it. A heat source placed in the path of the twisted yarn sets the twist before it is removed at the spindle. Textured yarns produced by this method are mostly nylon and polyester.

ii. Stuffer Box Method

The method is based on the principle of heat setting filaments, which are held in a confined space in a compressed state and then withdrawing them in a crimpled form. In practice, this is achieved by using a pair of feed rollers to feed a yarn into a tube where it is restrained at the exit. In this manner, the individual filaments are caused to bend or fold into a zigzag configuration. When the compressed filament is set by heating, deformation is set into the yarn. The product of this process has high bulk, soft handle and less stretch than yarn made by the false twist method.

iii. Crinkle-Type Textured Yarns

This type of yarn can be produced in two ways as follows; namely, *the Knit de Knit method* and *the Gear Crimpling method*. In the former method, a flat yarn is first knit and subsequently the knitted fabric is heated, and the fabric is then unraveled (de knit). It is possible to vary the crimp frequency and shape by changing the needle gauge and fabric structure. Fabric produced from such yarn has a pronounced sparkle, boucle-type texture, good stretch and recovery, and good handle. In the Gear Crimpling method crinkle type, textured yarns can also be produced by passing a yarn through closely meshed gears. In the process, bulk is generated in the structure. The gear head is heated so that the crinkle produced from the yarn is permanent.

iv. Air-Jet Texturing

Texturing using this method is achieved by blowing a jet of air stream into a yarn while it is being delivered at a higher rate than when it is being taken up. The air produces turbulence, which causes the formation of random loops in the overfed yarn. The loop structures are consolidated in the yarn by passing through a heat stabilization zone. Polyesters, polyamides, viscose, acetate, polypropylene and combinations of these materials are mostly used under this method to produce a wide variety of yarns. The process is capable of producing blended yarns. It is noteworthy that this is the only process that can be used to texture non-thermoplastic fibers. Another feature of this process is the possibility of mixing several filaments in a jet. This paves the way for mixtures of different basic materials such as color mixes, and structural effects. Fabrics produced from yarns processed by this method have a soft and a pleasant handle with an appearance similar to spurn products but with a high bulk. Yarns from polyesters and polyamides filaments are used in the sports and leisure wear industry to produce fabrics with good handle and appearance. Some of the areas of applications of air-textured yarns are outer wear, sportswear, furnishing fabrics, curtaining, wall covering, blankets, car seat fabrics and many more.

3.4.2.2 Tow-to Top Conversion

One other way of producing useful yarns from continuous flat filaments fibers is, as a first step, to turn them into a staple and subsequently spinning them on conventional staple spinning machines. Staple fiber is produced from the filaments by two methods namely by cutting the filaments into short lengths in a process called *crush cut* or by stretching them until they break into pieces in a process called *stretch-break*. The filaments processed by this method are invariable in a tow (a tow is a collection of thousands of parallel lying continuous filaments). The tow is the direct product of the man-made fiber processing. To cut production costs and time and to get a quality sliver and finer yarns, the cutting or breaking into staple of the tow is carried out such that the parallel arrangements of the individual filaments in the tow are maintained. In this manner, the resulting short staple is called a *top* (sliver), hence the name of the process *tow-to-top conversion*. In the crush cut method, the cutting machines shear the tows at an angle thereby ensuring that the continuity and parallel arrangement of the fibers are maintained. On the other hand, when the stretch break method is used, drafting rollers are used to stretch the tows beyond their breaking such that they will be broken. Since not all filaments will break in the same place, the sliver produced has variable length. The top from these processes is subsequently taken through similar process as described previously for the spinning of natural staple fiber. Man-made top can be processed to form 100% man-made fiber yarns; it can be blended with wool for a wool/man-made blended yarn.

3.5 Fabric Processing

Fabrics are formed from yarns. Once the yarn has been formed, the process of making a fabric is the same irrespective of the origin of the yarn. There are two major routes for producing fabrics. These are knitting and weaving. Other methods used for producing fabrics are crocheting and binding.

3.5.1 Knitting

Knitting is the process of making fabrics by forming loops (stitches) with yarns, which are interlaced in a variety of ways. In forming the loops, needles or pins are the principal tools used. Knitting may be done by hand or by machine. Knitting by use of machines is several times faster than knitting by hand. A piece of knitting begins with the process of *casting on* (also called *binding on*). Casting on involves the initial creation of stitches (loops) on the needle. By forming additional loops and passing them through the casting on, both lengthwise and in width, a fabric of a specified width and length may be produced. Once the knitting piece is finished, the remaining live stitches are 'cast off'. Casting off (or binding) loops the stitches across each other so they can be removed from the needle without unraveling the item. The process is made up of three basic tasks; (i) the active unsecured stitches must be held so they do not drop, (ii) these stitches must be released sometime after they are secured and (iii) new bights of yarn must be passed through the fabric, usually through active stitches, thus securing them.

There are two major ways of knitting, namely, *weft knitting* and *warp knitting*. In weft knitting, the loop and the weft thread move perpendicular to each other on the fabric. The loops are formed across the width of the fabric while the weft thread is fed, more or less, at right angles to the direction in which the fabric is produced. Weft knitting can take place with only one thread or cone of yarn. Circular weft knitting machines exist that can use up to 192 threads at a time. One of the more versatile methods of fabric making is weft knitting. It can be used to produce a wide variety of fabric structures and several yarn types can be utilized. It is the simplest method of converting yarn into fabric.

Warp knitting, on the other hand, is a method of producing fabrics by using needles similar to those used in weft knitting, but with the knitted loops made from each warp thread being formed down the length of the fabric. The loops (course or stitch) are formed vertically down the length of the fabric from one thread as opposed to across the width of the fabric, as is the case in weft knitting. While warp knitting is undertaken using machines, weft knitting can be undertaken by use of machines or by hand.

Knitting may produce a flat or circular fabric depending on whether flat or circular stitch was done. In circular knitting (also called knitting in the round), knitting is worked in rounds in a spiral. This is achieved by the use of a circular needle. Circular or tube-shaped products such as hats, socks, mittens, and sleeves are made using circular knitting. Flat knitting is done with two straight knitting needles and is worked in rows, horizontal lines of stitches. Flat knitting is used to produce such pieces as, scarves, blankets, afghans and the backs and fronts of sweaters.

A key factor in knitting is stitch definition, which defines how complicated stitch patterns can be seen when made from a given yarn.

3.5.2 Weaving

Weaving is the oldest method of making fabrics from yarns. Weaving is done using weaving machines (traditionally called looms). The loom is made up of lengthwise yarns called the warp, which form the skeleton of the fabric. The warp yarns usually require a higher degree of twist than the filing yarns (weft yarns) that are interlaced widthwise. To produce a woven fabric, the loom must make three primary motions namely; (i) shedding—the motion that moves the headframes up and down in order to separate the warp sheet into two layers and create a triangle in front of the reed through which the weft can be passed, (ii) weft insertion—it is the means by which the weft is projected through the shed. In modern machines, this is done by a projectile, rapier, air jet or water jet, (iii) beating-up—during this motion the reed, mounted in a reciprocating sley, pushes the weft into the fell of the cloth to form a fabric. In addition to the primary motions three secondary motions are critical in allowing weaving to take place and in controlling the quality of the final fabric being produced. These are: (i) left-off motion—this ensures that the warp ends are controlled at the optimum tension for the fabric woven; (ii) take-up motion—during this motion the fabric is withdrawn from the fell and stored at the front of the loom; (iii) weft selection motion—this motion is executed only when it is necessary to vary the weft being inserted.

In selecting a loom for a given task, a number of factors need to be taken into consideration. These include;

(1) The type of yarn to be woven. For instance, looms that are to weave continuous filament yarns will require many different features than those weaving staple fiber yarns.
(2) The width of the fabric to be woven. The loom must be able to weave the maximum desirable width required, so it will always be able to weave narrower fabrics.
(3) The method of weft insertion to be used. This is decided by the suitability of the weft insertion media to the type of yarn, the fabric structure, the patterning required from the shedding and the weft insertion systems, the type of fabric edge that will be produced and the amount of waste created.
(4) The shedding system. This is mainly decided by the range of weaves to be used and the versatility required from the system.
(5) Mechanisms to create specific features.
(6) The speed of the weaving machine and its ability to produce the required quality of fabric.

3.6 Wet Processing

Following the fabrication of the fabric, it is taken through a number of treatments before it is ready for the market. These processes are in the main meant to give the fabric aesthetic appeal, to prolong the life span of the product and to make the fabric suitable for certain applications. The processes are organized into three basic operations, namely: (i) preparation; (ii) coloration and (iii) finishing. All three operations are carried out in the form of wet or chemical treatment. This step of fabric/textile production is different from the hitherto considered processes of fabric production in that whilst all the processes until now have been purely mechanically based, the wet processes are mostly chemically based. The only exception is the formation of synthetic fibers, which is also a chemical process.

3.6.1   Preparation Processes

The main objective of the preparation processes is to ensure that the textile has the right physical and chemical properties to enable it to be colored or finished. Often, the fiber from which the yarns of the fabric are made contains impurities that are natural to the fiber or are picked up from earlier processing stages. It is important, in order to dye or print on the fabric, to first remove these impurities. It is the removal of the impurities that constitute the preparation of the fabric. Although there are some preparation steps that are general to all fibers, for instance scouring, most of the preparation procedures are applicable to specific groups of fibers, such as cellulosic fibers or man-made fibers. Even for those operations that are common to all fabrics, the material needed and the conditions under which the operation is carried out will differ for different fabrics from different fibers. Accordingly, these procedures will be discussed in subsequent sections based on the fiber from which the fabric was made.

3.6.1.1   Preparation of Cotton and Cellulosic Fabrics

A woven cotton fabric may be prepared according to the flowchart in Fig. 3.7.

Singe is the process whereby surface hair on woven fabric is removed. This is the first operation in the preparation of a fabric for dyeing or printing. This process is accomplished in a singeing machine. In the machine, the fabric passes at very great speed over a naked gas flame. In the process, the surface hairs are burnt away. Singeing may be done on one side or on both sides of the fabric. Unlike cellulosic fibers that form a dusty ash upon burning this method is not suitable for protein and other fibers that leaves out hard black residues upon burning. There are three main reasons for singeing a fabric. The first reason is to improve the wettability of the fabric. Surface hair helps to trap air into a fabric, as such when immersed in water it takes long for water to enter into the fabric. Second, the formation of pills on blended fabrics can be minimized with singeing. Pills are little balls of fiber that form on some fabrics as a result of abrasion that occurs during usage. Finally, singeing creates a smooth surface for printing and dyeing. Displaced or moved hair from the surface of a printed fabric will cause details of the print to be distorted or get fuzzy. This can be avoided by singeing the fabric surface. It is noteworthy that knitted fabrics are not singed.

Following the singeing process, the fabric is sized to facilitate weaving yarns by treatment with starch and/or other substances, for instance, polyvinyl alcohols and polyacrylic acids. For cotton and other cellulosic fibers, starch-based sizes are the ones often used. Sizes enable the yarn to withstand the stresses encountered in the loom. These substances, however, must be removed from the fabric before dyeing or printing. The starch and other substances applied for this purpose are called *sizes* and the process of removing them from the fabric is called *desizing*. Sizes are not applied to yarns used for knitting as the high stresses found in weaving are absent during knitting. Removal of starch-based sizes is carried out by the application of

Fig. 3.7 A generalized
flowchart for the preparation
of cotton fabric

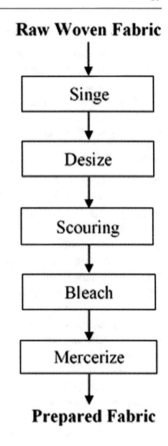

bacteria, acids, enzymes or oxidizing agents. Irrespective of the approach, the objective is to break down the insoluble starch into soluble components followed by dissolution and rinsing.

Desizing by use of bacteria

In this approach, bacteria are used to break down the starch in the fabric. This is achieved by soaking the fabric in water and placing it in a damped place for several hours. Bacteria from the atmosphere break down the starch in the fabric, which is then rinsed away. The process is very difficult to control as the bacteria could degrade the cellulose of the fabric itself if allowed to act too long on the fabric.

Desizing by use of acids

In this process, a 0.5–1.0% (wt) (on weight of fiber) cold solution of sulfuric acid or hydrochloric acid is used. The fabric is soaked in the acid and left for 2–3 h at room temperature. The method has the advantage that it removes any metal contamination from the fabric. However, the disadvantage is that the acid reacts with the cellulose of the fiber just as much as it does with the starch. This method is therefore not popular as a desizing agent.

Desizing with oxidizing agents

Hydrogen peroxide is an example of oxidizing agents used to desize starch from cellulosic fibers. The fabric is soaked in a hot solution of the oxidizer for several hours or steamed at 20 min or so at 100–150 °C. Unlike the previous methods, this one does not cause serious damage to the fiber. An advantage of this method is that the peroxide could be used also as a bleaching agent so that the desizing and the bleaching could be combined into one.

Desizing using enzymes

Amylases that are known to break down starch under the right conditions are used in this process. In this case, electrolytes are added to the medium containing the enzymes to improve on the activity of the enzymes. The enzyme solution is prepared using hard water. This method is the safest with regard to fiber damage as the enzymes act only on the starch.

After desizing, the fabric is ready for scouring. This is the process that removes wax, fat, partially broken down seeds or husks still remaining in the fabric and generally remove any remaining water soluble impurities from the fabric. Scouring may be done by the use of alkali solutions. An example is the use of sodium hydroxide solution at high temperatures. The alkali reacts with the fats and waxes and turns them into soap. The soap thus produced improves the detergency of the solution and facilitates the removal of dirt and the solubilization of the previously water-insoluble impurities. An 8.0% (w/v) alkali solution at 120 °C may take 2–20 min to complete the process. At temperatures above 100 °C, the process is carried out in the absence of air, otherwise the air would oxidize the fiber. Scouring may also be done by the use of organic solvents in which the fats and waxes are soluble. An example of such solvents is trichloroethylene. The disadvantage of scouring using solvents is that the solvents are unable to digest and remove seed and husk fragments in the fabrics.

Subsequent to the scouring, the fabric is bleached. Bleaching the fabric at this stage does two things. It destroys from the fabric any coloring matter and removes any residual seed or husk matter in the fabric. Bleaching is done by the use of oxidizing agents. The oxidizing agents commonly used are hydrogen peroxide (H_2O_2), sodium hypochlorite (NaOCl) and sodium chlorite ($NaClO_2$). The choice of the oxidizing agent depends on the nature of the fiber and the equipment to be used. For cotton, the most commonly used bleach is hydrogen peroxide. It is preferable to hypochlorite because it does not produce any toxic by-products and does not hurt the environment. It is applied at high temperatures and in the pH regime of 10–11. Its major disadvantage is its relatively poor stability. The use of sodium chlorite is not popular as it produces chlorine dioxide gas which is highly toxic.

The final step in the preparation of cotton fabrics is mercerization. It is the process of getting rid of little bundles (also called *neps*) that form on the surface of fabrics. The bundles are formed from immature fibers (commonly referred to as *dead fibers*) in the fabric. Mercerization is achieved by applying an appropriate medium to swell the immature fibers and give them a rounder shape and greater reflectivity. The method is only applicable to cellulose-based fabrics or fabric blends containing cotton. Woven as well as knitted fabrics may be mercerized. However, mercerizing knitted fabrics is much more difficult and requires specialized techniques. Mercerization changes the physical and chemical properties of a fabric by making it more lustrous, stronger, greater capacity to absorb more dyes, softer handle and more extensible. During mercerization the fabric, under tension, is soaked in 18–30% sodium hydroxide at 13–15 °C. Sufficient time is allowed for the sodium hydroxide to adequately absorb into the substrate. Thereafter, the alkali is washed out with the substrate still under tension. Both yarns and fabrics can be mercerized.

3.6.1.2 Preparation of Wool Fabrics

Wool fabrics are normally prepared according to the steps outlined in Fig. 3.8.

Cropping

The first step in the preparation of wool fabrics is removing surface hair from the fabric. This process is called *cropping* or *shearing*. Cropping is a mechanical process that is achieved in a cropping machine. The cropping machine is fitted with

Fig. 3.8 Flowchart for the preparation of wool fabric

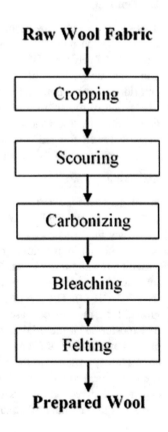

a series of helical blades, which rotate at very high speeds. As the fabric passes beneath the blades, the hair is removed by the cutting action created between the blade and a stationary ledger blade beneath the fabric. Cropping is much slower than singeing, but it makes possible a more controlled removal of hair. The method is applicable to any fiber type.

Scouring

Scouring of wool fabric is done to remove impurities from the fabric prior to printing or dyeing. Impurities such as lubricants, anti-static agents, oils and waxes applied to the yarn or fiber to improve processing and weaving, or knitting are all removed at this stage. Natural impurities such as wool grease (lanolin), suint (sweat), vegetable matter, parasites, branding fluids, pesticides and veterinary products are expected to have been removed by the time the fabric is formed. However, where they are not, the scouring process removes all of these. It must be noted that wool could also be scoured in loose stock or in yarn. Scouring of wool fabrics is done in scouring machines using soap or nonionic synthetic detergents and sodium carbonate solution (i.e. soap and soda scout). The temperature of the scouring medium is kept between 45 and 55 °C.

Carbonizing

This is a chemical treatment process during which any remaining vegetable impurity in the fabric is removed. To achieve this, the fabric is first impregnated with a 5–6% HCl or H_2SO_4 or a salt which will produce a strong acid when the goods are dried. Examples of such salts are aluminum chloride and magnesium chloride. After the application of the acid or acid salt, the fabric is dried and baked. The action of the acid or salt in the fabric is to hydrolyze the vegetable matter to dark, brittle hydrocellulose which is then removed mechanically by passing the fabric between franged rollers. As a final step of the process, the fabric is neutralized and rinsed. Carbonizing may also be carried out at the loose wool stage.

Bleaching

Wool may be bleached with both reducing and oxidizing agents, unlike cotton, which are bleached only with oxidizing agents. When reducing agents are used, the wool may be bleached by exposing the wool fabric to sulfur dioxide (SO_2) or by treating the fabric in a liquor containing reducing agents such as sodium bisulfate ($NaHSO_3$), sodium dithionite ($Na_2S_2O_4$), sodium metabisulfite ($Na_2S_2O_5$), or zinc formaldehyde-sulfoxylate $Zn(HSO_2CH_2O)_2$ or sodium formaldehyde-sulfoxylate ($NaHSO_2CH_2O$). The use of reducing agents in the bleaching of wool however is limited for a number of reasons. First, the whiteness produced is not permanent as on exposure to air the original color of the wool returns. This is due to an oxidation process that reverses the reducing action. Second, the use of sulfur dioxide results in an uneven appearance in the fabric. Hydrogen peroxide is the only oxidizing agent used to bleach wool. The other bleaching agents previously mentioned all have detrimental effects on the wool and as such are not used. Bleaching with hydrogen peroxide has a number of advantages. First, the whiteness it produces is permanent.

Second, it bleaches evenly. Bleaching with hydrogen peroxide is done at a pH of between 8.5 and 9.0 and at a temperature of 40–50 °C. Higher pH of the liquor will cause damage to the wool. It is important during the bleaching of wool that the pH is kept constant. This is achieved by adding sodium silicate to the liquor. The silicate acts as a buffer to keep the pH constant. An added advantage of sodium silicate is that it acts to stabilize the liquor against the catalyzing action of metals, which degrades the peroxide. After bleaching, the fabric is neutralized using acetic acid.

Felting

Felting is a characteristic of hair fibers only. It is the ability of fibers to form matted entanglements such that a woven fabric, for example, can become so matted that the warp and welt yarns are no longer visible. This results in a fabric that is apparently composed of a mangled web of fibers. This type of fabric is called a felt fabric. Felting occurs when the cuticle scales on two fibers lying alongside each other interlock during the relative motions of the fibers.

Felting is done for two reasons: (i) felt fabric has a different appearance to an unfelted one. Therefore, felting is sometimes carried out as a design consideration only; (ii) during felting the air spaces between fibers within the fabric get smaller. It, therefore, becomes more difficult for air to pass through the fabric. Since air is a good thermal insulator, fabric that has smaller air spaces will be warmer to wear than those with bigger spaces. Felting is always accompanied by shrinkage of the fabric because the fibers move closer to each other. Industrially, felting is carried out in milling machines in a process called *milling*. During this process, the fabric that is wetted by soap (the soap acts as a lubricant) is continually passed through the machine until the desired felting is achieved. The process is controlled by the liquor temperature and pH, the fabric speed through the machine and the pressure exerted on the fabric by the machine.

An important operation during the preparation of wool fabrics is setting. Setting is a process of giving a fabric a sustainable shape or form. It is done to avoid distortions in the fabric structure, such as waviness (cockling) during processing. The distortions are mostly brought about during the scouring process and if they are not resolved can adversely affect the subsequent processing of the fabric. The setting of the fabric during the preparation stage is non-permanent and is achieved through two techniques, namely, crabbing and blowing. During crabbing, the fabric is first soaked in hot water and subsequently wound around a roller under tension. A second roller presses down on the roll of cloth to apply the necessary smoothing pressure. During blowing, the fabric is wound under tension onto a hollow per-forated cylinder through which steam is then blown for about one or two minutes. Setting can be best achieved if the fabric pH is about 9. This can be performed by first giving the fabric an appropriate wet treatment. The set achieved would nor-mally remain in the fabric provided it is not wetted again at a temperature equal to or higher than that at which it was previously set.

3.6.1.3 Preparation of Man-Made Fiber Textiles

Man-made fibers by nature of their processing contain far less impurities than natural fibers. As a result, man-made fibers require far less preparation steps than those discussed in the previous sections. Most of the impurities accompanying fabrics made form man-made fibers are acquired more from the processes that follow the preparation of the fiber rather than from the fiber formation processes. Lubricants, anti-static agents, sizes that are added during yarn preparation and weaving/knitting must all be removed as impurities after the fabric has been formed. The principal preparation steps for fabrics made from man-made fibers are shown in Fig. 3.9.

Desizing and scouring

Desizing and scouring are combined in one operation for fabrics from man-made fibers. The operation is performed using a non-ionic synthetic detergent with the addition of a little alkali at a temperature of between 50 and 60 °C. Subsequent rinsing after the detergent treatment is sufficient to desize and scour fabrics from man-made fibers. The same process is applicable to both weaved and knitted fabrics.

Setting

Fabrics from man-made fibers, particularly, synthetics and modified regenerated cellulose fibers are set by a process called *heat setting*. Heat setting takes advantage of the ability of the fibers to soften and therefore can be molded when heated. During heat setting, the fabric is first exposed to dry heat or steam at a temperature above their softening temperature whilst they are held flat and to specified width.

Fig. 3.9 A generalized flowchart for the preparation fabrics from man-made fibers

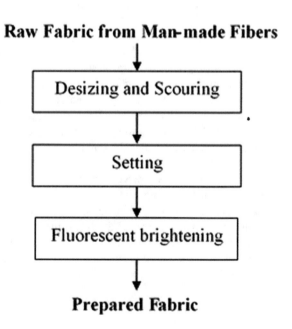

Subsequently, they are cooled whilst still held flat. The set conferred on man-made fibers has more permanency than that on wool brought about by crabbing or blowing processes.

Fluorescent Brightening

This is a process by which whiteness is added to a fabric to enhance it to a level above that which is achieved by bleaching alone. This is achieved with fluorescent brightening agents (FBA). The fluorescent brightening agents are substances that have the ability to absorb light (generally, radiation) and to re-radiate it at a longer wavelength. The FBA used in textiles often have the ability to absorb white light (typically with wavelengths of 300–400 nm in the non-visible area) plus the additional characteristics of being able to convert any ultra-violet radiation present into visible light at the end of the visible spectrum (at wavelengths of 400–500 nm). FBA may be added to bleaching liquor and applied to the fabric or applied by methods similar to those used to dye fabrics.

The most common class of brightness for textiles are the stilbenes (di- and tetra-sulfonated triazole-stilbenes and di-sulfonated stilbene-biphenyl). However, long-term exposure to UV caused them to fade as a result of the formation of optimally inactive stilbene cis-isomers found at the center of the molecule. Other classes of FBAs are coumarins, imidazolines, benzoxalines naphthalimide, pyrazine and triazine. Stilbene derivatives are FBAs used for cellulosics, while distyryl and triazolyl stilbene derivatives are used for polyamides, and benzimidazole derivatives for acrylic fabrics. FBAs for polyesters are benzoxazole and styryl derivations. Tinopal (a thiophenediyl benzoxazole class) is the most efficient brighteners for cotton and exhausts within a pH range of 8–11 or higher and work best when applied by padding method.

3.6.2 Coloration of Fabrics

Fabrics may be colored to enhance their aesthetic appeal or for functional reasons, for example, military camouflage and fluorescent jackets for road repair workers and the police. Coloration of fabrics is achieved with pigments or dyes. Dyes are water-soluble substances and are absorbed into the fabric when applied. Some dyes are fiber specific. On the other hand, pigments are water-insoluble substances that adhere to the surface of the fabric when applied. They are not fiber specific. Processes used to add color to fabrics are dyeing and printing.

3.6.2.1 Dyeing
Dyeing completely covers the fabric with color (i.e. the substrate is literally, completely soaked in color during dyeing). Dyeing is achieved using dyes that may be of natural or synthetic origin. Dyes are organic chemicals that are able to absorb and reflect visible light. Such molecules are called *chromophores*. Chromophores contain one or two functional groups on the molecule namely, chromogen-color providing group and auxogens-color intensifying or deepening group. The dye is

usually carried into the fabric in aqueous solutions from dye baths. The process may be carried out batch-wise or in a continuous manner. During batch dyeing, the fabric is first placed in a dyeing machine and the liquor pumped through the stationary fabric. Alternatively, the fabric is circulated through a 'stationary' dye liquor. In yet another version, both the fabric and liquor are circulated together. Four basic types of machines exist for carrying out a batch-wise dyeing operation. These are the jig, winch, beam and jet type machines. In the jig, the fabric is wound on a roll, which then passes through the dye liquor onto a second roll. The desired shade is achieved by repeating the process severally in a back-and-forth manner. Heating in the jig machine is achieved by the use of steam coils or by direct steam injection. The exhausted liquor is sent directly to the drain. Jigs are suitable for clothes, which cannot be creased during processing. In winches, the fabric circulates, in a continuous rope between 50 and 100 m, over reels and rollers and down into a dye-bath. This machine is useful for loosely woven cotton fabrics and for fabrics that can withstand creasing in rope form. Jigs and winches are operated under atmospheric pressures. For lightweight, delicate and knitted fabrics which suffer creasing, damage and poor dye uptake on jigs and winches, dyeing is carried out at high temperatures and pressures using beam or jet type machines. In beam machines, the fabric is wound around a perforated beam, which is placed horizontally in a dyeing vessel. The door of the vessel is then shut and bolted. Dye liquor heated about 120–130 °C is then circulated through the fabric. In jet-type machines, the fabric is circulated in a rope form with a vigorous circulation of the liquor, at temperatures between 120 and 130 °C.

The continuous dyeing process is used when a large amount of fabric must be dyed to the same shade. In continuous dyeing, the prepared fabric is first soaked with a suitable amount of dye liquor and then passed into a fixation chamber where dye fixation takes place in just a few minutes. Subsequently, the dyed fabric is washed to remove the unfixed dye and finally, the fabric is dried. In one version of the continuous process, the fabric is fed through a pad bath then nipped through rollers to remove excess liquor. It is then dried and the dye fixed. Fixation is done either at a high temperature using a steamer (saturated steam) or with a thermosol process. Dyeing can be done on fabrics as well as loose fibers, and yarns. Dyes may be classified according to how they are used in the dyeing process. They may also be classified by the nature of the chromophore carried on the molecule. Tables 3.5 lists the various classes of dyes.

3.6.2.2 Printing

Generally, printing is the localized application of dye or pigment in a thickened (often in paste form) onto a substrate, to generate a pattern or a design. There are four broad methods of printing. These are engraved printing, screen printing, transfer printing and hand block printing.

Engraved Roller Printing

In this method, the color is transferred continuously onto the fabric by engraved rollers. An advantage of engraved roller printing is that sharp lines and fine details

Table 3.5 Characteristics if dyes used to dye fabrics

Dye type	Characteristic of dye	Area of application
Acid	Water soluble, anionic	Silk, wool, nylon, modified acrylic
Alkaline (basic)	Water soluble, cationic	Acrylic fibers, wool and silk
Direct	Neutral or slightly alkaline medium	Cotton, paper, leather, wool, silk and nylon
Mordant	Requires a mordant, e.g. potassium dichromate	Wool
Vat	Insoluble in water	Cellulosic, polyester and protein
Reactive	Reacts directly with substrate	Cotton, other cellulose fibers, protein and polyamide
Disperse	Water insoluble, solid paste or powder	Nylon, cellulose acetate, polyester and acrylic
Azo	Produced directly onto fiber, water insoluble	Cellulosic, polyamide and polyester
Sulfur	–	Cellulosic

of the print can be obtained. The circular nature of the roller ensures that the designs are repeated at distances determined by the circumference of the roller. The dye/pigment paste is held in the engraved printing area on the roller. The design to be printed is cut into the surface of copper rollers. Each color in the pattern requires a separate roller. Large patterns will naturally require rollers with bigger circumferences to achieve the desired print. The depth of the shade of the print can be altered by varying the depth of the engraving on the roller.

Screen Printing

In screen printing, a screen that consists of essentially a synthetic fiber or metal gauze is stretched taut over a frame. Parts of the gauze have the hole blocked off (non-printing areas). The printing paste is forced through the open printing areas by a rubber or metal blade, called a squeegee, and onto the fabric beneath. Even though screen printing can and is actually done manually, this approach is not popular as it is slow and labor intensive. Increasingly, therefore, automatic machines are used. There are two main types of this method, namely, flat screen printing and rotary screen printing. In the rotary method, the color is applied to a fabric by passing it under a series of rollers of screens where a part of the pattern is exposed, allowing the color through to the fabric on the stroke of a squeeze. The printed part of the fabric is then dried in an oven. In flat screen printing, the color passes through a screen onto the cloth, the pattern being produced by masking out parts of the screen. Each color used in the design requires a separate screen. Well-designed, mechanized flatbed machines give precise prints and are suitable for wide fabrics or large pattern repeats.

Table 3.6 Relative merits of the fabric printing techniques

Parameter	Engraved roller	Hand block	Flat-bed screen	Rotary screen	Transfer printing
Productivity	High	Very low	Moderate	Very high	Moderate
Capital cost	Very high	Very low	Very high	High	Low
Space need	Low	Moderate	Very high	Moderate	Very low
Machine 'down time'	High	Low	Low	Low	Very low
Operative skill level	High	High	Semi-skilled	Semi-skilled	Unskilled
Dye efficiency	Duller prints	Bright prints	Bright prints	Bright prints	<100%
Fabric limitations	Width limitations	Wide widths possible	Wide widths possible	Wide widths possible	Confined to synthetics

Transfer Printing

This method differs from the first two in that the printed pattern is transferred from printed transfer paper onto the fabric by passing both over a heated roller. The special dyes used (selected disperse dyes) are first printed onto a cellulosic paper by conventional paper printing techniques. The fabric to be printed and this paper are then sandwiched and passed into a transfer calendar. In the calendar machine, the paper is heated for about 30 s at temperatures of up to 200 °C. At that temperature, the dyestuff transfers and diffuses into the fabric by sublimation. The relative merits of the various printing techniques are shown in Table 3.6.

Hand Block Printing

A template made from wood, sometimes with copper inserts, is designed with the requisite pattern. Color is applied to the block and the pattern is then stamped onto the fabric. The color and pattern are thus printed onto the fabric, which is subsequently dried. A separate block is required for each color to be printed.

Except for the transfer printing, in all the other methods of printing, the dye is applied to the fabric in a paste form. The viscosity of the dye is controlled by the use of appropriate thickening agents, which prevent dye migration from the surface of the fabric. Examples of thickening agents include starch, gum tragacanth, alginates, methyl cellulose, ether and sodium carboxymethyl cellulose.

An overview of the fabric printing methods is outlined in Table 3.6.

3.6.3   Fabric Finishing Processes

This is the final stage of the processes through which a fabric goes through before it reaches the market or the consumer. The finishing treatment may be performed to achieve a temporary effect or a permanent effect. The reasons for these processes are either to improve the attractiveness of the fabric; to increase the serviceability of the fabric; or to achieve a greater customer satisfaction with the product. Four methods are used to achieve these effects, namely, mechanical processes, chemical processes, heat setting and surface coating processes.

3.6.3.1 Mechanical Processes

In these processes, machines employ mechanical actions to achieve the desired effects on the fabrics. These include calendaring, raising, cropping or sueding and compressive shrinkage. These are explained below:

Calendaring

Calendaring is the modification of the surface of a fabric by the action of heat and pressure to give the fabric a smooth appearance. It is achieved by passing the fabric between two heavy, heated rotating rollers to give a flattened, smooth appearance to the surface of the fabric. The nature of the surface of the roller determines the kind of finish the fabric receives. In practice, the surface can be either smooth or engraved.

Raising

Raising is the process of teasing out the individual fibers from the yarn so that they stand out at the surface of the fabric to give it a brush or a napped appearance. Raising can be carried out manually or more commonly by the use of machines. The manual operation uses teasels (thistles nailed to a wood board) over which the fabric is drawn. This produces a fabric with a hairy surface, which has improved insulating effect. Raising machines, however, use a hooked or bent steel wire to tease the fibers from the surface of the fabric.

Cropping or Sueding

Cropping is the process whereby the fibers protruding out of the surface of a fabric are cut to an even height. The process has been previously described under Sect. 3.6.1.2 on the discussion of wool fabrics. A related process is called *sueding* (also called *emerising* or *sanding*) during which a sandpaper or emery paper is wrapped around rotating rollers over which the fabric passes. The fibers standing on the surface of the fabric are cut leaving an abraded appearance on the fabric. The difference between sueding and cropping is that, in the former process, the fibers on the surface of the fabric are cut whilst in the latter, the fibers are pulled off the surface of the fabric.

Compressive Shrinkage

Compressive shrinkage is carried out to replace the warp crimp, which has been pulled out during the preparation and coloration processes. The process uses compressive shrinking machines to effect the requisite change.

3.6.3.2 Chemical Finishing Processes

Chemicals are used to treat fabrics mainly to improve the serviceability of the fabrics. Chemical finishes to improve such fabric properties such as handle modification, water repellency, moth proof, antibacterial and antifungal behavior, anti-shrink, crease resistance, flame retardance, and antistatic behavior. The nature of the

Table 3.7 Chemicals and the properties they engender in fabrics

Fabric property	Chemical	Mode of application
Stiffening of fabric	Starch	Apply in solution to fabric followed by drying
Fabric softeners	Nonionic mineral oils Cationic surfactants	Apply in solution
Water repellent	Fluorocarbons Heavy metal soaps	Treat surface of textile
Oil repellent	Fluorocarbons	Treat surface of textile
Soil release	Polymers containing hydrophilic groups Block copolymers with hydrophilic and hydrophobic groups	Deposit on surface of textile
Moth proofing	Synthetic pyretheroids	Treat fabric with chemical
Crease recovery	Urea Formaldehyde	Treat fabric with chemical
Flame retardant	Proban process Pyrovatex process	Reaction of chemicals with fabric
Antistatic	Anionic and cationic agents	Treat fiber with chemical

treatment depends on the base fiber of the fabric. Chemical finishes hardly improve the attractiveness of fabrics. Table 3.7 lists the chemicals and the particular fabric property that they engender.

3.7 The Textile Industry and the Environment

Pollution of the environment is a major challenge of the textile industry. The processes used in the preparation of fiber and its subsequent conversion into finished fabric invariably involve the use of chemicals and production of by-products, some of which find themselves into the environment through process effluent waters and the surrounding air. Volatile solvents used during printing are released into the air alongside solids or dust particles, which are generated during fiber preparation. Most of the liquid effluent and air-borne chemicals or particles, for instance, halogens, reducing and oxidizing agents, metals, amines, etc. are not friendly to humans, flora and fauna. Therefore, it is important to monitor and curtail their release and to find measures to treat effluents containing these materials before discharge into the environment.

Process water from textile manufacturing is typically alkaline and has high biological oxygen demand (BOD) and chemical oxygen demand (COD) loads. Pollutants are found in textile effluents, which include suspended solids, anti-foaming agents, grease, spinning lubricants, surfactants, phenols, halogenated organics, heavy metals, pesticides and even microbial agents. Air emissions include dust associated with natural fiber and synthetic staple processing; carbon disulfide, hydrogen sulfide and nitric acid from processing of regenerated fibers; volatile organics and oils mists from printing and fabric cleaning processes; exhausts gases from power generation; and odors from dyeing and finishing processes. These and others are the pollutants that must be dealt with in the industry often by specific treatment processes. In view of the multiplicity of the processes involved in textile manufacture, and the specific treatment of these processes, there is a wide variation of chemicals or pollutants generated by the industry. This in turn demands several treatment protocols to deal with the specific pollutants. Table 3.8 lists the pollutants associated with specific unit operations of the textile industry and suggested measures to mitigate their release into the environment.

Table 3.8 Pollutants in the textile industry

Unit operations	Pollutants	Mitigation measures
Scouring	Detergents, lanolin, alkali, organic solvents, BOD and COD loads	Use biodegradable detergents, e.g. alcohol ethoxylates. Adsorption of volatile organic compounds emitting solvents. Mechanical removal of water prior to drying process
Desizing	Organic matters and solids. High BOD & COD loads	Use more biodegradable sizing agents. Use enzymatic or oxidative desizing with starch. Selection of raw materials with low add-on techniques
Bleaching	Sodium hypochlorite, sulfur dioxide, organic halogens, AOx, alkaline solutions, etc.	Use hydrogen peroxide instead of chlorine- and sulfur-based bleaches. Reduce use of sodium hypochlorite, Use biodegradable stabilizers where possible
Mercerizing	Highly alkaline effluents	Recovery and reuse of alkali
Dyeing	Halogens, color pigments metals (Cu, Zn, Co, and Ni), amines, antifoaming agents, alkali, reducing & oxidizing agents, high BOD & COD loads	Use of automatic systems for dosing and dispensing dyes. Use continuous or semi-continuous dyeing processes. Use of bleaching systems that reduce liquor-fabric ratios
Printing	Volatile organic compounds, mineral spirits, pigments& dyes, defoamers, resins	Reduce the volume of printing paste supply. Recycle the rinsing water used at the printing belt. Apply appropriate printing method depending on the job at hand, For instance, use transfer printing for synthetic fabrics and digital ink-jet printing for short runs of fabrics
Mothproofing	Biocides such as those based on permethrin & cyfluthrin	Minimize spillage of within the house of mothproofing agents during dispensing and transport. Avoid the use of dyeing auxiliaries that can retard the uptake of mothproofing agents

3.8   The Textile Industry in Ghana

The textile industry in Ghana can be divided into two broad groups. In the first group are those large-scale companies whose production capacity runs into thousands of yards of fabric per year. These companies use modern machinery and equipment and processes earlier described in the text for their production. The companies in this group manufacture mostly not only for the local market but also for export. These companies employ several hundred personnel in the production lines. In the second group are small scale, exclusively, indigenous enterprises that produce textiles mostly for the local market. This latter group can further be divided into two, namely, those that weave indigenous textiles such as kente or smock ('frugu'), and those that use homemade designs to print or decorate already woven fabrics. This latter group includes tie and dye (or tie-dye) manufacturers, batik cloth makers and 'Adinkra' cloth producers.

3.8.1   Commercial Textile Manufacturing in Ghana

The history of the textile industry in Ghana is a checked one. Originally developed as part of a grand vision of the post-independence governments to industrialize the nation, it employed at its peak in the early 1970s an estimated 25,000 skilled and non-skilled workers in about 40 textile manufacturing plants. The industry was and remains cotton based. Some of the companies had machinery that processed raw cotton into yarn and thence into fabrics and garment. Others only processed printed fabric from bleached fabric. The cotton was supplied mostly from local farms, which meant that a vibrant local cotton industry was developed alongside the textile industry. In the 1980s and thereafter, however, the industry began to decline, mainly as a result of a generally unfavorable economic conditions in the nation but increasingly as a result on unfair competition of cheaper products from China, Pakistan, Nigeria and Ivory Coast. By 2000, only 5000 workers were employed in the industry. Even so, the decline continued such that by 2005 only four major textile manufacturing plants remained in production which together employed only about 2961 persons. Presently, the only textile companies operating in the country are Ghana Textile Manufacturing Company (GTMC), Akosombo Textile Limited (ATL), Ghana Textile Product (GTP) and Printex. Among these, currently, only GTP process cotton into yarn and fabric. All the other import fabrics are then processed and printed/dyed into garments. These companies produce mostly cotton-based textile products, which include African prints (wax, java, fancy, bed sheets, and school uniforms) and household fabrics (curtain materials, kitchen napkins, diapers and towels). Some products from man-made fibers (synthetics, e.g. polyester) and their blends are also made. These include uniforms, knitted blouses and socks.

3.8.2 The Indigenous Textile Industry

Long before the advent of Europeans in Ghana, and before the setting up of commercial textiles industry in Ghana, indigenous people of Ghana made clothes using the loom and pattern printing with wood carved objects—locally called 'Adinkra'. In the North of the country smock (frugu), weaving was an already established industry. Recently, tie and dye and batik cloth making have also been introduced into the country. These, mostly craft industries, are carried out on a small scale by indigenous Ghanaians.

3.8.2.1 Kente Weaving

The product of Kente weaving is the Kente cloth, a brilliant colorful cloth made from cotton, rayon, sheen, or silk and woven on the traditional loom. This loom device is built with wood and operated manually. The design and operation of the loom are many respects similar to the loom described earlier in Sect. 3.5.2. In particular, features such as heddles, pulleys with spools, shuttles with bobbins, beaters, sword stick combined with other structures such as skein winder, bobbin winder, bobbin holder and a heddle-making frame are all used with the traditional loom. Kente is usually woven into four inches wide strips, which are then sown together into various sizes, which are further added together to manufacture a larger Kente cloth. The main weaving communities of Kente cloth are in the Ashanti region where legend has it that Kente weaving originated and Volta region, which also has a very strong tradition of Kente weaving. Even though data are scanty on the extent of the Kente weaving industry in the country, it is a thriving business in Bonwire, Adanwomase and surrounding towns in the Ashanti region, and also in the Agortime and Sogakope areas of the Volta Region. Pockets of Kente manufacturers may also be found in some other southern communities in Ghana.

3.8.2.2 The Adinkra Cloth

Adinkra is the name of an indigenous hand printed fabric made from unbleached 'cotton' fabric by stamping colored patterns (designs) into the fabric using wood-carved stamps. During Adinkra making, the cotton fabric is, first, completely soaked in a base colored dye (often black but sometimes red) following which the fabric is dried often in the sun. Subsequently, the dyed fabric is spread out on a flat surface and patterns (wood bearing designs) are pressed into the base fabric to give it a designed appearance. The patterns are first dipped into dye solutions which may have a color different from the color of the base fabric. Adinkra like Kente is a ceremonial cloth except that Adinkra is worn mostly during funerals and during occasions that demand sober reflections. In Ghana Adinkra is almost exclusively made in Ntonso and surrounding towns in the Ashanti Region.

3.8.2.3 Tie and Dye Cloth

Tie and dye is yet another technique used to decorate fabrics. It derives its name from the processes by which the fabric is colored or decorated. During tie and dye, the fabric is, literally, first tied into folds and subsequently dyed. The process uses bleached cotton fabric as the base cloth which is first folded and tied depending on the color design/patterns envisaged for the fabric. Portions of the folded fabric are thereafter soaked completely in a base dye. When the fold is spread out, the dye will spread across the fabric making beautiful colors and patterns. Finally, the fabric is dried in order to drive out moisture. Sodium chloride, sodium hydroxide and chrolex are usually dissolved in the dye solution in order to improve the fastness of the dye into the fabric. Tie and dye, like the Kente and Adinkra, is carried out on small scale in Ghana. It is, however, an industry that can be found in most communities throughout the country, mostly carried out by women folks and community groups.

3.8.2.4 Batik Cloth

The batik process is yet another approach to decorating fabrics with colors and designs. In batik making, the fabric design is achieved by dipping patterns/design on wood or foam into melt wax and applying the wax into the fabric. Subsequently, the waxed fabric is dipped into a dye whereupon those areas of the cloth without wax are soaked with the dye. After drying the fabric, the application of wax and dipping into a dye solution may be repeated in order to achieve the desired design in the cloth. As a final step, the wax is removed by boiling the fabric in a hot water, and finally drying the fabric. Unlike, tie and dye or Adinkra making, the batik process is tedious, lengthy and more complex. A flow chart of the batik-making process is shown in Fig. 3.10.

3.8.2.5 Frugu Making

In Ghana, smock wear is a distinctively of Northern Ghana origin. In recent times however, it is worn throughout the country by all manner of Ghanaians. Some use it as every day wear—particularly in the North, others use it as an office wear, yet others wear it on special occasions. The process of making frugu is in many respects like Kente making. Like Kente, strips of the fabric are first woven on the loom. Subsequently, the strips are sown together into large sizes. Furgu is made exclusively from cotton yarn, which may be dyed prior to weaving into fabric or not.

Fig. 3.10 A simplified schematic of the batik making process

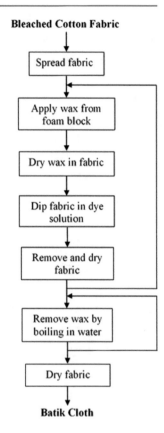

Bleached Cotton Fabric

Spread fabric

Apply wax from foam block

Dry wax in fabric

Dip fabric in dye solution

Remove and dry fabric

Remove wax by boiling in water

Dry fabric

Batik Cloth

Further Readings

Babu, K. M. (2018). *Silk: Processing, properties and applications* (2nd ed.). Woodhead Publishing Ltd.

Cobman, B. P. (1985). *Fiber to fabric* (6th ed.). McGraw Hill Higher Education.

Hearle, J. W. S. (2008). *Physical properties of textile fibres* (4th ed.). Woodhead Publishing Ltd.

Johnson, I., Cohen, A. C., &Sarkar, A. K.(2015).J.J. Pizzuto's fabric science swatch kit (11th ed.). London: Fairchild Books

Kadolph, S. J. (2016). *Textiles* (12th ed.). Pearson.

Morton, W. E. (2008). *Physical properties of textile fibres* (4th ed.). Woodhead Publishing Ltd.

Thompson, R. (2014). *Manufacturing processes for textile and fashion design professionals.* Thames & Hudson Ltd.

Woodings, C. (Ed.). (2001). *Regenerated cellulose fibres.* Woodhead Publishing Ltd.

Fertilizer Technology

4

Abstract

Fertilizers are substances that are added to the media where plants are grown to stimulate plant growth and promote crop yield. Fertilizers added to the soil for uptake of nutrients through the roots but may also be applied by foliage feeding for update through the leaves. Fertilizers may occur in nature (e.g. peat or mineral deposits, saltpeter) or produced through natural process such as composting, or may be manufactured by chemical processes (e.g. the Harber Process). Each nutrient/fertilizer has specific roles on plant growth. For instance, nitrogen is required by plants for protein metabolism and promotes leaves and stem growth. Potassium facilitates the workings of the nervous system of the plants. Potassium deficiency causes yellowing of plant, readiness to wilt, falling of leaves and poor development of fruits. Phosphorous provides the energy requirement of plant reactions. Sufficient amount in soils promotes good root growth and the production of flowers and fruits. In this chapter, the methods for manufacturing the various fertilizers and nutrients have been covered in detail. Also discussed in this chapter is the fertilizer industry in Ghana.

4.1 Introduction

Plants require nutrients to grow and to yield crops to meet the expectations of farmers. In fact, in tilling the land, i.e. growing plants on the land, the main aim of all farmers both commercial and subsistence is that the plant will yield crops in acceptable quantities. Plants achieve this by feeding on nutrients in the soil, which are essential both for plant growth and crop yield. In a natural cycle, the nutrients return to the soil when the plant dies and decays, and thus are released for another batch of plant growth. However, commercialization of farming and stress on land because of the increased need for food for a growing world population has meant that this natural cycle of plant nutrient uptake and return to the soil is hardly any longer sustainable. For instance, in commercial farming, the harvested crops are often transported far away from the point of harvest so that the nutrients carried by the crops do not return to the soil from where they were harvested. Also, increased population and therefore increased need for food have meant that the same soil has to be tilled severally within the natural cycle of plant growth and decay. In other words, the nutrients extracted from the soil by a plant which in the normal cycle will return to the soil do not get to the soil at the time another batch of crops are cultivated. These and other causes, for example, loss of soil nutrients from soil as a result of erosion born of over exploitation of land has meant that for the soil to continually support plant growth, its nutrients have to be replenished from sources other than the natural cycle for it to continually support plant growth. Substances that are added to various media in which plants are grown in order to promote plant growth and foster crop yield are called *fertilizers*. The elemental nutrients required by plants may be grouped into four, depending on the extent of their presence in plants, as structural elements, primary elements, secondary elements and micronutrients. The relative amounts of specific elements in the various groups found in a healthy plant are listed in Table 4.1. In Table 4.2, are the relative

Table 4.1 Elemental requirement of a healthy plant

Element	Amount in whole plant (% dry weight)
Structural elements	
Oxygen	45.0
Carbon	44.0
Hydrogen	6.0
Primary nutrients	
Nitrogen	2.0
Phosphorus	0.5
Potassium	1.0
Secondary nutrients	
Calcium	0.6
Magnesium	0.3
Sulfur	0.4
Micronutrients	
Boron	0.005
Chlorine	0.015
Copper	0.0001
Iron	0.020
Manganese	0.050
Molybdenum	0.0001
Zinc	0.0100
Total	99.9011

Table 4.2 Comparing elements (1) normally in soils; (2) required to be in the soil in assimilable form for healthy plant growth and (3) present in the plant

Elements	In the soil (kg/ha)		Presence in plant (dry matter basis) (µmol/g)
	Normally present	Requirement in assimilable	Plants
Carbon	$40.0–55.0 \times 10^3$	$1.0–1.5 \times 10^3$	35.0×10^3
Hydrogen		$1.5–2.5 \times 10^3$	60.0×10^3
Oxygen			30.0×10^3
Nitrogen	$2.5–3.5 \times 10^3$	$1.0–2.0 \times 10^2$	10.0×10^3
Phosphorous	$0.9–1.6 \times 10^3$	$0.5–1.5 \times 10^2$	60.0
Potassium	$3.8–4.5 \times 10^4$	$0.5–1.5 \times 10^2$	250.0
Calcium	$9.0–23.0 \times 10^3$	$0.3–0.5 \times 10^2$	125.0
Magnesium	$6.5–13 \times 10^3$	$0.2–0.6 \times 10^2$	80.0
Sulfur	$9.0–18.0 \times 10^2$	$0.1–0.5 \times 10^2$	30.0
Chlorine	$0.2–1.0 \times 10^2$		3.0
Boron	$0.1–0.3 \times 10^2$	0.1–1.0	2.0
Iron	$10.0–100.0 \times 10^3$	3.0–11.0	2.0
Manganese	$0.4–20 \times 10^3$	5.0–22.0	1.0
Zinc	$0.2–5.0 \times 10^2$	5.0–11.0	0.3
Copper	$0.1–0.3 \times 10^2$	5.0–22.0	0.1
Molybdenum	0.4–10	0.01–0.07	0.001

amounts of these elements generally found in soil, required in soil to support plant growth and the concentration in a healthy plant.

Due to the large abundance of carbon, oxygen and hydrogen in nature, their supply is considered outside the responsibilities of the normal fertilizer manufacturer. With this in view, fertilizers are grouped into three broad categories based on the requirement of plant for growth. Primary fertilizers comprise nitrogen, phosphorus and potassium fertilizers. Plant requirements for calcium, sulfur and magnesium are an order lower than for primary fertilizers. Fertilizers of these elements are therefore called secondary fertilizers. The requirements of plants for the other elements including copper, iron, manganese, zinc, boron and molybdenum are of an order lower than secondary fertilizers. Fertilizers based on thise last group of elements are therefore called tertiary fertilizers. Fertilizers are often administered to plants via the soil for uptake through the roots, but they may also be applied by folia feeding for uptake through the leaves. They may be naturally occurring, such as, for instance, peat or mineral deposits, saltpeter or manufactured through natural processes—such as composting, or chemical processes—such as the Harber Process.

4.2   The Role of Specific Nutrients (Elements) in Fertilizer on Plant Growth

Each element on which a fertilizer is based plays a specific role in a healthy plant growth and crop yield. These roles are discussed in the following sections.

4.2.1   Nitrogen

Nitrogen is required by plants for protein metabolism. They are needed mostly in the leaves where amino acids and proteins are produced from carbohydrates. The leaves increase in size because cells are able to grow from additional protoplasm produced. A plentiful nitrogen supply may cause relatively more protoplasm to be produced than cell wall (the cell wall is essentially made of carbohydrates and cellulose) and the leaves become soft and juicy thereby becoming attractive to animals and bacteria. Lack of nitrogen leads to yellowish, small, thin and premature leaves. Crops that are not grown for their leaves require careful control of nitrogen. Nitrogen is introduced into the soil in one of two forms, namely, through organic matter, for instance, leguminous plants or through nitrogen compounds introduced into the soil.

4.2.1.1   Nitrogenous Organic Matter
Nitrogen-containing organic matter in the soil releases its nitrogen in assimilable form through the activities of microorganisms and animals devouring the organic matter and excreting nitrogen-containing compounds. The form of nitrogen-containing compound excreted depends on the nature of the organism devouring the organic matter.

4.2.1.2   Other Nitrogenous Matter
Nitrogen is also introduced directly into the soil in the form of nitrogen-containing compounds such as nitric acid, which is formed by the combination of nitrogen and oxygen in the atmosphere during thunderstorm and subsequently dissolved in rain. Other nitrogenous compounds introduced into the soil are nitrates, ammonium compounds, urea, cyanamide, cellulose nitrate and urea–formaldehyde resins. In the soil, all nitrogen-containing compounds are first converted into nitrates or ammonia before assimilation by plants. Before the nitrogen is made use of by the plant, it must first be able to assimilate the substance when it is in soil solution. Subsequently, once the substance has entered the plant, it must be able to be absorbed into the metabolic processes in the plant.

4.2.2   Potassium

Potassium plays important role in plants of facilitating the workings of the nervous system of plants. In plants, the role of the nervous system has to do with communication mechanism among the various parts of the plant. Potassium deficiency manifests in the yellowing of the plant, readiness to wilt, a falling of the leaves, 'leaf scotch' and poor development of fruit. Plants growing on soil with sufficient potash do not wilt or scotch so easily in warm dry weather.

4.2.3   Phosphorous

Inorganic orthophosphate provides the energy requirement of most plant reactions. Plants take up phosphorous in the form of H_2PO_4. Adequate supply of phosphorous in the soil is associated with good root growth and production of flowers and fruits as compared with leave and stem growth, which is associated with nitrogen. Phosphate deficiency is generally difficult to diagnose just from the appearance of the plant because the effect is of general debility in anthropomorphic terms as though the plants were generally lacking in energy and needed tonic.

4.2.4   Calcium

The presence of calcium in the soil is identified with controlling the uptake of ions by the plant. It also plays a role in the strengthening of cell membrane and in nitrification processes in the soil. In the presence of other fertilizers, it assists in the disintegration of animal and vegetable remains in the soil and in releasing ammonia from substances in the soil.

4.2.5   Magnesium

Magnesium is a constituent of chlorophyll, the compound known to play a very important function of photosynthesis in plants. It is known that when cattle feed on grass that lack chlorophyll, they develop the disease called staggers. Magnesium is taken up by plants in the form of magnesium ions, Mg^{2+}.

4.2.6   Sulfur

Sulfur is present in many important constituents of plants, for instance in proteins, in the vitamins thiamin and biotin, in mustard oils and in allene sulfides. Deficiency of sulfur causes chlorosis in maize—a condition where old leaves are green while new ones are paler. When sulfur is abundant in the soil, chlorophyll, vitamins and proteins of legume plants all increase. It appears that sulfur is taken up by plants in the form of sulfate ions, SO_4^{2-}.

4.2.7   Tertiary Plant Foods

These elements occur in trace quantities in plants, yet the importance of their role in plants growth cannot be over-emphasized. These elements must be available to plants before such classes of compounds as enzymes, vitamins, hormones may be produced in plants. Metabolism in plants would be hampered or changed if these elements are absent in plants. These elements include boron, iron, copper, manganese, zinc, molybdenum, gallium, vanadium, chlorine, iodine, cobalt, silicon and selenium. The isolated effect of individual nutrients is not always conspicuously evident as most of the elements act in conjunction with others. Nonetheless, in certain cases, these have been studied and the results are evidently clear. Table 4.3 lists the role some of these elements play in plant growth.

4.3   Determining the Nutrient Need of Plants

Before any fertilizer is applied to the soil, it is necessary to first figure out the nutrient need for a particular crop or plant. The need will depend on the specific crop or plant that is being cultivated and the state of the soil on which the cultivation will be done. Other factors such as the climate, the time of the year for planting and the fertilizer and/or manure addition history of the soil are also considered. The best way to determine which nutrients are added and in what amounts is to find out through a soil test or through a field experiment.

4.3.1   Soil Test

To do a soil test, a representative sample of the soil is first collected from the area or field where the cultivation is to be done. In collecting a representative sample from the intended field of cultivation, it must be born in mind that differences in soil in various parts of a field as small as an acre can sometime be significant. A test is done for the presence of essential nutrients such as useful nitrogen, soluble phosphorus and potassium, the soil acidity, soil lime requirement, soil texture, soil drainage and the presence or otherwise of other essential plant food elements.

4.3.2   Field Experiment

During a field experiment, a series of plants of different species from seeds in the samples of the soil intended for those species are grown on the field. Subsequently,

Table 4.3 The role of tertiary elements in plant growth

Trace element	Effect when present in plants	Effect when absent in plants
Iron	Necessary for the synthesis of chlorophyll. Involved in the transport of oxygen in respirator reactions. It is present in the enzymes, which catalyze nitrogen fixation by legumes	
Boron	It plays a role in the dividing process of the cell nucleus	General inability of the plant to develop, to form roots, to form flowers and to form growing points
Copper	It is involved in the photosynthesis and respiratory processes of plants. It is able to precipitate toxins in the soil, and thus prevent their uptake by the plant	Lack of copper can make cereals prone to 'lodging'
Manganese	It is required in protein synthesis and influences the selection of amino acids entering the protein molecule	Lack of it causes 'grey speck' in oats, 'speckled yellow in sugar-beet, 'marsh spot' in beans, and dark-brown spots along the veins of potato leaves
Zinc	Helps in the elimination of carbon dioxide from plants. It is involved in the production of chlorophyll. Helps in good seed formation of plants. It is also involved in the reproductive system of plants	
Molybdenum	It is associated with enzymes, which reduce nitrates to ammonium ions for the formation of amino acids. It is also present in the enzymes necessary for the fixation of nitrogen by legumes	
Cobalt	Seems to be necessary for some microorganisms	
Silicon	It gives turgidity to plants. It increases the yield of rice and sugar cane	

the uptake of food elements in the grown plants is measured. Also, bacteria and fungi that are known to have the ability to extract nutrients of all known kinds from the soil are grown on the field and to analyze their growth for nutrients.

Based on the results and observations of these tests, conclusions are drawn and recommendations are made on the nutrient needs of the soil and the kind and quantities of particular fertilizers or manures that could be applied to the soil.

4.4   Types of Fertilizers

Fertilizers may be classified broadly as organic or inorganic and naturally occurring or manufactured. These grouping and examples in each group are shown in Fig. 4.1.

4.4.1   Organic Fertilizers

These are fertilizers made of natural materials that undergo little or no processing and include plant, animal and mineral materials. There are two groups of these fertilizers; naturally occurring organic fertilizers and manufactured organic fertilizers. In the former group are manure, slurry, worm castings, peat, sewage and green manure among others. This group also includes naturally occurring minerals such as mined rock phosphate, sulfate of potash and limestone. Materials in this group are applied to the soil without any further processing save sometimes size reduction of the material. The manufactured organic fertilizers group includes compost, blood meal, bone meal and seaweed extracts. Others are natural enzyme digested proteins, fish meal and feather meal. This later group of fertilizers, even though of organic origin, is subjected to one kind or the other of physical processing before application. Some ambiguity in the usage of the term 'organic' exists because some of synthetic fertilizers, such as urea and urea formaldehyde, are fully organic in the sense of organic chemistry. Indeed, it is difficult to make distinction between the chemistry of urea from nature and that synthesized by man. It is equally noteworthy that some fertilizer materials commonly approved for organic agriculture are actually inorganic compounds. Examples are powdered limestone and mined "rock phosphate".

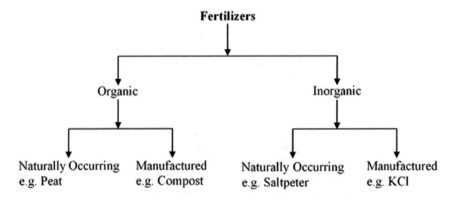

Fig. 4.1   Types of fertilizers available

Generally, the density of nutrients in organic material is comparatively modest. Yet their use as a source of plant nutrients continues to attract interest because their use has several advantages. These include:

i. They can be produced on-site and thus lowering transport costs.
ii. Most nitrogen supplying organic fertilizers contain insoluble nitrogen and act as a slow-release fertilizer.
iii. They mobilize existing soil nutrients, so that good growth can be achieved with lower nutrient densities while wasting less.
iv. They release nutrients at a slower and more consistent rate, helping to avoid boom-and-bust pattern.
v. They help to retain soil moisture and hence reducing the stress associated with temporary moisture stress.
vi. They improve the soil structure.
vii. Application of natural fertilizers avoids the possibility of 'burning' of plants sometimes accompanying the concentrated use of chemicals.
viii. Their use avoids the progressive decrease of real or perceived 'soil health', apparent in loss of structure, reduced ability to absorb precipitation, and lightening of soil color.
ix. Their use avoids the necessity of reapplying artificial fertilizers regularly.
x. Helps to reduce the cost of fertilizer application and affords the farmer a great deal of independence.

The foregoing advantages notwithstanding, the continued use of organic fertilizers comes with its own costs, such as;

i. The composition of organic fertilizers tends to be more complex and variable than a standard organic product.
ii. When improperly processed, organic fertilizers may contain pathogens from plant or animal matter that may be harmful to humans or plants.
iii. When significant amounts of nutrients are needed in the soil, large volumes of organic fertilizer would have to be transported to be applied (because they are diluted nutrient source compared to inorganic fertilizers). This results in an increased cost of transportation.

4.4.2 Inorganic Fertilizers

Plant nutrients are most often supplied to the soil as chemical fertilizers. These chemicals occur naturally in the earth's crust as, for instance, in the case of saltpeter or are synthesized through chemical reactions as in the case of ammonia. The chemicals commonly used in this group are of inorganic origin and hence the name inorganic fertilizers. The inorganic fertilizers are applied to the soil as single elements, though often as a compound with other elements. Examples of elements applied in this manner are nitrogen in ammonia or potassium in potassium chloride. The inorganic fertilizers may also be applied as a complex mixture of several plant nutrients as in the case of potassium nitrate and NPK. In these fertilizers, the plant nutrients are in a concentrated form and as such are not applied in the same bulk as are organic fertilizers. This feature of the inorganic fertilizers offers them a major advantage over their organic counterparts.

The bulk of inorganic fertilizers applied to the soil are primary fertilizers since these are the ones used up in greater quantities by plants. They are, therefore, the ones often depleted from the soil and hence require regular replenishment. Of course, other plant nutrients, namely, secondary and tertiary nutrients, as the need may be, are also applied to the soil as inorganic fertilizers either as single elements of other compounds or as a mixture in conjunction with other plant nutrients. Depending on their nutrient content, fertilizers can be classified into straight and complex fertilizer. Straight fertilizers contain only one plant nutrient whereas complex fertilizers contain more than one primary or major nutrient element.

4.4.2.1 Straight Fertilizers

Sometimes also called *single nutrient fertilizers*, these are mostly the primary plant nutrients i.e. nitrogen, phosphorous and potassium (NPK). In these fertilizers, the nutrients in the form of compounds with other elements are applied. However, they are straight because in the compounds in which they feature they are often the one plant nutrient, as for instance nitrogen in ammonia or potassium in potassium chloride. The straight fertilizers of the primary nutrients are discussed below.

i. Manufacture of Nitrogen Fertilizers

Nitrogen is an important plant nutrient. It has, perhaps, the quickest and the most pronounced effect on plants. Even though it comprises 79% by volume of the gases in the earth's atmosphere, only a few plants, chiefly legumes, can utilize the atmospheric nitrogen directly. For most plants including such important plants as rice, maize, wheat, the nitrogen uptake is from 'fixed nitrogen'—nitrogen in chemical form dissolved in soil solution. The bulk of the nitrogen in the soil (soil solution) gets there as a result of deliberate application of nitrogen-containing substances to the soil by man. Most of this is in the form of inorganic nitrogen fertilizers. However, a small percentage of the nitrogen in the soil gets there as a result of natural phenomena. For example, lightning discharges convert small amounts of atmospheric nitrogen to nitrogen oxides, which are then carried into the soil by rainwater. Nitrogen is found in the ammoniacal, nitrate or amide forms in nitrogenous fertilizers. Table 4.4 lists the fertilizers of nitrogen that fall in the category of straight fertilizers. The mass percentage of elemental nitrogen (N) in the fertilizers is listed in the second column of the table.

The bulk of the nitrogen fertilizers listed in Table 4.4 is synthetically manufactured. The exceptions are sodium nitrate and ammonium sulfate. While sodium nitrate is mined as an ore, ammonium sulfate is produced as a by-product of a number of chemical processes.

Table 4.4 Straight fertilizers of nitrogen

Name of fertilizers	Percentage of nitrogen (% wt)
Ammonium sulfate	20.6–21.0
Urea	44.0–46.0
Ammonium chloride	25.0+
Ammonium nitrate	32.0–35.0
Ammonium sulfate nitrate	2.6
Calcium ammonium nitrate (CAN)	25.0
Sodium nitrate	16.0
Calcium nitrate	15.6–21.6
Potassium nitrate	13.0
Calcium cyanamide	212.0
Ammonia	82.4

Sodium Nitrate

An impure form of sodium nitrate known as 'caliche' is mined as a mineral from rocklike deposits in a relatively dry desert area of Chile. The mined ore is first sized, and the sodium nitrate is leached with water. The resultant solution is evaporated yielding crystalline sodium nitrate, which is applied directly as a fertilizer. The fertilizer grade is called 'Chilean nitrate', 'nitrate of soda' or 'sody'.

Ammonium Sulfate

The first chemical process from which ammonium sulfate is produced as a by-product is the *coking process*. During the production of coke from coal, ammonia is released as a by-product. The emerging ammonia is recovered when the emerging gases of the coking process are passed through a scrubber-crystallizer system containing sulfuric acid. In the scrubber, the ammonia in the effluent gas reacts with the sulfuric acid to produce ammonium sulfate crystals, which is an excellent fertilizer providing both the primary nutrient nitrogen and the secondary nutrient sulfur to plants. Another process that yields ammonium sulfate as a by-product is during the production of caprolactam and acrylonitrile. These latter processes provide larger volumes of ammonium sulfate than the cooking process. The last of the processes from which ammonium sulfate is produced is from the reaction of spent sulfuric acid from other processes with ammonium. This reaction is used in industry as a means of disposing of spent sulfuric acid.

ii. Manufacture of Phosphorus Fertilizers

Phosphorus is the second most important fertilizer element after nitrogen. It is an essential constituent of every living cell and for the nutrition of plant and animal. It is an active component of all metabolisms of plants. Plant uptake of phosphorous is entirely from the soil via root absorption. Phosphorous is taken up by plants in the form of H_2PO^{4-}, or $HPO4^{2-}$, $PO4^{3-}$ ions. Native soils have a very low phosphorous content. Again, a portion of the soil phosphorous is present in forms that are not assimilable or at best slowly available to crops. The phosphate in fertilizers is present in two forms, namely, water soluble and citrate soluble forms. The former is soluble in water while the latter is not soluble in water but rather is soluble in 2% neutral ammonium citrate solution. The value of the sum of the two forms of phosphate in phosphate fertilizers is termed as *available phosphates*. Cropping of native soils without phosphate fertilization soon depletes the supply of phosphorous from the soil and renders the soil barren. Modern, high-yield agriculture depends on regular fertilization with phosphorous compounds that either is immediately soluble in the soil or that becomes soluble at a rate sufficient to supply the crop. An important parameter in phosphorous fertilization is the soil 'fixation' of phosphorous. This parameter has to do with the characteristic of the soil to convert the applied phosphorous fertilizer into a chemical form available to plants. This parameter depends on the type of soil onto which the fertilizer was applied. It is highest in soils of high clay content. On some soils, as much as 60% of the applied phosphorous fertilizer may be lost to fixation. Table 4.5 lists the major phosphorous fertilizers and the weight percentage of phosphorous (calculated as P_2O_5 equivalent) in them.

All phosphorous fertilizers are derived from mined ores. A flowchart of the synthesis routes of various phosphorous fertilizers is shown in Fig. 4.2.

iii. **Manufacture of Potassium Fertilizers**

Potassium is the third most important fertilizer element. Potassium acts as a chemical traffic policeman, root booster, stalk strengthener, food former, sugar and starch transporter, protein builder, breathing regulator, water stretcher and as a disease retarder. It is, however, not effective without its co-nutrients such as nitrogen and phosphorus. Plants affected by a deficiency of potassium usually are stunted and spindly and exhibit a rather distinctive chlorosis and necrosis that is most prominent in the lower leaves. This may lead to a severe reduction in crop yield. Potassium is present in the earth crust in abundant quantities. It is widely distributed geographically and is commonly found in association with sodium compounds. However, they are difficult and costly to convert to forms suitable for use as fertilizers. The important forms of potassium in the fertilizer industry are potassium chloride (muriate of potash), potassium sulfate, potassium nitrate, and potassium-magnesium sulfates of varying potassium/magnesium ratios. In the fertilizer industry, the potassium content of fertilizer salts is expressed in terms of

Table 4.5 Major phosphorous fertilizers	Name of fertilizers	Percentage of P_2O_5
	Ordinary superphosphate	16.1–20.0
	Double superphosphate	30.1–35.0
	Triple superphosphate	45.0–50.0
	Monoammonium phosphate	48.0–55.0
	Diammonium phosphate	46.0–53.0
	Rock phosphate	20.0–25.0

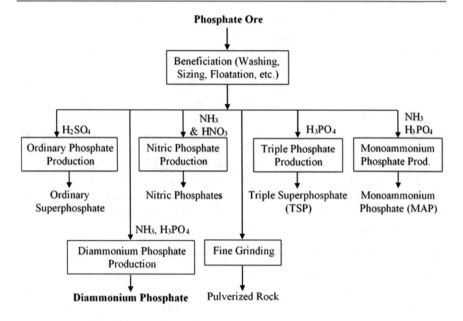

Fig. 4.2 Process routes for the production of phosphate fertilizers

potassium oxide (K_2O) content. Potassium-containing fertilizers are manufactured from potassium ore mined from the crust of the earth. The main sources of potassium for use as fertilizers are listed in Table 4.6.

Muriate of Potash (Potassium Chloride)

The chemical is refined from sylvite ore, a mechanical mixture of potassium chloride and sodium chloride. The sodium chloride component of the ore is harmful to plants and as such is separated from the KCl. The separation of the potassium salt from the sodium salt is achieved by physical methods. In one such method, a heavy medium, for example, a solution of magnetite in brine is mixed with the crushed ore in a floatation vessel. The density of the magnetite-brine solution lies between that of the potassium chloride and that of the sodium chloride, as such upon suspending

Table 4.6 Sources of potassium for use as fertilizer	Name of fertilizers	Chemical formula	Percentage of (K_2O)
	Muriate of potash	KCl	50.0–60.0
	Potassium sulfate	K_2SO_4	48.0–52.0
	Kainite	$KCl \cdot MgSO_4 \cdot 3H_2O$	19.26
	Langbeinite	$K_2SO_4 \cdot 2MgSO_4$	22.7
	Potassium Nitrate	KNO_3	

the crushed ore in magnetite-brine solution the KCl floats off, and the contaminating NaCl sinks and are drawn off and rejected from the bottom of the floatation vessel. Entrapped magnetite from both the product and waste stream are recovered magnetically and recycled.

In yet another physical method called the froth floatation method (explained earlier in Sect. 2.4.2.5), the crushed ore that has been previously ridden of clay particles in a scrubber is first treated with an aliphatic amine acetate and a froth-promoting alcohol. The sylvite particles are preferentially attracted to the amine acetate phase. In a subsequent step, the ore is maintained in suspension, and the coated sylvite particles are attracted to and entrained on rising air bubbles generated by the floatation agitator. The ore particles rise to the surface and are collected hydraulically or mechanically, centrifuged to remove the hydraulic medium, dried, screened into various particle sizes and sent for storage. A flowchart of the froth floatation method is shown in Fig. 4.3.

Kainite (KClMgSO₄·3H₂O) and Langbeinite (K₂SO₄·MgSO₄) Fertilizers

Kainite ($KClMgSO_4·3H_2O$) and Langbeinite ($K_2SO_4·MgSO_4$) Fertilizers

These salts are important sources of potassium fertilizers. Additionally, they supply significant quantities of magnesium and sulfur to plants. Kainite contains 9.94% magnesium and a sulfur content of 13.11%. Langbeinite contains 11.71 and 23.18% of magnesium and sulfur, respectively. Both salts are recovered from potash mines. During the recovery of Langbeinite, the ore is first crushed and the contaminating

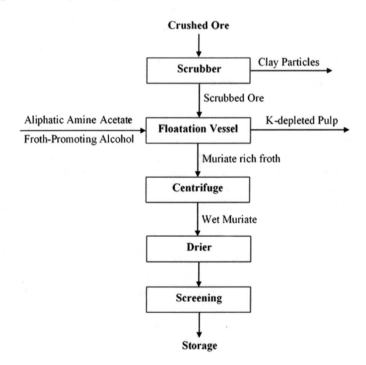

Fig. 4.3 Flowchart for the physical separation of muriate from sylvite ore

KCl and NaCl are removed by contacting the ore with water. Subsequently, the ore is centrifuged, dried and lastly screened for storage.

Potassium Sulfate Fertilizers

Potassium sulfate, even though is relatively costly to produce, is a preferred source of potassium for some vegetables, fruit crops and tobacco (i.e. those crops that are sensitive to chloride forms of potassium). The chemical is produced by a number of methods. In the first method, it is produced by fractional crystallization from natural brines. In the second method, it is produced from the controlled decomposition reaction of a complex sulfate salt or the reaction of the salt with KCl. An example is the reaction of KCl with langbeinite to form sulfate of potash and magnesium chloride, which proceeds as follows:

$$K_2SO_4 \cdot 2MgSO_4 \; + \; 4KCl \; \rightarrow \; 3K_2SO_4 \; + \; 2MgCl_2 \qquad (4.1)$$

Approximately half of the world's production of potassium sulfate is derived from the reaction of KCl with either langbeinite or kainite. A third method of producing potassium sulfate is from the reaction of KCl with H_2SO_4. The reaction proceeds as follows:

$$H_2SO_4 + KCl \; \rightarrow \; K_2SO_4 \; + \; 2HCl \qquad (4.2)$$

The disadvantage of this latter reaction is that both the reactant sulfuric acid and the product hydrochloric acid are corrosive substances.

Potassium Nitrate

Potassium nitrate is a very attractive fertilizer material because it contains two fertilizer elements, namely, potassium and nitrogen in a form that is available to plants. However, it is costly to produce. This chemical is produced by the reaction between nitric acid and potassium chloride as follows;

$$HNO_3 + KCl \; \rightarrow \; KNO_3 \; + \; HCl \qquad (4.3)$$

A crude form of the chemical is marketed from natural deposits of caliche ($NaNO_3$) in Chile. The quantity marketed in this form is, however, very small.

4.4.2.2 Complex Fertilizers

These are fertilizers that contain more than one of the primary soil nutrients of N, P and K. Complex NPK fertilizers have the advantage of having each nutrient in each granule. Weight for weight, they are more expensive than the primary nutrient from which they were made. However, the product quality is guaranteed. If needed, other nutrients can easily be incorporated into complex NPK fertilizers. By applying a single dose of a complex fertilizer, farmers are able to achieve the presence of each required nutrient in the soil. Mixed fertilizers are, therefore, efficient products but trends during the past few years in the types of fertilizers used have been determined by their economics, and not by considerations of agricultural efficiency or

Table 4.7 Examples of complex fertilizers

Material	Total nitrogen (N)	Neutral ammonium citrate soluble phosphate (P_2O_5)	Water-soluble phosphate (P_2O_5)	Water-soluble potash (K_2O)
Ammonium phosphate				
11-52-0	11.0	52.0	44.2	
18-46-0	18.0	46.0	41.0	
Ammonium phosphate sulfate				
16-20-0	16.0	20.0	19.5	
20-20-0	20.0	20.0	17.0	
18-9-0	18.0	9.0	8.5	
Ammonium phosphate sulfate nitrate				
20-20-0	20.0	20.0	17.0	
Nitro phosphate				
20-20-0	20.0	23.0	12.0	
23-23-0	23.0	23.0	18.5	
Ammonium nitrate phosphate				
23-23-0	23.0	23.0	20.5	
Urea ammonium phosphate				
28-28-0	28.0	28.0	25.2	
24-24-0	24.0	24.0	20.4	
20-20-0	20.0	20.0	17.0	
Potassium nitrate (crystalline/prilled)				
13-0-45	13.0			45.0
Mono potassium phosphate				
0-52-34			52.0	34.0
NPK fertilizers				
15-15-15	15.0	15.0	4.0	15.0
10-26-26	10.0	26.0	22.1	26.0
12-32-16	12.0	32.0	27.2	16.0
22-22-11	22.0	22.0	18.7	11.0
14-35-14		35.0	29.0	14.0
17-17-17	17.0	17.0	14.5	17.0
14-28-14	14.0	28.0	23.8	14.0
19-19-19	19.0	19.0	16.2	19.0
17-17-17	17.0	17.0	13.6	17.0
20-10-10	20.0	10.0	8.5	10.0

sustainability. Examples of mixed fertilizers are listed in Table 4.7. In mixed fertilizers, the amount of each component in the mixture is expressed as a percentage of the total mass of materials in the fertilizer. For example, for a mixed fertilizer of composition 15-15-15, the weight percentages of the three primary nutrients N, P, K are 15%, 15% and 15%, respectively. The remaining 55% of the fertilizer is ballast material (i.e. other materials that facilitate the uptake of the fertilizer by the plant). Table 4.7 is a list of the important complex fertilizers. Mixed fertilizers can be divided, by physical characteristics and production methods into four groups. The groups are nongranular mixtures, compound granular, bulk blends and fluid mixtures.

i. Nongranular Mixtures

These mixtures are produced by weighing batches of the N, P and K containing chemicals and blending them in drum mixers. The resulting products are stored and later applied to the field. In these products, the source of phosphate is almost invariably ordinary superphosphate (OSP), the source of nitrogen is fine-crystal by-product ammonium sulfate and the source of potassium is nongranular potassium chloride. A disadvantage of nongranular fertilizers is that they form cakes. Another disadvantage is that these products have poor handling and storage properties. This type of fertilizer is no longer in vogue. Only a small quantity is still produced and marketed.

ii. Compound Granulars

Granular fertilizers were introduced to improve the handling, transportation and the storage properties of fertilizers by increasing the particle sizes of the chemicals and eliminating fines. A number of approaches for producing granular fertilizers exist. Among these are the TVA continuous ammonia-granulator process, slurry granulation, melt granulation and steam granulation. Central to all the processes is the granulator in which the feed materials are turned into granules. The differences in the methods are due in part to the treatment of the raw materials prior to entering the granulator and also due to the handling of materials in the granulator. In the TVA continuous ammonia-granulator, all the raw materials namely nongranular ordinary sulfate, nongranular potassium chloride and ammonium sulfate crystals are fed to the granulator while sulfuric acid and anhydrous ammonia or an ammonia containing nitrogen solution are sparged under the rolling bed of solids in the drum. In the slurry granulation, the ammonia is first reacted with the acid (in this case phosphoric acid) in a preneutralizer. The reaction product, a slurry, is thereafter fed to the granulator drum where it mixes with the other components and is granulated. In the melt granulation process, a pipe reactor is used in place of the preneutralizer vessel. The product of the ammonia-phosphate reaction is a melt of very low moisture content, which is added to the rolling bed of dry solids for granulation. The steam granulation process is a much simpler granulation procedure. In this process, the feed materials are finely ground, mixed and moistened in a rotary drum or pan granulator with either water spray, steam or both. The rolling action of the granulator causes granules to form, which are hardened by a subsequent drying operation. A flowchart of the steam granulation process is shown in Fig. 4.4.

iii. Bulk Blends

If all the N, P and K materials are available in granular form, granular mixed fertilizers of almost any nutrient grade can be made by simple proportioning and blending (mixing). Blended fertilizers are very popular and now dominate the

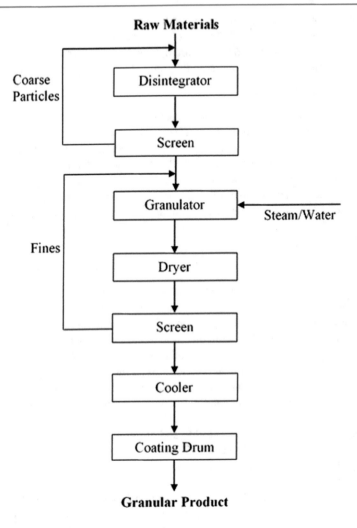

Fig. 4.4 A flowchart for fertilizer granulation by the steam granulation process

fertilizer market. Fluid fertilizers are next in popularity followed by compound granular fertilizers. This method has the advantage that it provides the farmer a wide choice of plant-food ration or grades on a 'prescription' basis. Also blending products are economical as bulk quantities can be handled without bagging.

iv. Fluid Mixtures

Fertilizers, especially mixed fertilizers (NP and NPK) may also be applied to the soil in the form of solutions or fluids. A good example is the application of urea-ammonium nitrate solution and aqua ammonia. The use of fertilizers as fluid mixtures has been promoted by two factors, namely, ease and precision with which fluids can be applied to the soil, and second the adaptability of fluids to the homogeneous incorporation of micronutrients, herbicides and insecticides. There are two major categories of fluid fertilizers. The first is fertilizer solutions, in which the plant food content is entirely in solution. The solvent may be water as in KCl solution (KCl granules are dissolved in water) or the solvent itself may be a plant nutrient as in ammonium orthophosphate solution fertilizer where raw materials (both plant nutrients) that are put together to form the product are fluids just as the product. The second category of fluid fertilizers is the fertilizer suspensions in which the nutrient solubility is exceeded in the solution, and hence the excess is held in suspension. The nutrient is kept in suspension by a suspension agent. These were developed chiefly as a means of breaking the grade barriers imposed on solution fertilizers by solubility limitations. There is a large variety in the types of fertilizer raw materials used in the preparation of suspension fertilizers. Successful preparation of this type of fertilizers is predicated on three factors: (i) small particle size of the suspended particles; (ii) the addition of gelling-type clay as a suspension agent and (iii) sufficient high-shear agitation to properly disperse the clay. The clay often used is hydrated magnesium aluminum silicate (commonly called attapulgite clay). For most fertilizer suspensions, clay concentration of 2–3% is sufficient. In one approach to preparing the fertilizer suspensions, a pre-gelling operation is carried out in which high-shear agitation is applied only to the clay plus a limited amount of water or fertilizer solution. The pre-gelled mixture then is added to the major body of fertilizer with only mild mixing. Pre-gelled clay is also marketed as liquid clay, which can be bought and used. In most operations, however, metered quantities of the main fertilizer nutrients, clay and water are fed to a mixing tank where high shear mixing is applied. The provision of high shear mixing, in addition to gelling of the suspending clay, is useful for disintegrating the granular raw materials such as diammonium phosphate or ammonium phosphate.

4.4.2.3 Secondary Nutrients

Calcium, magnesium and sulfur are called the secondary nutrients. When present in soil, calcium and magnesium are, to a large extent, responsible for the acidity or alkalinity of the soil. They are required for plant growth in relatively smaller quantities than primary nutrients. A deficiency in these nutrients and other essential elements adversely affects the efficiency of primary nutrients. Their presence in crops is necessary alongside primary nutrients to yield optimum results. They are moderately soluble, so in regions of high rainfall they are washed from the soil and carried away. Soils in these regions are usually acid, and lime (calcium carbonate) is often added every year to make the soil less acid. In areas of low rainfall, and where the soil is derived from limestone, the soil is high in calcium and magnesium and is usually alkaline. Sulfur is another secondary nutrient present in many commercial fertilizers. As sulfate, it is highly acidic and is often used to make alkaline soils acidic. The principal sources of the secondary nutrient are listed in Tables 4.8, 4.9 and 4.10.

Table 4.8 Common calcium sources

Material	Percent Ca	Relative neutralizing value[a]
Calcitic limestone ($CaCO_3$)	32	95–100
Dolomitic limestone	29	50–70
Basic slag	22	None
Gypsum ($CaSO_4 \cdot 2H_2O$)	22	15–85
Mari	24	120–135
Hydrated lime ($Ca(OH)_2$)	46	150–175
Burned lime (CaO)	60	–

[a]Based on pure calcium carbonate at 100%

Table 4.9 Common magnesium sources

Material	Percent Mg
Dolomitic limestone (Mg Carbonate)	3.0–12.0
Magnesia (Mg oxide)	55.0–60.0
Basic slag	3.0
Magnesium sulfate	9.0–20.0
Potassium-magnesium sulfate	11.0
Magnesium chloride	7.5

Table 4.10 Common sulfur sources

Material	Chemical formula	Percent S
Ammonium sulfate	$(NH_4)_2\ SO_4$	24
Ammonium thiosulfate	$(NH_4)S_2O_3 \cdot 5H_2$	26
Ammonium polysulfide	$(NH_4)_2S_x$	40–50
Potassium sulfate	K_2SO_4	18
Potassium magnesium sulfate	$K_2SO_4 \cdot 2MgSO_4$	22
Elemental sulfur	S	>85
Gypsum	$CaSO_4 \cdot 2H_2O$	12–18
Magnesium sulfate	$MgSO_4 \cdot 7H_2O$	14

i. Calcium

Calcium chloride is the source of calcium when in soil. The other form in which calcium may be applied into the soil is as a component of foliar spray. Calcium is mostly applied to correct soil acidity.

ii. **Magnesium**

Magnesium is mostly found in dolomitic limestones in soils. Other sources include potassium-magnesium sulfate, magnesium oxide and basic slag. Crops require magnesium in substantial quantities.

The common Mg sources along with their percentage Mg content are given in Table 4.9. The Mg sulfate forms are more soluble than dolomitic lime and may be the preferred Mg source on those soils where a quick crop response is required.

iii. **Sulfur**

Soil organic matter is the primary soil source of sulfur. More than 95% of the sulfur found in the soil is tied up in organic matter. Sulfur may also be found in animal manures, irrigation water and the atmosphere. Animal manures contain sulfur levels ranging from less than 0.02 to 0.3% depending on species, method of storing and application. Sulfur dioxide and other atmospheric gases dissolved in rain and snow can contribute up to 22 kg of sulfur per hectare per year or even more in some industrialized areas. Irrigation water can contain fairly high levels of sulfur. When the sulfate-sulfur content of irrigation water exceeds 5 ppm, a sulfur deficiency is highly unlikely. Still, starter fertilizer applications of sulfur for new crops may be beneficial in certain instances. Common sulfur sources are given in Table 4.10.

Most fertilizer sulfur sources are sulfates and are moderately to highly water soluble. Soluble forms also include bisulfites, thiosulfates and polysulfites. The most important water-insoluble sulfur fertilizer is elemental sulfur, which must be oxidized to the sulfate-sulfur form before plants can use it. Bacterial oxidation of sulfur in the soil is favored by warm soil temperature, adequate soil moisture, soil aeration, and fine particle sizes. The water-soluble sulfates are immediately available to plants and should be used when sulfur is needed quickly. They are commonly used in dry fertilizers, although ammonium sulfate solutions are also common. Ammonium thiosulfate (ATS; 12-0-026) is a clear liquid well-suited for use in fluid fertilizers or addition to irrigation water. It should not be seed placed and, if band applied, should be at least 2.5 cm from the seed. Ammonium polysulfide (APS) is a red fluid sulfur source with strong odor of ammonia, commonly applied in irrigation water. Sulfur in APS must be oxidized to the sulfate form to be available to plants. Although gypsum (calcium sulfate) is less water soluble than the other sulfates, it is an effective and inexpensive sulfur source.

4.4.2.4 Micronutrients

Micronutrients are required by plants in small quantities. However, they are very essential for plant growth. Examples of micronutrients are iron, zinc, copper, molybdenum, boron, manganese and chlorine. Micronutrients may be applied in several ways. They can be directly to the soil, incorporated in NPK fertilizers, applied in solution or sprayed on bulk blends. Salts of micronutrients dissolved in water are sprayed onto crop foliage for quick action.

Table 4.11   Boron fertilizer sources, boron contents and water solubility

Source	wt % Boron (B)	Water solubility
Borax	11.3	Yes
Sodium pentaborate	18.0	Yes
Fertilizer borate 46	14.0	Yes
Fertilizer borate 65	20.0	Yes
Boric acid	17.0	Yes
Colemanite	10.0	Low
Solubor	20.0	Yes
Boronated single superphosphate	0.18	Yes
Ulexite	10.0–15.0	Half is in water-soluble form

i. **Boron Sources**

Soil application rates for responsive crops may be as high as 3 kg B/ha and for low and medium responsive crops, 0.5 to 1.0 kg B/ha. Methods of application of boron include direct application to the soil as a broadcast or band application or applied foliar as a spray or dust. Common boron fertilizer sources, their boron contents and water solubility are given in Table 4.11.

ii. **Copper Sources**

Table 4.12 contains the list of common copper sources, their copper contents, water solubility and application methods.

iii. **Iron Sources**

Common iron fertilizer sources and their iron contents are given in Table 4.13. Iron formulations may be directly to the soil or as a foliar spray. Applying soluble materials such as ferrous sulfate to the soil is not very efficient, because the iron converts rapidly to unavailable forms. Foliar sprays are much more effective. Most iron fertilizers are therefore often applied as foliar spray. Injections of dry iron salts directly into trunks and limbs have controlled iron chlorosis in fruit trees. This method uses smaller amounts of the iron material than when applying to the soil.

Iron deficiencies in the soil can also be corrected by altering the soil pH in a narrow band in the root zone. Oxidizable sulfur (S) products will lower soil pH and convert insoluble soil iron to forms the plant can use.

Table 4.12   Sources of copper available for use in fertilizers

Source	Percent Cu	Water soluble	Application methods
Copper sulfate ($CuSO_4$)	22.5–24.0	Yes	Foliar, soil
Copper ammonium Phosphate	30.0	Slight	Foliar, soil
Copper chelates	Variable	Yes	Foliar, soil
Other organic	Variable	Yes	Foliar, soil

Table 4.13 Sources of iron available for use in fertilizers

Source	Percent Fe
Iron sulfate ($FeSO_4$)	19.0–23.0
Iron oxide (FeO)	69.0–73.0
Iron ammonium sulfate	14.0
Iron ammonium polyphosphate	22.0
Iron chelates	5.0–14.0
Other organics	5.0–10.0

iv. Manganese Sources

Manganese deficiencies can be corrected by adding soluble manganese salts, such as manganese sulfate, with starter fertilizer and applied in a band. High phosphorous starter fertilizer helps mobilize manganese into the plant. The manganese-containing material may also be applied as a foliar spray. Spraying with 10 kg/ha $MnSO_4$ in water is a common treatment form for manganese deficient soybeans. In extremely acid soils (pH < 5), manganese toxicity may occur. Liming will eliminate this problem. Table 4.14 shows common manganese fertilizer sources and their manganese contents.

v. Molybdenum Sources

The common molybdenum fertilizer sources, their molybdenum contents and water solubility are given in Table 4.15.

vi. Zinc Sources

Zinc deficiencies in field crops can be corrected by applying a zinc fertilizer broadcast or in a starter fertilizer beside the row. Application rates range from 1 kg Zn/ha to as high as 10 kg Zn/ha, depending on soil test levels. Zinc has excellent residual effects and high application rates may be sufficient for 3–4 years. Zinc fertilizer sources are listed in Table 4.16. When soil tie-up of zinc is expected under pH conditions or where an emergency situation exists on an established crop, zinc may be applied as foliar spray. Foliar sprays usually require about 0.5–1.0 kg Zn/ha.

Table 4.14 Common manganese sources and their manganese content

Source	%(wt) Manganese
Manganese sulfate	26.0–30.5
Manganese oxide	41.0–68.0
Manganese chelate	12.0
Manganese carbonate	31.0
Manganese chloride	17.0

Table 4.15 Common molybdenum sources, their molybdenum content and their solubility

Source	%(wt) Molybdenum	Water solubility
Ammonium molybdate	54.0	Yes
Sodium molybdate	39.0–41.0	Yes
Molybdic acid	47.5	Slight

Table 4.16 Sources of zinc available for use in fertilizers

Source	%(wt) Zinc
Zinc sulfate (hydrated)	23.0–36.0
Zinc oxide (ZnO)	78.0
Basic zinc sulfate (ZnSO$_4$)	55.0
Zinc-ammonia complexes	10.0
Zinc chelates	9.0–14.0
Other organics	5.0–10.0

4.5 Fertilizer Application Methods

There are generally three methods of applying fertilizers for intake by plants. These are: (i) they may be applied into the soil; (ii) they may be applied onto the soil and (iii) they may be applied directly to the plants. The method to use depends on the nature of the site and the plant condition. The most important point to remember is to apply them at the proper rate, as over-application can result in plant damage or death. It is important always to follow soil test recommendations or manufacturer's directions. Regardless of whichever method is used, the aim is to apply the nutrients in the fertilizers uniformly, effectively and in a cost-effective manner.

The method of application used depends on the type of fertilizer material used. Consideration must be given to whether the fertilizer is water-soluble granules or powder, water-insoluble granules or powder, liquid or gaseous.

i. Solid water-soluble fertilizers are distributed on to the soil surface. In this case, penetration into the root zone takes place during leaching after dissolution by water. They may also be placed directly into the root zone, for instance, beneath the seed at sowing time.
ii. Solid water-insoluble fertilizers are first distributed onto the soil surface and subsequently, mechanically mixed into the arable layer. Where this is not practical, for example, on grassland, there is a natural but slow penetration by soil organisms.
iii. Liquid fertilizers, either in solution or suspension form, are either sprayed onto the soil surface in their original concentration and left to penetrate into the soil (this approach is not recommended if gaseous losses occur). Or they may be

mixed into the soil immediately after application to prevent gaseous losses. Another way to apply liquid fertilizers is to inject them directly into the soil. Finally, liquid fertilizers may be applied by spraying them in a diluted form directly onto the plants. Special spraying machines are used for this purpose.

iv. Gaseous fertilizers are injected into the top layer of the soil.

v. Organic manures are evenly distributed onto the soil surface to the extent possible and mixed into the arable layer by harrow (disc or plough harrow). Organic manures with low nutrient content may be used as mulch (i.e. as surface protection layer).

Applying fertilizers directly to the leaves (foliar application) has the advantage of quick action. However, the amount of fertilizer that can be distributed in a foliar dressing is limited by the sensitivity of the leaves to osmotic agents such as dissolved salts or organic chemicals such as urea. As a result, foliar application can supply only a limited amount of the primary nutrients needs of plants. The exception is some nitrogen fertilizers, where due to high tolerance of urea, suitable results can be achieved with foliar application. Foliar application is most suitable for applying micronutrients because a relatively large proportion of the total requirement can be supplied in a single spraying. In the case of secondary nutrients, only partial results can be attained by the use of foliar application. Spraying is the most effective method of folia application, and the risk of scorch is minimized, if the spray droplets do not dry too rapidly.

4.6 The Fertilizer Industry in Ghana

According to Ministry of Food and Agriculture of Ghana, Ghana has a fertilizer requirement of 600,000 metric tonnes per annum (equivalent to fertilizer consumption of about 38.4 kg/ha.). In the years 2016 and 2017, 239,883 metric tonnes and 444,236 metric tonnes, respectively, were imported into the country. Thus, the nation fell short of supplying its fertilizer needs in those years. Indeed, the rate of fertilizer consumption in Ghana for the period 2006–2017 as shown in Table 4.17 indicates that the country has almost always fallen short of its fertilizer need.

Ghana imports all its inorganic fertilizer need from other countries, namely, Morocco, Finland, China, Estonia, Latvia and Italy. Table 4.18 lists the types and quantities of fertilizers imported into the country in the period 2013–2019.

There is no plant in Ghana that produces primary inorganic fertilizer. Fertilizer plants in the country are predominantly blending plants (secondary production units) that import fertilizers in compounds and bulk which are blended into various formulations. Notable among the companies operating secondary fertilizer production pants are Yara Ghana and Chemico Limited, which are all located in the port city of Tema. Also, operating a secondary fertilizer production plant is MicroFertil, which is located in Kpong in the Greater Accra Region. Since the introduction of the government's flagship program of Planting for Food and Jobs

Table 4.17 Consumption and importation of fertilizer in Ghana from 2006–2017

Year	Fertilizer import (metric tonnes)	Fertilizer consumption (kg/ha.)
2017	444,236	31.2
2016	239,883	20.9
2015	290,156	23.5
2014	111,083	15.7
2013	296,086	25.3
2012	523,171	34.8
2011	57,144	13.2
2010	175,809	18.7
2009	182,281	19.0
2008	85,1923	14.5
2007	156,386	17.8
2006	206,014	20.1

Table 4.18 Fertilizer imports into Ghana in the period 2013–2019 (reproduced with permission from AfricaFertilizer.org. @2020)

Fertilizer name	2013	2014	2015	2016	2017	2018	2019
NPK	117,047	44,880	138,440	132,632	213,887	224,176	217,024
Urea	36,104	202	18,348	23,268	88,259	42,005	77,011
Ammonium sulfate	54,863	6,282	64,015	23,268	43,865	15,993	42,235
Organic fertilizers	6,465	5,523	7,818	8,772	37,643	9,460	29,300
TSP	47,173	21,258	32,052	13,802	26,766	10,084	17,326
MOP	19,849	22,715	18,707	13,842	24,325	5,875	4,673
Other fertilizers	16,587	10,223	11,077	8,532	9,582	7,564	37,542
Total (mt)	298,086	111,083	290,156	239,883	444,236	315,157	425,110

(PFJ) in 2017 and the attendant increased demand for fertilizers in the country, a number of fertilizer blending plants have been set up to manufacture fertilizer for the local market. These include **OmniFert Fertilizer Factory**, a wholly-owned Ghanaian fertilizer manufacturing company located at Dawhenya in the Greater Accra Region. The company operates a total of five warehouses on a 12-acre land, with a total storage capacity of 100,000 metric tonnes. It specializes in blending its own variety of fertilizers and sources its stock from international suppliers. The company operates two blending plants and also sacks printing plant. The blending and bagging production rate of the plant is about 50 tonnes per hour. The largest fertilizer blending plant in the country is Glofert, which is situated at Asuboi in the Eastern Region of Ghana. Glofert produces fertilizer comprising, urea, ammonium, sulfate as well as various types of nitrogen, phosphate and potassium (NPK), which are widely used by farmers in Ghana and across the West African sub-region. The company has a production capacity of 120 metric tonne per hour (over 800,000 metric tonnes per year). Thus, when operated at full capacity the company can supply all the fertilizer need of the country.

Other factories producing fertilizers in the country include Jekora Ventures, which produces organic fertilizer from fecal sludge and agro-industrial waste. The plant is a joint venture between International Water Management Institute (IWMI) and Jekora Ventures Limited. It has a production capacity of about 200 metric tonnes of compost annually. It is located at Somanya in the Eastern Region of Ghana. Safi Sana Ghana Limited is yet another plant producing organic fertilizer in Ghana. It is located at Ashaiman in the Greater Accra Region. The Safi Sana plant processes 25 tonnes per day of local organic waste and the output of the Communal Service Blocks into fertilizer and biogas.

Further Readings

AfricaFertilizer.org. (2020). Fertilizer statistics overview—Ghana, 2015–2019, 2020 edn.

Engelstad, O. P. (Ed.). (1985). *Fertilizer technology and use* (3rd ed.). Soil Science Society of America, Madison.

Fertilizer Association of India. (1977). Handbook of fertilizer technology, 2nd edn. Fertilizer Association of India, New Delhi.

Gowariker, V., Krishnamurthy, V. N., Gowariker, S., Dhanorkar, M., & Paranjape, K. (2009). The fertilizer encyclopedia, 1st edn. Wiley-Blackwell, Hoboken.

https://africafertilizer.org/blog-post/fertilizer-statistics-overview-ghana-2015-2019/. Accessed 07 Jun 2020.

Industrial Development Organization, U. M. (Ed.). (1998). *Fertilizer manual* (3rd ed.). Springer.

Lowrison, G. C. (1989). *Fertilizer technology*. E. Horwood.

Mishra, B. (2012). *Technology and management* (1st ed.). IK International Publishing House Pvt. Ltd.

Mortvedt, J. J. (1999). *Fertilizer technology and application*. Meister Pub. Co.

Munn, V. (Ed.). (2018). *Fertilizer technology and soil fertility*. Larsen and Keller Education.

Palgrave, D. A. (Ed.). (1991). *Fertilizer science and technology*, 1st edn.CRC Press, Boca Raton.

Park, M. (2001). *The fertilizer industry* (1st ed.). Woodhead Publishing.

Shishir, S., Pant, K. K., & Bajpai, S. (2015). *Fertilizer technology I: Synthesis*, 2 vol. Stadium Press, LLC, Houston.

Cement and Clay Products Technology

5

Abstract

Cement is used mostly in the construction, plumbing and masonry industries where it finds its application in the production of concrete, preparation of mortar, marking grout, and the preparation of stucco. Two types of cement are used in the construction industry, namely hydraulic cements (e.g. lime, gypsum plasters, oxychloride) and non-hydraulic cement (e.g. Portland cement). There are special cements that have Portland as the base but formulated for special applications, such as expansive cement, super high strength, alinite and special high C_2S cements. In this chapter, the chemistry of cement manufacture and the production processes have been covered. Refractories that are ceramic materials with high strength and fracture toughness have also been covered. They are inert to most chemical attacks, abrasion-resistant and can withstand high temperatures above 1100 °C. The techniques used to manufacture refractory materials have been presented in this chapter, namely the mold and burn process, electric fusion process and ceramic fiber formation process. There are safety and environmental concerns associated with the production and use of cement such as noise pollution, emissions of dust particles, chromium content and the release of large amounts of CO_2, NO_x and SO_x emissions. Methods applied to minimize and control the safety and environmental problems have been considered. Finally, an overview of the cement industry in Ghana has been presented.

5.1 Introduction

Cement is any material that when in contact with moisture in the right amount sets and hardens independently and, in the process, binds other materials together. During setting, mixtures containing cement acquire form and become solid. Hardening, on the other hand, relates to the ability of a mixture containing cement to bear increasing load (force). Cement finds its use mostly in the construction, plumbing and masonry industries as follows:

i. It is used in the manufacture of concrete. Concrete is an artificial stone made from a carefully controlled mixture of cement, water, fine and coarse aggregates (usually fine and coarse rock or sand). The exact composition of concrete depends on the use to which it is to be put and the strength required of it. In the construction industry, concrete finds its use in the making of buildings, walls, tunnels, dams, bridges, roads and other structural materials.

ii. Cement is also used in the preparation of mortar. Mortar is a mixture of sand, cement and water in correct proportions. It is used in masonry to bind construction materials together and to fill the gaps between them. Mortar is applied in its paste form which then sets and hardens. It can also be used to fix masonry where the original mortar has washed away. Lime can be used as a binder in mortar in place of cement.

iii. Cement finds its application in the preparation of stucco. Stucco is a mixture of an aggregate, a binder (e.g. cement) and water in correct proportions. It is used as a coating for walls, ceilings and decorations. It is applied wet and hardens when dried.

iv. Finally, it is used in making grout. It is made up of water, cement, sand and occasionally fine gravel. Grout is used to bind tiles to walls, embed rebars in masonry walls, connect sections of pre-cast concrete, fill voids and seal joints.

The widespread application of cement is in the construction industry where it finds its use mostly in the making of concrete and mortar. Cements used in the construction industry are characterized as hydraulic or non-hydraulic. Hydraulic cements set and harden after being combined with water, as a result of chemical reactions with water, and that after hardening retains strength and stability even under the water. The reaction with water produces hydrated compounds which are insoluble in water and which therefore provides the product of the mixture with strength and stability. The most common of this kind of cement is Portland cement. Non-hydraulic cements, unlike hydraulic ones, must be kept dry in order to gain strength. Examples of such cement are lime (fat lime), gypsum plasters and oxychloride cements. Lime mortars, for example, set only by drying out, and gain strength only very slowly and on the absorption of carbon dioxide from the atmosphere to re-form calcium carbonate through carbonation.

5.2   Hydraulic Cements

The central constituent of all hydraulic cements is lime. It is the compound responsible for the setting and hardening of cements. The common types of this kind of cements are discussed below.

i. Natural cements

They are made from impure limestone. Raw limestone is first calcined at moderate temperatures and subsequently cooled, ground and bagged for use as cement. Calcination is the process whereby calcium carbonate (limestone) is heated to produce calcium oxide and carbon dioxide according to the reaction as follows:

$$CaCO_3 \quad \xrightarrow{\Delta} \quad CaO \quad + \quad CO_2 \tag{5.1}$$

Natural cements require very stringent compositions, and their manufacture requires strict control measures.

ii. Hydraulic limes (Argillaceous limestone)

These are also made from impure limestone which has high clay matter. The clay content could be as high as 25%. Hydraulic lime is formed from the limestone by burning the limestone in an ordinary stack kiln, then cooling and hydrating or slaking it. The hydraulic properties of this cement are due to the silicates and aluminates which form when the lime reacts with the silica, alumina and iron oxide present in the clay.

iii. Portland cements

Portland cement is made by grinding together Portland clinker and gypsum. The percentage of gypsum is often about 5% (wt). Clinker itself is made by burning a mixture of limestone and clay matter at very high temperatures near 1450 °C, then cooling the product and grinding it into powder. The main constituents of Portland clinker are calcium silicates, calcium aluminates and calcium aluminoferrite. The name Portland cement is derived from a stone in Portland (Portland stone), an island on the coast of England, which has a color similar to Portland clinker. There are several types of Portland cement based on the composition of the main constituents.

iv. Granulated blast furnace slag cement

Blast furnace slag is a product of pig iron-making processes. The residue dumped from the furnace after extracting the pig iron when cooled quickly or quenched and then ground has cementitious properties. The quenching process is called *granulation* and the resultant product is called *granulated blast furnace slag*. This product when mixed with lime or Portland clinker forms cement called *slag cement*. Slag from other industries can also be used to make Portland slag cement. Only granulated slag is effective as a cementing agent. Two types of cement are commonly made from this slag, namely:

(a) Portland blast furnace slag cement (PBFSC): This contains up to 70% (wt) blast furnace slag, 5–6% (wt) gypsum and the remaining Portland cement clinker. This kind of cement has all the physical properties of ordinary Portland cement (OPC). It is therefore used as an economic alternative to OPC. It is more sulfate-resistant and has a low heat of hydration. It develops strength slowly and has very high ultimate strength. As the slag content is increased, early strength is reduced. However, all compositions produce high ultimate strength.

(b) Super-sulfated cement (SSC): This is more sulfate-resistant than PBFSC and SRC (sulfated slag cement). SSC is prepared by grinding together granulated slag, anhydrite (or gypsum) and clinker (or lime) in a proportion of approximately 80:15:5. It exhibits good resistance to aggressive agents in general. Again, they produce high strength by the formation of ettringite with strength growth similar to slow Portland cement.

v. Portland pozzolana cements (PPC)

Pozzolanas are natural or synthetic clay matter, which, when ground with lime or clinker and mixed with water, produce cementitious compounds. Portland pozzolana cement is made by mixing and grinding highly reactive pozzolana or fly ash, Portland cement clinker and 5–6% (wt) gypsum. The Pozzolana or fly ash content should not be more than 25% (w). PPC has a much lower heat of hydration and is also fairly sulfate-resistant. It develops strength slowly and has very high ultimate strength. Its physical properties are the same as that of OPC, except that it has lower shrinkage. It can, therefore, be used for all construction work for which OPC is used. Fly ash is a fine powder residue resulting from the burning of pulverized coal. It is generally finer than cement and consists mainly of glassy-spherical particles as well as residues of hematite and magnetite, char, and some crystalline phases formed during cooling.

vi. Masonry cement (MC)

It is made by inter-grinding Portland cement clinker with limestone, sandstone or granulated slag in the proportion of 1:1. Hydrated lime and/or plasticizer are often added to give better plasticity. Masonry cement, unlike the other types of cement, gives a more plastic mortar. It is used mostly for masonry work such as laying bricks (blocks) and for plastering. The other cements when mixed with sand and water to form mortar give a somewhat harsh mortar that does not retain water very well.

vii. Calcium aluminate cements

These are made by grinding limestone and bauxite together. Monocalcium aluminate ($CaAl_2O_4$) and Mayenite ($Ca_{12}Al_{14}O_{33}$) are the active ingredients of this type of cement. Strength is formed by hydration to calcium aluminate hydrates. They are well adapted for use in refractory concretes, e.g. furnace linings.

viii. Calcium sulfo-aluminate cements

These are produced from a mixture of clinkers that has Ye'elimite ($Ca_4(AlO_2)_6SO_4$) as the base material. During hydration, ettringite is formed and unique properties are developed by adjusting the concentration of calcium and sulfate ions in the medium. Energy requirements for its production are lower because of the lower kiln temperatures required for reaction, and the lower amount of limestone in the mix.

They are used in expansive cements in ultra-high early strength cements and low-energy cements.

ix. **Geopolymer cements**

These are made from mixtures of water-soluble alkali metal silicates and aluminosilicate mineral powders such as fly ash and metakaolin. These are very low-energy materials; hence the production of geopolymer cement results in very large energy savings. The properties of geopolymer cement, when used to make concrete, are equivalent to other cements in terms of the structural qualities of the resulting concrete. The properties of geopolymers are significantly dependent on the silica/alumina ratio. For example, a geopolymer cement with Si/Al ratio of 1 is used in bricks and ceramics, while that with Si/Al ratio of 2 is used in low CO_2 concretes. To obtain a geopolymer cement with acceptable mechanical properties, it is necessary to enhance the activity and solubility of Al-Si source materials in the alkali solution. One way of meeting this condition effects on the final properties is thermal activation of the source material.

5.2.1 Variations of Portland Cement

The main constituents of the different variations of cement are the same. They vary only in the composition of the constituents and in the end-use to which they are put. However, the method of manufacture is the same for all of them. The range of composition of constituents of Portland cement types is shown in Table 5.1.

In practice, it is the presence and relative amounts of the compounds tricalcium silicate (C_3S), dicalcium silicate (C_2S), tricalcium aluminate (C_3A) and tetracalcium aluminoferrite (C_3AF) in cement that imparts the different properties to the different types of Portland cement. The probable composition of these compounds in Portland cement has the ranges shown in Table 5.2. The specific composition of the compounds and the unique properties they confer of the cement products are listed in Table 5.3.

Table 5.1 Constituents of Portland cement

Compound	Range of composition (%wt)
SiO_2	19.0–25.0
Al_2O_3	2.0–8.0
Fe_3O_2	0.3–6.0
CaO	60.0–65.0
MgO	1.0–6.0
SO_3	1.0–3.0
Alkalis	0.5–1.5

Table 5.2 Compounds in Portland cement that imparts its properties

Compound	Composition (%wt)
C_2S	20.0–60.0
C_3S	20.0–60.0
C_3A	0.0–16.0
C_4AF	1.0–16.0

Table 5.3 Types of Portland cement

Type	Composition	Unique properties
Ordinary Portland cement (OPC)	55% C_3S, 19% C_2S, 10% C_3A, 7% C_4AF	Develops enough compressive strength in 3 days, 7 days and 28 days
Moderate heat Portland cement (MHPC)	51% C_3S, 24% C_2S, 6% C_3A, 11% C_4AF	They release less heat when hydrated than OPC
Rapid hardening cement (RHC)	56% C_3S, 19% C_2S, 10% C_3A, 7% C_4AF	These give higher strength in 24 h than OPC will give in 3 days
Low heat Portland cement (LHC)	28% C_3S, 49% C_2S, 4% C_3A, 12% C_4AF	These give very low heats of hydration. Used in large concrete works such as in dam construction when contraction cracks must be avoided
Sulfate-resisting cement (SRC)	38% C_3S, 43% C_2S, 4% C_3A, 9% C_4AF	Concretes from such cements resist sulfate attacks. Used in areas where soils and water with high concentrations of sulfates
Oil well cement (OWC)	Portland cement with C_3A content of less than 3%	Grout from such cement does not set for 90–120 min at high well bottom temperatures. Once the grout is in place it is expected to set and harden within 24 h. Used in lining the annulus of oil wells
White cement	Proportion of iron oxide is reduced to below 0.4%	It does not burn well during clinker formation. It has all the physical properties of OPC

5.2.2 Special Cements

Special cements have Portland as the base but are formulated to impart on them special properties for special use. Cements falling under this category are listed in Table 5.4.

Table 5.4 Constituents of special cements

Type	Composition	Unique properties
Expansive cement	Extra gypsum and aluminous cement clinker added to Portland cement clinker	When measured properly, the composition gives a measured expansion. Concrete from OPC shrinks when set and hardened. Expansive cements solve this problem when used to make concrete
Super high strength cement	Has high C_3S content. Portland cement clinker ground to very high fineness. The C_3S should be about 60% and the fineness should be 6000 cm^2/gm or more	These are used for either urgent repairs of some important concrete structures or for very massive foundation
Alinite cement	Raw mix plus 5% calcium chloride prior to grinding	Good compressive strength. The burning temperature of the raw mix is lowered considerably
Special high C_2S cements	Portland cement clinkers with 60% or more C_2S content	Have very low early strengths and dust due to betta-gamma conversion

5.3   The Chemistry of Portland Cement Manufacture

The central component of Portland cement is Portland clinker, the chemistry of whose formation is the subject of this section. Clinker itself is a composite of a number of complex compounds, which form during the reactions between the raw materials used in cement manufacture. Generally, the nature of the reaction products and their relative amounts during clinker formation affect the properties of the eventual cement product and also the clinker formation process, particularly, the burning of the raw materials. Portland cements are made by burning a finely ground mixture of lime bearing materials such as calcium carbonate or calcium sulfate—generally called *calcareous materials*—and aluminum silicate, also called *argillaceous materials*, such as clay or laterite in measured proportions at high temperatures. Temperatures in the range 1400–1450 °C are commonly employed during clinker formation. The main product of the reaction is called *Portland cement clinker*. This product when cooled and ground with gypsum forms Portland cement.

At the high temperatures of the clinker formation reaction, a succession of reactions takes place producing the main substances responsible for cement behavior, namely tetracalcium aluminoferrite (C_4AF), dicalcium aluminate (C_2A), tricalcium silicate (C_3S), dicalcium silicate (C_2S), free lime and MgO, among others. The mixture of these products together constitutes a clinker. In Table 5.5 the major constituents of Portland clinker are listed. The exact chemical composition of the clinker depends on the type of raw materials and the fuel used. The calcareous raw material may be limestone, marl, calcareous sea sand, seashells, coral or

Table 5.5 Clinker compounds

Formula	Name	Abbreviation
$2CaO \cdot SiO_2$	Dicalcium silicate (Allite)	C_2S
$3CaO \cdot SiO_2$	Tricalcium silicate (Bellite)	C_3S
$3CaO \cdot Al_2O_3$	Tricalcium aluminate	C_3A
$4CaO \cdot Al_2O_3 \cdot Fe_2O_3$	Tetracalcium aluminoferrite	C_4AF
MgO	Magnesium oxide	M

by-product calcium carbonate. The argillaceous raw materials, on the other hand, can be clay, shale or laterite. Where necessary, small quantities of iron oxide or sandstone may be added to bring up the composition of the raw mix to a predetermined level. All Portland cements are made from limestone. In few instances, however, calcium carbonate sludge (a by-product) or gypsum or anhydrite may be used as the main raw material. Where coal is used as fuel, part of the ash present in the burnt coal is added to the raw mix during burning. In this instance, the composition of the coal ash should be taken into account while designing the raw mix.

5.3.1 Role of the Individual Compounds in Cement

Tricalcium silicate (C_3S) is the main strength-giving compound in Portland cement. It contributes toward early strength. Dicalcium silicate (C_2S) contributes toward later strength. Clinker with very high potential C_3S, however, it is difficult to burn. Therefore, in OPC, the proportion of C_3S is kept between 40 and 50% and C_2S between 25 and 35%. In rapid hardening cement (RHC) and in oil well cement (OWC), the C_3S must be kept more than 50%, at about 50–60%, so that high early strength can be achieved. To enable the clinker to be burnt properly, alumina (Al_2O_3) and iron oxide (Fe_2O_3) which act as fluxing agents are added to the raw mix. On burning, the Al_2O_3 and Fe_2O_3 present in the raw mix first form C_4AF and the remaining forms C_3A. C_3S and C_3A on hydration generate more heat. The presence of higher proportions of C_3A in cement makes it less resistant to sulfate action. In moderate heat cement (MHC), the C_3S content is kept a little lower and the C_3A content is kept below 5%. In low heat cement (LHC), both C_3S and C_3A must be kept as low as possible, at about 30% and 6%, respectively. In sulfate-resistant cement (SRC), C_2S must be kept as high as in OPC but C_3A must be kept below 5%. In OWC, the C_3S content must be kept higher than 50%, the C_3A content should be kept below 3% (preferably at 0%), but the combined C_4AF and C_2A content should be below 20%. In white cement (WC), the Fe_2O_3 must be below 0.4%. The specific roles of the individual constituents of clinker are listed in Table 5.6.

The cement industry and research use a unique nomenclature to describe compounds found in cement clinker. This is mostly to simplify the rather complex and long names associated with the clinker compounds. According to this notation,

Table 5.6 Specific functions of individual clinker compounds in cement

Compound	Function
C_3A	Causes setting of the reaction mixture
C_3S	Responsible for early strength (at 7 or 8 days)
C_2S and C_3S	Responsible for final strength (1 year)
Fe_2O_3, Al_2O_3, Mg, and Alkalis	Lowers the clinkering temperature

Table 5.7 Nomenclature of compounds in cement clinker

Oxide	Notation	Oxide	Notation
CaO	C	SO_3	$\hat{S}$
SiO_2	S	Na_2O	N
Al_2O_3	A	K_2O	K
Fe_2O_3	F	CO_2	$\hat{C}$
MgO	M	H_2O	H
TiO_2	T		

oxides are referred to by their first letter: 'C' represents CaO; 'M' is MgO and so on, for all the oxides encountered in cementitious systems, as shown in Table 5.7.

The notations in Table 5.7 are used in the literature, commerce and industry to communicate information on cement.

5.4   Production of Portland Cement

There are five important stages in the manufacture of Portland cement, namely the selection of raw materials, preparation of raw materials, production of clinker, adding additives to the clinker and finally grinding the mixture. A flowchart for the process of cement manufacture is shown in Fig. 5.1.

5.4.1   Raw Materials Selection

In selecting raw materials for the manufacture of Portland cement, great care must be exercised as they affect the quality of the cement product, the clinker formation process and finally the fuel to use to do the burning. The physical condition, the composition of the substantive minerals and the levels of impurities in the raw material will all have an impact on the selection of processing route and on the quality of the cement product. Raw materials used in the manufacture of cement may be classified into two broad groups. These are the calcareous materials which are rich in lime, and the argillaceous materials which are rich in silica. The most common form of calcium carbonate is limestone. It is used extensively, worldwide,

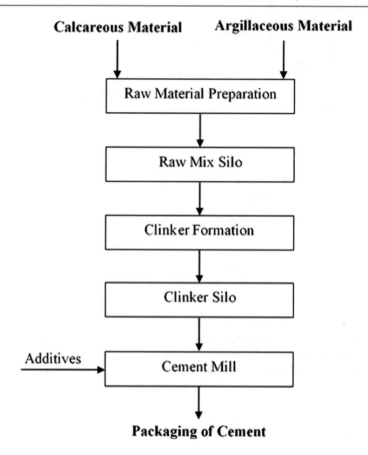

Fig. 5.1 A generalized flowchart for cement manufacture

for the manufacture of cement. Other sources of limestone are marls, chalk, sea sand, seashells and coral reef. Also, by-product calcium carbonate sludge from the paper and pulp industry or from the process of manufacturing ammonium sulfate from gypsum are sometimes used. The argillaceous raw materials may be clay, shale or laterite. Where necessary, the silica modulus of the raw mix may be boosted by the addition of small quantities of bauxite, sandstone and iron ore. The compositions of these raw materials are listed in Tables 5.8 and 5.9. Another raw material of Portland cement is gypsum. This material, when ground with cement clinker in about 5–6 wt%, acts as a retarder of the setting time of cement. (Note, cement when mixed with water sets immediately, which situation is not wanted in the use of cement for several applications.)

Table 5.8 Calcareous raw materials

Composition	Raw materials			
	Limestone	Seasand	Seashell	Marl
SiO_2	1.0–15.0	1.0–10.0	0.5–2.0	10.0–20.0
Al_2O_3	1.0–6.0	1.0–3.0	0.5–1.0	2.0–6.0
Fe_2O_3	0.2–5.0	0.5–2.0	0.1–0.5	1.0–5.0
CaO	40.0–55.0	45–55.0	53.0–55.0	35.0–45.0
MgO	0.2–4.0	0.2–2.0	0.5–1.0	1.0–4.0
Alkalis	0.2–1.0	1.0–1.5	0.3–0.5	0.5.0–1.0
Chlorides	≤ 0.1	1.0–2.0	0.2–0.5	Trace
Sulfates	≤ 3.0	0.5–10.0	1.0–0.3	Trace
L.O.I	35.0–44.0	38.0–44.0	42.0–45.0	30.0–38.0

L.O.I = Loss on ignition. The compositions are in % (wt)

Table 5.9 Argillaceous materials

Composition	Raw materials						
	Clay	Shale	Laterite	Bauxite	Iron-ore	Sandstone	Coal-ash
SiO_2	40–70	40–80	10–30	3–15	5–10	85–95	50–70
Al_2O_3	15–30	15–30	20–40	40–54	2–5	2–5	15–30
Fe_2O_3	3–10	3–10	20–40	2–10	85–95	1–3	5–10
CaO	1–10	1–10	2–4	2–4	Trace	1–3	2–5
MgO	1–5	1–5	1–2	1–2	Trace	1–3	1–3
Alkalis	1–4	1–4	Trace	Trace	Trace	1–2	2–4
SO_2	≤ 2	≤ 2	Trace	Trace	Trace	Trace	1–2
L.O.I	5–15	2–5	15–25	20–30	≤ 5	2–5	≤ 2
TiO_2	–	–	2–5	2–6	–	–	–

L.O.I = Loss on ignition. The compositions are in % (wt)

5.4.1.1 Impurities in the Raw Materials and Their Effects

Several impurities also accompany the raw materials and care must be taken as some of these materials are harmful to the cement burning process and the quality of the final cement product. Table 5.10 lists some of the impurities found in the raw materials and the manner they affect the cement and the cement manufacturing process.

5.4.1.2 Clinker Fuel

The fuel used in burning the clinker is also an important raw material for making cement. Coal, oil or gas is the fuel usually used to burn clinker. The raw materials, fuel and the process selected for the burning must be compatible. For instance, in the manufacture of white cement only oil can be used because the Fe_2O_3 in the ash present in coal will contaminate the white cement. Where coal is used as a fuel,

Table 5.10 Impurities in raw materials and their effect on cement

Impurities in raw mix	Effect in cement
SO_3	Its presence in the raw material is undesirable. It is difficult to dissociate and drive off the SO_3. Sometimes a carbonaceous material has to be added to help the dissociation. SO_3 is often present in by-product sludge from synthetic ammonium sulfate plants. (As a gas, it forms air pockets in the cement product and weakens the structural strength of the product).
Alkalis	Large amounts of alkalis present in by-product sludge from paper and pulp industry is also undesirable. The alkalis react with silica in alkali-silica reaction (ASR), which leads to serious expansion and cracking in concrete, resulting in critical structural problems that can even force the demolition of a particular structure.
P_2O_5	If it is in raw mix in excess of 0.5%, the resultant cement may be slow setting and have very low early strengths.
Fluoride	Up to 0.5% fluoride is helpful as a flux. However, higher quantities in the raw mix attack the refractory lining of the kiln and increase balling action in the kiln.
Mn_2O_5	It is undesirable when present in the raw mix in more than 0.5%. It makes the color of the resultant cement very dark. It also affects the early strength of the cement.
TiO_2	In raw mix, it acts in the same way as SiO_2. In small quantities, it is not harmful to the cement.
MgO	When present to the tune of 4% or more in the raw mix, it causes delayed expansion of the resultant cement product with attendant cracks in the products. Because of this, the use of dolomite or dolomitic limestone as a raw material should be avoided.

sometimes small quantities of oil are used to increase the flame temperature and hence the efficiency of the process and quality of the cement.

5.4.2 Raw Material Preparation

The calcareous and argillaceous materials, which are the main raw materials for cement manufacture, when mined are often in the form of large lumps ranging in size from about 200 mm to 1 m and beyond depending on the mode of mining and nature of the raw material. For instance, when mined mechanically limestone comes in lumps of sizes 1 m or more, while when manually mined the lumps sizes are about 200 mm. On the other hand, the mined clay, gypsum, coal, sandstone, etc. have sizes of about 300 mm. In these forms, they present a physical barrier for use in the burning equipment where they must necessarily be fed to be reacted to form the cement clinker. Such large lumps do not also make efficient use of space in the burning equipment neither do they facilitate intimate mixing of the reactants and heat transfer in the equipment. To make for efficient operation of the clinker formation process, the raw materials brought from the mines are first worked on before feeding into the burning equipment. Generally, the raw material is prepared with six purposes in mind, namely:

i.   to reduce the large lumps of the raw limestone and clay into fine particles to facilitate feeding into the burning equipment;
ii. to facilitate intimate mixing of the limestone and the clay (the reactants);
iii. to improve energy transfer in the equipment;
iv. to prepare the raw mix in a suitable physical state for the subsequent burning in the kiln;
v.   to improve on the efficiency of the burning process;
vi. to optimize the use of kiln space.

The preparation of the raw materials constitutes two steps. In the first step, the lumps are reduced into particulate sizes of about 20 mm using crushers. In the second stage, the products of the crushing operations are mixed and ground in mills to a size of about 90 μm, which are then fed to the kiln as a slurry or raw dry meal depending on whether a wet burning or dry burning will be the next operation. A flowchart of the raw material preparation process is shown in Fig. 5.2.

The type of crusher to use depends on the size of the lumps of raw materials and the nature of the material. For limestones of lump sizes of 200 mm or less and argillaceous materials generally, hammer crushers are used in a single-stage crushing operation. For large factories where the raw limestone is mechanically mined, a two-stage crushing operation is often used. During the first stage, jaw crushers, gyratory crushers or impact crushers crush lump sizes of about 1000 mm

Fig. 5.2 A generalized flowchart for the preparation of raw material for cement manufacture

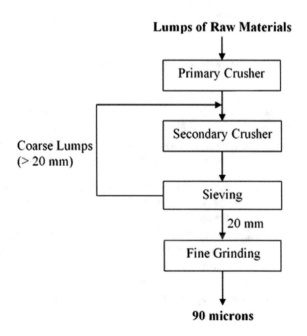

to products of sizes 200 mm or less. The product from the first stage is subsequently crushed in a second crusher using a hammer crusher, impact crusher or cone crusher. The final product has particle sizes of 20 mm. Larger particles are screened and returned to the second stage to be recrushed. Hammer mills are used to crush shale, gypsum, coal and other argillaceous materials. The crushing processes generate a lot of dust and air-borne particles. Consequently, dust collectors and fans must be installed at the plant to deal with dust pollution.

The products from the crushing operations are usually mixed in steel ball mills where they are ground to particle sizes of 90 microns before being fed in an appropriate form into the burning equipment. Two types of mills are used in the grinding process, namely compartment mills where the mill is divided into two or three chambers with perforated diaphragms in between them or in tube mills. Either of the two employs steel media for the grinding action. Again, the milling may be a closed-circuit or open-circuit mill. In open-circuit mills, the feed enters at one end and the finished product comes out from the other end. In a closed-circuit mill, the product exiting from the mill is fed to a separator where the coarse particles are separated and recycled to the mill and the fine material sent for storage. The degree of grinding depends on the composition of limestone used. If the limestone is of cement grade composition, the argillaceous constituents are intimately mixed.

The physical state, in which the raw mix is fed to the kiln, depends on the burning process to be used. In the wet process, the raw materials are ground with the addition of sufficient water so that the resultant slurry contains about 30–40% water. It is this slurry that is fed to the kiln. In the dry process, the raw materials are ground in the dry state and fed in that state to the burning equipment. In the semi-dry process, the ground raw meal is first nodulized by the addition of 10–12% water before feeding to the burning equipment.

5.4.3   Storage of Raw Materials

Limestone in lump form is generally not stored in the factory. Only crushed limestone is stored in the factory. The raw lumps are therefore crushed immediately on reaching the factory and then stored. The argillaceous and other corrective materials, however, may be stored as received or in crushed form. Three types of storage devices are used in the cement industry, which are:

i. Crane gantry with electric overhead traveling (EOT) cranes.
ii. Storage bins with overhead feed belt conveyers and underground reclaiming belt conveyers. This type is more economical than the first.
iii. Bed blending system: This is usually used for factories with high capacities. In this facility, crushed limestone from different faces of the mines and correctives are spread in pre-determined layers at one end and the material reclaimed with the help of cutter blades at the other end. In this manner, a pre-blended raw mix of approximately the desired chemical composition is fed to the raw mill hopper.

Ground raw materials must also be stored before use. The mode of storage of the ground materials depends on whether the product of the grinding process is a slurry or dry raw meal. The ground slurry is stored either in silos or in basins with continuous mixing. In the dry process, tall silos are used for the storage of dry raw meals. Mixing silos or homogenizing silos are placed above the storage silos so that ready-mixed and corrected raw mix are fed to the storage silos by gravity. This practice saves power.

5.4.4 Clinker Formation

Burning is the unit operation where limestone and the corrective materials are heated and reacted to form the cement clinker. This operation is the most important step in the cement-making process. Burning is mostly carried out in rotary and shaft kilns, but fluidized bed reactors and sinter grate kilns are also sometimes used. During burning, the limestone breaks down to produce lime which then reacts with the alumina, silica and ferrite in the clay matter to produce the clinker. The burning takes place in kilns. There are three types of burning processes. First, there is a dry burning process where the raw mix is fed in the dry state to the kiln. Then, there is a wet burning process during which the raw mix is fed to the kiln in the form of a slurry. Finally, there is a semi-dry burning process in which wet, nodulized raw meal is fed to the kiln.

The sequence changes that occur during burning in the long and short rotary kilns are the same irrespective of whether the dry, wet or semi-dry process is at stake. The raw mix is first heated to 100 °C at which temperature the moisture present in the raw mix is removed. At about 400–500 °C, the combined moisture present in the clay matter of the raw mix is dissociated and removed and magnesium carbonate, if present, is also dissociated. At 900 °C, the calcium carbonate (the main ingredient of the clinker formation process) is dissociated into carbon dioxide and lime. When the temperature reaches about 1200 °C the lime, the silica, alumina and iron oxide in the raw mix begin to react with each other as follows; first, all the iron oxide present in the raw mix and a part of the alumina combine with lime to form C_4AF which is in a liquidus state. At the same time, a part of the silica present starts combining with lime to form C_2S. The remaining alumina present combines with lime to form C_3A. At this point, the material would have become more liquidus and proper clinker nodule formation begins. Further down the kiln, in a zone called the actual clinkering zone, all the silica present combines with lime to form C_3S and the remaining lime combines with C_2S to form C_3S. Very little free lime is left at the end of this process. From the clinkering zone, the clinker leaves the kiln and is immediately cooled.

Lime silicates (C_2S and C_3S) are the true cementations materials. C_3S contributes toward early strength and C_2S contributes toward late strength of cement products. In making clinker, therefore, the objective is to achieve a clinker with the highest possible C_3S and lowest possible free lime. This is achieved by judicious

Table 5.11 Reactions during clinker formation

Temperature (°C)	Reactions	Heat change
100	Evaporation of free water	Endothermic
500 and above	Evolution of bound water	Endothermic
900 and above	Crystallization of amorphous dehydration products of clay	Exothermic
	Evolution of carbon dioxide from calcium carbonate	Exothermic
900–1200	Main Reaction between lime and clay	Endothermic
1200–1250	Commencement of liquid formation	Endothermic
1250 and above	Further formation of liquid and completion of formation of cement compounds	Endothermic

selection of the raw materials available and the proper control of the burning process.

The composition of the clinker and the final burning temperature depend on the composition of the raw mix. For instance, a mixture of pure lime and silica in a proportion of 2.8:1 produces a clinker, which is 100% of tricalcium silicate (Ca_3SiO_5). However, it will require prolong heating at 1600 °C or more to obtain a perfect combination. Such prolonged heating conditions are in practice not realistic. What is done in practice, therefore, is to add some flushing materials to bring about a lower temperature and a shorter reaction time. Compounds like alumina, iron oxide, magnesia and alkalis already present in the raw mix act as flushing agents. Therefore, by measured inclusion of these oxides in the raw mix it is possible to shorten the reaction time to reasonable limits at the prevailing temperatures in actual practice. The reactions that occur during clinker formation are summarized in Table 5.11.

The burning process and the quality of the cement product produced are affected by oxides other than CaO, SiO_2, Al_2O_3 and Fe_2O_3. These other oxides include MgO, Mn_2O_3, TiO_2, K_2O, Na_2O, SO_2 and P_2O_5. The presence of the oxides in certain percentages could be harmful to both the burning process and the quality of the resultant product. Their specific effects have been previously discussed. The fluoride and chloride compounds when present in higher quantities also affect the burning and the product quality.

5.4.4.1 Changes Occurring During Clinker Formation

In a correctly designed raw mix composed of CaO, SiO_2, Al_2O_3, Fe_2O_3 and MgO, during the process of burning, all the Fe_2O_3 and part of the Al_2O_3 first combine with CaO to form C_4AF; the remaining Al_2O_3 combines with CaO to form C_3A. Then the C_4AF and C_3A combine to form silicates. All the SiO_2 in the mixture merges with a portion of the CaO to produce C_2S. The CaO still left (uncombined) then combines with a part of the C_2S already formed to form C_3S, which is the ultimate desired product. In a correctly designed raw mix, it is expected that no CaO (free) should be left in the clinker. In practice, however, about 0.5–2% free CaO is always left in the clinker.

5.4.4.2 Calculating Raw Mix and Clinker Composition

Formulae exist that can be used to design the composition of the raw mix if the expected clinker composition is known. For a typical Portland cement clinker of the following composition, see Table 5.12.

The oxide composition of the raw mix is calculated using Eqs. (5.2) to (5.5) as follows (the chemical formulae represent weight percentages):

$$Fe_2O_3 = C_4AF \times 0.326 \tag{5.2}$$

$$Al_2O_3 = C_3A \times 0.377 + Fe_2O_3 \times 0.64 \tag{5.3}$$

$$CaO = C_3S \times 0.737 + C_2S \times 0.651 + C_3A \times 0.621 + C_4AF \times 0.456 + Free\ CaO \tag{5.4}$$

$$SiO_2 = C_2S \times 0.26 + C_3S \times 0.35 \tag{5.5}$$

Bogue calculations (shown in Eqs. 5.5–5.8) enable estimation of the approximate amounts of the four main constituents (i.e. C_3S, C_2S, C_3A and C_4AF) in clinker based on the oxide composition as indicated by analysis of the clinker. Note, the chemical formulae represent weight percentages.

$$C_2S = 8.6024SiO_2 + 1.0785Fe_2O_3 + 5.0683Al_2O_3 - 3.0710CaO \tag{5.6}$$

$$C_3A = 2.6504Al_2O_3 - 1.6920Fe_2O_3 \tag{5.7}$$

$$C_4AF = 3.0432Fe_2O_3 \tag{5.8}$$

$$Silica\ Modulus = \frac{SiO_2}{Al_2O_3 + Fe_2O_3} \tag{5.9}$$

Bogue's calculations assume that the four main clinker minerals are all pure minerals. Further, it must be noted that clinker is made by combining lime with silica and also lime with alumina and iron. If some of the lime remains uncombined,

Table 5.12 Composition of a typical Portland clinker

Compound	Composition (%wt)
C₄AF	10.0
C₃A	10.0
C₃S	47.0
C₂S	28.0
MgO	4.0
Free lime	1.0

which is often the case, then the uncombined amount must be subtracted from the total lime content before the calculation is done in order to get the best estimate of the proportions of the four main clinker minerals present. Consequently, a figure for the combined free lime must be indicated in any clinker analysis.

Clinker characteristics are often described using the four parameters as follows:

$$\text{Silica Modulus} = \frac{SiO_2}{Al_2O_3 + Fe_2O_3} \tag{5.10}$$

The silica modulus is also referred to as the *silica ratio*. It measures the relative presence in the clinker of calcium silicates compared to the aluminates and ferrites. A high silica modulus means there are more silicates in the clinker and less of the aluminates and the ferrites. The value of silica modulus is typically between 2 and 3 for clinker.

$$\text{Alumina Ratio} = \frac{Al_2O_3}{Fe_2O_3} \tag{5.11}$$

The alumina ratio (AR) determines the potential relative proportions of aluminate and ferrite in the clinker. A high AR implies more aluminate and less ferrite in the clinker. In ordinary Portland cement the value of AR is usually between 1 and 4.

$$\text{Lime Saturation Factor (LSF)} = \frac{CaO}{2.8SiO_2 + 1.2Al_2O_3 + 0.65Fe_2O_3} \tag{5.12}$$

The LSF is a ratio of CaO to the other three main oxides, namely, Al_2O_3, SiO_2 and Fe_2O_3. The LSF accounts for the potential combining power of SiO_2, Al_2O_3 and Fe_2O_3 with CaO. In Portland cement, the LSF should be between 0.66 and 1.02. Up to a limit, the higher the LSF, the better. If it is potentially near 1.0 in a clinker, it is difficult to burn. The silica modulus is a measure of the fluxes in the clinker. Up to a limit, the lower it is in the raw mix, the better for burning. It should be maintained between 2.0 and 3.0. The optimum values for these ratios are: LSF of 0.91; silica modulus of 2.41; and alumina ratio of 1.80. MgO is necessarily present in the raw mix and hence in cement. It has some fluxing action in the clinker. Its presence can be disadvantageous sometimes. For instance, when burnt at very high temperatures it forms hard burning periclase crystals in the clinker. These crystals hydrate very slowly (over a year) and expand during hydration. Thus, if a clinker contains MgO higher than a maximum specified value, it will be unsound on hydration. In some standards, the MgO content in clinker is specified at 6%.

Ideally, in the manufacture of clinker for any type of cement, e.g. OPC, MHC, LHC, SRC, OWC etc., it will be necessary first to decide upon the compound composition of the cement; from this, the oxide composition of the raw mix can be calculated, and finally, the quantities of the various raw materials to be used can be estimated. It is, however, noteworthy that beyond the composition of the raw mix other factors such as the conditions of burning and particulate size of the raw mix also affect the eventual composition of the clinker.

5.4.5 The Role of Gypsum in Cement

Gypsum is ground with clinker to produce cement. Gypsum is a set retarder. It plays an important part in the setting and hardening process of cement. It serves as a pacemaker for the hydration of the different lime compounds in cement paste by forming calcium sulfo-aluminates. Set and hardened cement paste is composed, essentially, of a tobermorite like calcium silicate hydrate (C_2SH) phase, and a calcium sulfo-aluminate ($Ca_3 \cdot Al_2O_3 \cdot CaSO_4 \cdot 31H_2O$) phase, both phases of varying composition depending on the chemical composition of the cement. The presence of more or less of these compounds in the hardened cement gives different properties to the cement paste, like moderate strength, high early strength, high ultimate strength, low heat hydration, better sulfate resistance, etc. It must be noted that most types of Portland cement when ground alone and then mixed with water will set immediately (flash set). This is, however, not desirable as the initial set should not occur in less than 30 min or so. It is to retard the setting time that gypsum is used. To achieve good results as a retarder, pure gypsum must be used. A purity level as low as 60% is applicable.

5.4.6 Finish Grinding

The last unit operation in the chain of operations during the manufacture of Portland cement is finish grinding. During this operation, the clinker granules formed in the kilns are ground into a fine powder before packaging and transportation. The grinding process is a dry one and gypsum (a setting retarder) in about 3–6% (wt) is added to the clinker and ground together. Compartment mills are used for grinding and the process may be either open or closed circuit. The energy requirement for this operation is the highest of all the unit operations used in the manufacture of cement. It varies from 35 kWh/ton to 45 kWh/ton. To curtail energy expenditure, it is important to operate the mills under the right conditions, such as:

i. uniform feed size and uniform feed rate of clinker to the mill must be used;
ii. the diaphragm openings of the mills must be of proper dimensions and must be kept clean;
iii. the loadings of the individual chambers of the mills must be trimmed size-wise and load-wise;
iv. necessary and sufficient draft must be maintained through the mill;
v. grinding aids must be used as much as possible;
vi. efficient dust collecting system must be used.

5.5 Chemistry of Cement Use

The main reactions occurring when cement is mixed with water are hydration and hydrolysis. These reactions lead to the formation of several hydration and hydrolysis products and results in the setting and hardening of cement products. When cement is mixed with water, brief and intense hydration starts. During this period, called the *pre-induction period*, the calcium sulfate in the cement dissolves completely and the alkali sulfates almost completely. The following reaction thereafter occurs between the various compounds in the mixture;

    i. Alite (tricalcium silicate) and water react as follows:

$$2Ca_3OSiO_4 + 6H_2O \rightarrow 3CaO \cdot 2SiO_2 \cdot 3H_2O + 3Ca(OH)_2 \qquad (5.13)$$

This is a relatively fast reaction, causing setting and strength development in the first weeks of the product.

The reaction of water with belite (Dicalcium silicate) proceeds as follows:

$$2Ca_2OSiO_4 + 4H_2O \rightarrow 3CaO \cdot 2SiO_2 \cdot 3H_2O + Ca(OH)_2 \qquad (5.14)$$

This is a slow reaction that is mainly responsible for the strength of cement products.

Tricalcium aluminate hydration is controlled by the presence of calcium sulfate, which is in solution. It causes the rapid formation of ettringite, thereby slowing down the hydration of the tricalcium aluminate. The ettringite reaction takes place as follows:

$$Ca_3(AlO_3)_2 + 3CaSO_4 + 32H_2O \rightarrow Ca_6(AlO_3)_2(SO_4)_3 \cdot 32H_2O \qquad (5.15)$$

The ettringite subsequently reacts slowly with further tricalcium aluminate to form 'monosulfate'—an 'AFm phase' (an 'alumina, ferric oxide, monosulfate' phase). The reaction (in Eq. 5.15) is completed after 1–2 days.

$$\begin{aligned} Ca_6(AlO_3)_2(SO_4)_3 \cdot 32H_2O + Ca_3(AlO_3)_2 + 4H_2O \\ \rightarrow 3Ca_4(AlO_3)_2(SO_4) \cdot 12H_2O \end{aligned} \qquad (5.16)$$

Calcium aluminoferrite reacts slowly due to the precipitation of hydrated iron oxide as follows:

$$\begin{aligned} 2Ca_2AlFeO_5 + CaSO_4 + 16H_2O \rightarrow Ca_4(AlO_3)_2(SO_4) \\ \cdot 12H_2O + Ca(OH)_2 + 2Fe(OH)_3 \end{aligned} \qquad (5.17)$$

The formation of a microstructure of the hydration products of varying rigidity is the cause of the setting and hardening of the cement products. The products set in a few hours and harden over a period of several weeks. The processes vary depending on the composition of the cement mix, the water/cement ratio and the conditions of curing of the product. Usually, the mixture is in a plastic state at the water/cement ratios of 0.25 and 0.75. Typically, concrete sets. That is, it becomes rigid in about 6 h and develops a compressive strength of 8 MPa in 24 h. The strength rises to 15 MPa in 3 days, 23 MPa in one week, 35 MPa in 4 weeks and 41 MPa in three months. All things being equal, the strength should continue to rise as long as the water is available for hydration to occur. However, concrete is usually allowed to dry out after a few weeks, and this causes strength growth to stop.

The reaction products and processes cause the setting and hardening to occur in the following manner. Upon the dissolution of the sulfates and inception of the reactions, short hexagonal needle-like ettringite crystals form at the surface of the clinker particles as a result of the reaction between calcium and sulfate ions with tricalcium aluminate. Further, originating from tricalcium silicate, first, calcium silicate hydrates (C-S-H) in colloidal shape can be observed. Caused by the formation of a thin layer of hydration products on the clinker surface, this first hydration period ceases and the induction period starts during which almost no reaction occurs. The first hydration products are unable to form a consolidated microstructure. This is because the hydration products are too small to bridge the gap between the clinker particles. Consequently, the mobility of the cement particles in relation to one another is only slightly affected, i.e. the consistency of the cement paste turns only slightly thicker. The setting starts after approximately 1–3 h when first the calcium silicate hydrates and forms on the surface of the clinker particles, which is very fine-grained in the beginning. After completion of the induction period, further intense hydration of the clinker phase takes place. This third period (accelerated period) starts after approximately four hours and ends after 12–24 h. During this period a basic microstructure forms, consisting of C-S–H needles and C-S–H leaves, platy calcium hydroxide and ettringite crystals growing in longitudinal shape. As the crystals grow, the gap between the cement particles is bridged. During further hydration, the hardening steadily increases, but with decreasing speed. The pores of the microstructure fill and its density increases. The filling process causes strength gain. The hydration reaction is exothermic and causes an increase in the temperature of the reaction mixture. The heats of hydration of the various clinker compounds at various stages of cure of the cement product are given in Table 5.13.

Table 5.13 Heat of hydration of clinker compounds (J/g)

Compound			Days		
	3	7	28	90	365
C_4AF	290	495	495	416	377
C_3A	888	1557	1378	1302	1168
C_2S	50	42	106	176	222
C_3S	243	222	377	435	490

5.6 Cement Additives

Besides gypsum which is added and ground together with clinker to regulate the setting time of cement, other additives may be added to cement to acquire cement products specific properties. Amongst such additives are air-entraining agents, de-icers, accelerators, extenders, fluid loss additives, friction reducers, retarders and corrosion inhibitors.

5.6.1 Air-Entraining Agents

These increase the resistance of hardened concrete to scaling as a result of alternate freezing and thawing. Their use is very important in cold climates. They are applied in tiny quantities. Examples of air entrainers are resinous materials, tallows and greases.

5.6.2 De-icers

De-icers help to prevent the formation of ice from concrete. An example is $CaCl_2$.

5.6.3 Accelerators

These are materials, which when added to Portland cement or Portland cement blends, shorten the pumping time, setting time and time to develop compressive strength. Accelerators are primarily used in wells having BHCT less than 49 °C, for example, calcium chloride flake used at concentrations from 1 to 4% BWOC, or NaCl, used at concentrations below 10% BWOW, KCl, used below 5% BWOC, and non-damaging to sensitive formations.

5.6.4 Extenders

These are used to increase the yield of a sack of cement and, at the same time, decrease slurry density to reduce hydrostatic pressure. This depends on the application for a low-density cement slurry, i.e. scavenger for mud removal, filler (lead) slurry for reduced or balanced hydrostatics, or as a low density, cost-effective primary or remedial cementing system. Extenders may react or non-react with the cement to produce the desired properties. Examples include cement-grade clay extender, 2–16% BWOC effective concentration; diatomaceous earth extender, 10–40% BWOC effective concentration, and inert, fine aggregate and volcanic glass extender.

5.6.5 Fluid Loss Additives

These are used in cement slurries to keep the mix water from being forced out of the slurry into porous media (formation rock) or in narrow annuli when the slurry is subjected to downhole pressures and temperatures. Some of the benefits from using Messina fluid loss additives in a cement system design are that it: (1) maintains slurry homogeneity and desired flow properties; (2) minimizes formation damage caused by cement filtrate; (3) ensures cement column uniformity and complete annular fill; and (4) allows higher success rates for remedial (squeeze cementing). Examples include powdered, organic polymer blend for fresh or saltwater slurries up to 18% NaCl (BWOW), BHCT 125–250 °F (52–121 °C); powdered, organic polymer for salt slurries up to 37.2% NaCl (BWOW), BHCT up to 250 °F (121 °C).

5.6.6 Friction Reducers

Dispersants are added to cement slurries to facilitate blending at densities greater than the American Petroleum Institute (API) recommended densities for classes of cement without the need for excess water. Some of the features and benefits for using dispersants are: (1) they reduce friction pressures; (2) they reduce critical pump rates for turbulent flow placement; (3) they enhance fluid loss additives; (4) they enhance retarder response; and (5) they improve rheological properties, for example, granular organic acid dispersant for use in salt slurries from 18 to 37.2% NaCl (BWOW).

5.6.7 Retarders

Retarders are classified as any reactive chemical which will extend the pumping time and/or thickening time for a cement slurry.

5.6.8 Corrosion Inhibitors

These when present in cement do inhibit the corrosion of steel reinforcing bars in concrete. An example of such a compound is calcium nitrite.

5.7   Safety and Environmental Effects of Cement Manufacture and Use

5.7.1   Safety Issues

In cement manufacture and use, safety relates to the effect the processes and materials involved in cement manufacture have on humans and measures that must be taken to mitigate the effects when they occur. For instance, when cement is mixed with water a high alkali solution (pH of about 13) is formed by the dissolution of calcium, sodium and potassium hydroxides. When this solution comes into contact with any part of the body it stands to damage it. Again, the cement technology is associated with lots of dust particles which when inhaled stands to damage the respiratory tracts. Excessive noise due to vibration and operation of machine parts could damage the eardrums if measures are not taken to protect them. To avoid any damage occurring, gloves, goggles, filter masks and ear covers are used for protection. Cement can cause serious burns on the skin if contact is prolonged or if the skin is not washed properly.

In several countries, chromium levels in cement are not allowed to exceed 2 ppm as chromium is thought to be toxic and a major skin irritant. The major safety concerns associated with cement use and production and measures to deal with the problems are listed in Table 5.14.

5.7.2   Environmental Effects

The environment can be adversely affected at every stage during the manufacture of cement. Materials and processes that adversely affect the environment during cement manufacture and ways to mitigate their effects are:

i. Emission of air-borne pollution in the form of dust particles and gases during quarrying and manufacture. This can be mitigated by the installation of

Table 5.14 Potential safety concerns of cement use and mitigation measures

Problem area	Cause	Mitigation
Noise pollution	Machine/engine operation	Wear ear mufflers
Respiratory tract	Dust particles	Wear filter mask
Skin and body damage	Cement/water mixture	Wear goggles, gloves and protective clothing
Skin irritation	Chromium in cement	Keep chromium ≤ 2 ppm
Body toxicity	Chromium in cement	Keep chromium ≤ 2 ppm
Accidents	Faulty equipment/carelessness	Well-maintained equipment/well-trained workforce

equipment to reduce dust during quarrying and manufacture. Also, equipment to trap and separate emission gases and dust particles may be installed.

ii. Damage to flora, fauna and land during quarrying. This is dealt with by the institution of land/soil reclamation and restoration measures to re-integrate quarries into the countryside after they have been closed down.

iii. Release of large quantities of CO_2 into the atmosphere as a result of burning large quantities of fossil fuel. The CO_2 contributes to the greenhouse effect. The CO_2 associated with Portland cement manufacture falls into four categories. In the first instance, there is the CO_2 derived from decarbonation of limestone. Secondly, there is CO_2 from kiln fuel combustion. Thirdly, there is the CO_2 released from electricity generated for use during cement production, especially during grinding of raw materials and cement, and lastly, CO_2 is released into the environment from the vehicles used in cement plants and distribution.

The first source of CO_2 is fairly constant, from a minimum of around 0.47 kg CO_2/kg of cement to a maximum of 0.54 kg CO_2/kg cement. The second source varies with the efficiency of the plant. An efficient pre-calciner plant emits 0.24 kg CO_2/kg cement; a low-efficiency wet process plant emits 0.565 kg CO_2/kg cement; a typical modern plant emits about 0.30 kg CO_2/kg cement. The last source is almost insignificant at 0.002–0.005 kg CO_2/kg cement. In sum, therefore, typically, 0.80 kg CO_2/kg finished cement is released into the environment. This value does not include CO_2 associated with electric power generation, where the power generated by coal is about 0.09–0.15 kg CO_2/kg finished cement. CO_2 emission can be mitigated by modification of the chemistry of cement by use of wastes, and by adopting more efficient processes.

5.7.3 Other Pollutants in Cement Production

Other pollutants released into the environment during cement manufacture and are therefore controlled include nitrogen oxides, sulfur dioxide and dust. Carbon monoxide, volatile organic compounds (VOCs), polychlorinated dibenzodioxins (PCDD), dibenzofurans (PCDF) and heavy metals are also released in trace amounts.

i. Dust Emissions and Control

Large quantities of dust are released into the environment during the mining, transportation and preparation of raw materials used for making cement. Dust is also generated in kilns used in the production of clinker, and finally during the grinding and bagging of cement.

The best available techniques (BATs) for reducing dust emissions in the cement industry are electrostatic precipitators and bag filters. Electrostatic precipitators (ESPs) use electrostatic forces to separate dust from gaseous streams. An electrostatic field is generated across the path of a particulate matter in the air stream. The particles acquire negative charges and move in the direction of positively charged plates where they are collected. Dust accumulated on the collection plates is dislodged and collected by electrode rapping. Bag filters for de-dusting gaseous waste streams comprise a fabric membrane that is permeable to gas flowing toward and through it but which will retain particulates. As dust accumulates on the fabric membranes of the 'bag filters' the resistance to gas flow increases. Therefore, regular cleaning is required to maintain performance. The most common cleaning methods include reverse airflow, mechanical shaking, vibration and compressed air pulsing.

ii. NO_x Emissions and Control

NO_x is a general term used, collectively, to refer to all the gaseous oxides of nitrogen, namely nitrogen monoxide (NO), nitrogen dioxide (NO_2), nitrous oxide (N_2O) and nitrogen tetroxide (N_2O_4).

NO_x is formed by the oxidization of the molecular nitrogen of the combustion air as well as the nitrogen compounds in the fuels and raw materials during the combustion of fuel. The NO_x formed in the kiln burning zone at temperatures above 1200 °C is called the thermal NOx. The amount of thermal NO_x produced in the burning zone is related to both burning zone temperature and oxygen content. The NO_x generated by the combustion of the nitrogen present in the fuel is called fuel NOx. Fuel NOx is formed in the pre-calciner where the prevailing temperature is in the range 850–900 °C, which is not high enough to form thermal NO_x.

NO_x emissions can be reduced via primary process optimization measures, including optimization of clinker burning process, computer-based expert system for kiln operation, optimization of the main burner and multi-stage combustion for in-line pre-calciners. NO_x emissions in the kiln flue gas can also be abated using either the selective non-catalytic reduction (SNCR) or the selective catalytic reduction (SCR) methods. During SNCR, NO_x in flue gases is converted into N_2 and water (H_2O) by reaction with NH_4–X compounds injected into the flue gas. The ammonia-containing solution may be supplied in the form of anhydrous ammonia, aqueous ammonia or urea. The SCR method involves the injection of ammonia-based reagents into the flue gas stream in the presence of a catalyst. The common reagents used include anhydrous ammonia, aqueous ammonia and urea. NO_x compounds are converted to N_2 and water and the catalyst allows these reactions to occur at lower temperatures, between 300 °C and 400 °C, than those required for SNCR.

NO_x emissions can also be reduced through primary process optimization measures, including optimization of the clinker burning process, computer-based expert system for kiln operation, optimization of the main burner (low NO_x burner) and multi-stage combustion for in-line pre-calciners.

iii. SO_x Emissions and Control

SO_2 is the most common form of the oxides of sulfur released into the environment during cement manufacture. Sulfur dioxide is formed during the combustion of fuel in the burning zone of the kiln and the pre-calciner. Sulfur dioxide is also formed as a result of the oxidization of pyrite and organic sulfur in raw feed materials in the pre-heater or the kiln inlet of long wet or dry kilns. The SO_2 formed by the combustion of fuel is absorbed in the basic feed material in the pre-heaters of modern cement kilns. Therefore, in modern kilns, any SO_2 emissions from the cement plant are predominantly due to oxidation of pyrite in the feed materials in the upper stages of the pre-heater.

The methods applied in the cement industry to reduce SO_2 emission are:

Use of absorbents

The addition of dry or wet absorbents, such as slaked lime, quicklime or activated fly ash with high CaO content to the exhaust gas of a kiln can absorb SO_2. Where emissions are below 1,200 mg/m^3 the most efficient way of utilizing slaked lime is to add it to the kiln feed prior to clinker burning and let the SO_2 absorption occur in the kiln, rather than inject it into the flue gases after sintering. Where SO_2 concentrations are higher this is not economically feasible.

Dry scrubbers

At emissions levels in excess of 1,500 mg/m^3, a separate scrubber may be required to reduce emissions below 500 mg/m^3. A dry scrubber can be installed to precede the raw meal entering the kiln.

Wet scrubbers

These are the most commonly used method for flue gas desulfurization in coal-fired power plants. A slurry or an aqueous solution of calcium carbonate, hydroxide or oxide is used to absorb SO_2 from the flue gases. The slurry is sprayed in a counter-current direction to the exhaust gas and collected in a recycle tank at the bottom of the scrubber where the formed sulfite is oxidized with air to sulfate and forms gypsum, $CaSO_4 \cdot 2H_2O$. The gypsum is usually separated and used as a set controlling agent during cement milling, while the water is returned to the scrubber.

Activated carbon

SO_2 can be adsorbed from flue gases using activated carbon. The flue gases pass through an activated carbon filter constructed as a packed bed and the used activated coke is periodically replaced with fresh adsorbent.

Primary optimization processes

SO_2 emissions can also be reduced via primary process optimization measures such as reduction in sulfur content in fuels and kiln feed, optimization of the clinker burning process, and addition of slaked lime to the kiln feed with the use of a dry scrubber if necessary.

iv. Trace Emissions

Other emissions from the cement industry which occur in trace quantities include carbon monoxide, volatile organic compounds (VOCs), polychlorinated dibenzo-dioxins (PCDDs), dibenzofurans (PCDFs) and heavy metals.

5.8   Quality Control Indicators

The quality requirements of Portland cements are specified based on their physical or chemical properties.

5.8.1   Chemical Specification and Analysis

The chemical composition of cement determines its physical properties. Therefore, chemical analyses are performed to ascertain that the cement product meets standard or customer specifications. The analyses also give an idea of how much of each raw material should be ground to achieve the expected quality. Generally, quality tests would be done on the raw materials, processes and product. Some chemical specifications and analysis of raw materials and cement products are listed as follow:

i. The ratio of the percentage of lime to the sum of percentages of silica, alumina and iron oxide should not exceed 1.02 and should neither be less than 0.66. Mathematically the ratio is expressed as:

$$\frac{CaO - 0.7SO_3}{2.8SiO_2 + 1.2Al_2O_3 + 0.65Fe_2O_3} \tag{5.18}$$

The ratio ensures that there are sufficient cementitious compounds (i.e. lime silicates) in the cement. A value of 1.02 would mean 100% C_3S in the cement and 0.66 would mean 100% C_2S.

ii. The ratio of alumina to iron oxide should not be less than 0.66. In white cement this ratio can be as high as 15. Note, white cement contains less ferrite compounds.

iii. The weight of insoluble residue in the cement should be less than 2.0%. This value checks the adulteration of cement with rock dust or ground coal ash. It must be noted that all the clinker compounds and gypsum are soluble in hydrochloric acid.

iv. The weight of magnesia should not be more than 6%. Excess magnesium oxide present in the raw mix forms periclase and is burnt at a high temperature. The over burnt periclase hydrates very slowly over the years and expands producing expansion cracks in concrete in the late ages. One way to neutralize the effect of the periclase is to cool the clinker very rapidly from the burning temperature to about 800 °C so that the periclase does not crystallize but is in a glassy state. For PBSF cement, the limit of MgO is 8%.

v. Total sulfur content, calculated as sulfuric anhydrite (SO_3), should not exceed 2.75%. The SO_3 is derived from the gypsum ($CaSO_4 \cdot 2H_2O$), which is added to moderate the setting of cement. High proportions of gypsum in cement, however, cause the set cement to expand, which is not desirable. The gypsum content of the cement measured in terms of percentage SO_3 in the cement is therefore set at 2.75% to avoid the expansion of cement products and at the same time achieve the correct setting of the product. To determine the SO_3 in cement, it is first precipitated from the cement using sodium hydroxide with Bromo-crystal as an indicator. The precipitate is subsequently filtered, charred and the residue weighed.

vi. Total loss on ignition (LOI) should not be more than 4%.

5.8.2 Physical Requirements

The physical requirements and analysis of cement products are presented in the following:

i. Setting Time Test

Setting time is the time required for cement to change from a plastic state to a stiff solid state. The initial set refers to that state when the paste begins to stiffen, and the final set refers to that state when the cement has hardened. The test is done in a Vicar apparatus. Exactly 500 g of cement is mixed with water to obtain a standard consistency in a mixer. The paste is then positioned under the needle of the Vicar apparatus. The needle is lowered gently until it is in contact with the paste. Subsequent automatic penetration is affected by the device. The setting time is the time it takes the needle to penetrate the paste to a set distance. When tested by the standard method the initial setting time of low heat cement should not be less than 60 min, and for all other types of cements, it should not be less than 30 min. The final setting time should not be more than 10 h for all types of Portland and blended cements.

ii. Compressive Strength

An important quality indicator of cement is its compressive strength development in mortar and concrete. The compressive strength gives a measure of how much load the mortar can carry before failure. To test the compressive strength, a mortar is prepared by mixing 450 g of cement, 1350 g of sand and 225 ml of water in a mixer and later poured into a 40 × 40 × 160 mm mortar prism. It is then cured in succession for 2 days and 28 days. The cured mortars are then subjected to compressive loading until failure in a compression machine. The values obtained are then compared with standard values to assess quality.

The compressive strength of 50 mm cubes made of 1:3 cement:sand mix (using standard graded sand) expressed as kg/cm^2 should not be less than the values given in Table 5.15.

iii. Heat of Hydration

For low heat cement, the heat of hydration test is compulsory and the value should not be more 65 cal/g at 7 days and 75 cal/g at 28 days.

Table 5.15 Compressive
strengths of different cements
at different durations

Curing period	OPC	RHC	LHC	PBFSC	PPC	MC	SRC
	Compressive strength kg/cm^2						
Day 1	160	–	–	–	–	–	–
Day 3	160	275	100	160	–	–	150
Day 7	220	–	160	220	220	25	220
Day 28	–	–	350	–	310	50	300

OPC = Ordinary Portland cement; RHC = Rapid hardening
cement; LHC = Low heat cement; PBFSC = Portland blast
furnace slag cement; SRC = Sulfate-resistant cement;
PPC = Portland pozzolana cement; MC = Masonry cement

iv. **Fineness**

The specific surface by the air permeability method should not be less than 3250
cm^2/g in the case of rapid hardening cement, 3200 cm^2/g in the case of low heat
cement, 3000 cm^2/g in the case of PPC and 2250 cm^2/g in the case of all other types
of Portland cement and PBFS.

v. **Soundness**

When tested by the boiling test, the expansion should not be more than 10 mm in
all types of cement or when tested by the autoclave test, the expansion should not
be more than 0.8% in all types of cement. A cement containing more than 3% MgO
must pass the autoclave test.

5.9 Refractory Materials

Refractories are ceramic materials that are characterized by high strength and
fracture toughness. They are also inert to most chemical attacks, are abrasion
resistant and can withstand high temperatures. Generally, refractory materials can
withstand temperatures above 1100 °C without softening, and this property is
sometimes used to differentiate refractory materials from non-refractory materials.
The strength of refractory materials is due in part to the relatively reduced number
and distribution of microcracks in the finished product. They are used in applications where heat resistance, hardness, fracture toughness and strength are essential.
Specific applications of refractory materials are in incinerators, furnace and kilns
where they are used as linings, in engine bearings, engine and turbine parts, cru-
cibles and cutting tools. They are also used in electronic and structural applications.

5.9.1 Types of Refractory Materials

Refractories may be classified based on their behavior in a chemical environment, their composition or simply as clay and non-clay. Based on their behavior in a chemical environment, they are classified as acidic, basic or neutral. Acidic refractories cannot be used in a basic environment as they would be corroded. Examples of such refractories include zircon, fireclay and silica. Similarly, basic refractories cannot be used in an acidic environment. Dolomite, magnesite are examples of basic refractories. About 95% of refractories manufactured are non-basic. Among the neutral refractories are alumina, chromite, silicon carbide, carbon and mullite. These latter ones can be used in both acidic and basic environments.

Based on their composition, refractories may be grouped as oxides, nitrides, borides or carbides. The oxides are compounds of oxygen. Among the oxides is aluminum oxide or alumina (Al_2O_3), beryllium oxide or beryllia (BeO), zirconium oxide or zirconia (ZrO_2), and mullite, which is a mixture of alumina and silica (SiO_2/Al_2O_3)—71.8% alumina and 28.2% silica. The nitrides invariably have nitrogen as a prominent element in their compound. Examples are aluminum nitride (AlN), silica nitride (Si_3N_4)) and two forms of boron nitride (BN)—the cubic and hexagonal forms. Prominent among the carbides used in the preparation of refractories are silicon carbides. They are two distinct forms of silicon carbides: the beta silicon carbide (a cubic form) and alpha silicon carbide, which is a generalized term for the non-cubic form of silicon carbide.

Clay refractories are produced from fireclay (hydrous silicates of alumina) and alumina (57–87.5%). Kaolin, bentonite, ball clay and common clay are some of the other clay minerals used in the production of refractories. The non-clay refractories are made from a composition of alumina (<87.5%), mullite, chromite, magnesite, silica, silicon carbide, zircon and other non-clays.

5.9.2 Manufacture of Refractory Materials

Refractories are produced in two basic forms, namely formed objects and unformed granulated or plastic compositions. The formed objects are called bricks and shapes. These are used to construct the walls, arches and floor tiles of high-temperature process equipment. The unformed compositions include mortars, gunning mixes, castables (refractory concretes), ramming mixes and plastics. These latter products are cured in place to form a monolithic, internal structure after application. The single-most important property to produce during refractory manufacture is high bulk density. The density affects such other properties as strength, volume stability, slag and spalling resistance, and heat capacity. For insulating refractories, a porous structure is required, which translates into a low-density material. Refractory products are manufactured by one of three approaches. These are the mold and burn process, electric fusion process and ceramic fiber formation process.

5.9.2.1   Mold and Burn Process

The four major unit operations involved in this process are raw material processing, product forming, firing and final processing. A generalized flowchart for the mold and burn process is shown in Fig. 5.3.

    The first step in this process is the preparation of the raw material. The object of this step is to reduce the size of the lumps and the moisture content in the raw materials to the level that will facilitate the subsequent operations of forming and firing. This step involves crushing and grinding of the crude clay or ore followed by

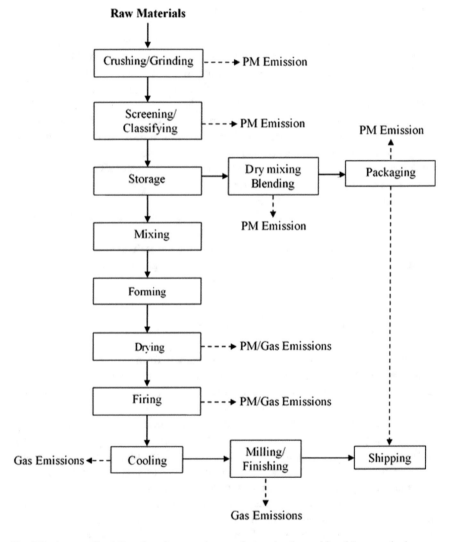

Fig. 5.3 A generalized flowchart for ceramic manufacture by the mold and burn method

size classification and calcining or drying. The size of the particles in a batch is one of the important factors during refractory formation. It is known that a mixture in which the proportion of coarse and fine particles is about 55:45, with only a few intermediate particles gives the densest mixtures. Achieving this requires careful screening, separation and recycling. The method works well on highly crystalline materials but is difficult to obtain in mixes of high plasticity. The processed raw material may be dry-mixed with other minerals and chemical compounds, packaged and shipped as a product. The mixing is done with the sole aim to distribute the plastic materials so as to coat thoroughly the non-plastic constituents. This serves the purpose of lubricating the materials during molding and allows the bonding of the mass with a minimum number of voids.

Forming consists of mixing the raw materials and molding them into desired shapes. The operation occurs under wet or moist conditions. The molding is a dry-press process with mechanically separated presses. This method is particularly well suited for batches that consist mostly of non-plastic constituents. It has been found necessary to de-air the bricks during pressing. This avoids laminations and cracking when the pressure is released. When present, the air is absorbed by the clay or condensed during pressing. De-airing is achieved by applying vacuum through vents in the molds box. The mold is **dried** before firing. The drying removes the moisture added before molding to develop plasticity. The elimination of water leaves voids in the mold and causes extensive shrinkage and internal strains. To avoid this in some cases, drying is completely skipped and the mold fired immediately after forming.

Firing involves heating the formed shapes to very high temperatures in a batch or continuous kiln to form the ceramic bond that gives the product its refractory properties. Two important things happen during firing. First, permanent bonds are developed in the mixture by partial vitrification. Secondly, stable mineral forms are developed. The changes that take place are the removal of water of hydration, followed by the calcination of carbonates and the oxidation of ferrous iron. The final volume of the product may be as much as 30% less than the non-fired material. The reduction in volume causes severe strains in the final product. The shrinkage may be avoided by pre-stabilization of the materials used. The final processing stage involves **milling, grinding** and **sandblasting** the product from the kiln to give it the correct shape, size and aesthetic appeal. Sometimes, during this stage, the product is impregnated with tar and pitch and finally packaged.

5.9.2.2   Electric Fusion Process
In this process, the raw material, often a mixture of diaspore clays of high alumina (to furnish a $3Al_2O_3/2SiO_2$), is introduced into the top of an electric arc furnace. Molten alumina silicate at 1800 °C is subsequently tapped from the furnace at intervals and run into molds built from sand slabs. The molds containing the blocks are annealed for about 6–10 days before the blocks are usable. The products from this process have a vitreous, non-porous body that has a linear coefficient of expansion of one and half times that of good firebrick. The voids content is only

about 0.5% in contrast to the usual 17–29% for fire-clay blocks. Refractory blocks cannot be cut or molded. They can, however, be ground on Alundum wheels. The products have a long life and minimum wear but have a high initial cost. They are used in glass furnaces, in linings of hot zones of rotary kilns, in modern boiler furnaces exposed to severe duty and in metallurgical equipment such as in forging furnaces.

5.9.2.3   Ceramic Fiber Formation

This process requires that the feed material, for instance, the calcined kaolin, is first melted in an electric arc furnace. Subsequently, the molten clay is made into fiber in a blow chamber with a centrifuge device. An alternative route to fiberizing the molten clay is by dropping the melt into an air jet and immediately blowing it into fine strands. The last stage of this process is curing the fiber in an oven. The curing adds structural rigidity to the fibers. Oils are added to lubricate both the fibers and the machinery used to handle and form the fibers during the curing process. This process is very similar to that used for the production of mineral wool. Table 5.16 lists some varieties of refractory materials, their application and composition.

5.9.3   The Cement Industry in Ghana

As of the end of 2018, 11 cement plants operated in Ghana. Out of these, 10 were grinding plants. The Savannah Cement Company plant at Buipe in northern Ghana is an integrated plant that has the capacity to produce clinker. However, for a long while the plant only operated as a grinding plant. As of 2018, no plant in Ghana produced clinker. What this means is that the clinker used by cement plants in Ghana are all imported into the country. The 11 plants and their location in Ghana are listed in Table 5.17. These plants have an installed capacity of 12 million tons as of the end of 2018. The annual demand for cement in Ghana stood at about 8 million tons in 2018. Of this amount, 500,000 tons are bagged cement imported into the country annually.

5.9.3.1   Sources of Raw Materials in Ghana for the Production of Cement

There are sufficient deposits of lime and clay, key ingredients for the production of cement clinker, to support the production of cement in the country for several years. However, as of the end of 2018, with the exception of the Buipe plant, no other clinker plant existed in the country.

Limestone deposits in the country are estimated to be over 500 million tons. The four major limestone deposits in Ghana are found in Buipe in the Savannah region, Nauli in the western region, Oterkpolu in the eastern region and Bongo-Da in the northeastern region. Smaller deposits can also be found at Wenchi, Abeasi and Kintampo (all in Bono East Region), Daboya in the northern region, and Salega-Yeji in the Savannah region, Sadan-Abetifi in Ashanti region, Anyaboni in the eastern region, and Du area in the upper eastern region.

Table 5.16 Varieties of refractory materials

Refractory	Application	Composition
Fire-clay brick	Most widely used refractory material used in linings furnace, foundries kilns, boilers, gas-generating sets	SiO_2/Al_2O_3 mixtures of varied compositions
Silica brick	Used for making arches in large furnace, are also used in coke ovens and in gas retorts	95–96% SiO_2 and 2% line
High-alumina	Used in lining of glass furnace, oil fired furnaces, high pressure oil stills, roof of lead softening furnaces and in regenerator checkers of blast furnaces	Made from clay rich in bauxite and diaspore
Magnesia	Used in open-hearth and electric-furnace walls, in the burning zones of cement kilns, and in roofs of non-ferrous reverberatory furnaces	Made from domestic magnesites or magnesia extracted from brines
Insulating brick	Used as backing for refractory bricks or in place of regular refractory bricks depending on the type. Fire-clay ground, mixed and molded	There are two types. They are made from natural porous diatomaceous earth or from waste cork
Silicon carbide	Used for making rocket nozzles, furnace and radiant heater tubes, combustion chambers for ceramic-based gas turbine engines, and in iron-making blast furnace	Made from a mixture of crude silicon carbide clay or finely ground silicon carbide
Electrocast refractory	Used in glass furnace, in linings of hot zones of rotary kilns, in boiler furnaces exposed to severe duty, and in metallurgical equipment such as forging furnace	Made from an electrically fused mixture of $2SiO_2/3Al_2O_3$ ratio
Pure oxide refractory	They have been developed for light refractory products	They are made from alumina, magnesia, beryllia, thoria and zirconia

The limestone deposits in Oterkpolu and Nauli are already being mined by Ghacem to supplement its limestone import into the country. Ghacem uses limestone as a clinker filler for the production of cement in Ghana. About 24% of the clinker weight in cement is replaced with limestone by the company.

The distribution of notable limestone deposits in the country is as follows: Nauli in the western region has over 400 million tons which could be used for the production of cement clinker to substitute all imports of clinker. Buipe in the northern region has limestone and mudstone deposits estimated at over 6.03 million tons. It also has limestone and dolomite deposits estimated at over 44 million tons. Limestone deposits at Bango-Da are estimated to be over 15 million tons in addition to 30 million tons of dolomite deposits. In Daboya over 162,000 tons of limestone and over 500,000 tons of dolomite can be found.

Like limestone, the other major raw material in clinker production, clay is also abundant in Ghana. It is estimated that clay deposits in Ghana amount to over 600

Table 5.17 Cement manufacturers in Ghana as of 2019

Name of company	Location	Type of plant	Type of cement	Plant capacity 10^6 tons/yr
Wan Heng Ghana Ltd	Accra	–	–	0.5
Xin An Safe Cement	Konongo	–	–	0.5
Ciments de L'Afrique Ghana Ltd	Tema	Grinding	–	1.0
CBI Ghana Ltd	Tema	Grinding	32.5R	0.8
Dangote Cement Ghana Ltd	Tema	Bagging	Portland	1.5
Dangote Cement Ghana Ltd	Takoradi	Grinding	Portland	1.5
Ghacem Ltd	Tema	Grinding	32.5R/42.4R	2.2
Ghacem Ltd	Takoradi	Grinding	32.5R/42.5R	2.2
Diamond Cement Ghana Ltd	Aflao	Grinding	42.5R/32.5R	1.8
Savannah Cement Company Ltd	Buipe	Integrated	42.5R/32.5R	0.44
Western Diamond Cement	Takoradi	Grinding	42.5R/32.5R	1.0

million tons and can be found in all the regions in Ghana. However, there is no known deposit of gypsum, the other key component of clinker making in Ghana.

Further Readings

Alsop, P. A. (2019). *The cement plant operations handbook* (7th ed.). Tradeship Publications Ltd.

Bye, G. C. (1999). *Portland cement composition, production and properties*. Thomas Telford.

Bhatty, J. I., Miller, F. M., & Boahn, R. P. (2011). Innovations in Portland cement manufacturing. Skokie, IL: Portland Cement Association.

Chatterjee, A. K. (2018). *Cement production technology: Principles and practice*. CRC Press.

Deolalkar, S. P. (2009). *Handbook for designing cement plants*. CRC Press.

Deolalkar, S. P., Shah, A., & Davergave, N. (2013). *Cement designing green plants*. Butterworth-Heinemann.

Johansen, V., Taylor, P. C., & Tennis, P. D. (2006). Effects of cement characteristics on concrete properties. Skokie, IL: Portland Cement Association.

Kohlbaas, B., & Labahn, O. (eds.). (1982). Cement engineer's handbook. Intl Public Service.

Locher, F. W. (2005). *Cement: Principles of production and use*. Verlag Bau + Technik.

Taylor, H. F. W. (1997). *Cement chemistry*. Thomas Telford.

Pulp and Paper Technology

6

Abstract

Paper has many uses such as in writing, printing, packaging, paper money, cheques voucher, cleaning, as construction materials, etc. Based on its applications, paper may be characterized as bank paper, book paper, bond paper, construction paper, cotton paper, electronic paper, photo paper, wallpaper, etc. The popular raw materials used to make paper are esparto-grass, straw, wood, flax, hemp, jute and rags-cotton and linen. The paper manufacturing process consists of two main stages: pulp making and conversion of pulp into paper. The process of producing pulp may be done by chemical or mechanical means, or by a combination of both depending on the nature of raw material used and the end application of the produced paper. Three operations are required to manufacture paper that are beating, conversion of pulp to paper and finishing. This chapter comprehensively covers the types of papers, their uses, raw materials and their manufacturing processes. Pulp and paper mills have adverse effects on the environment such as depletion of forest covers, and chemical pollution caused by liquid effluent discharges, solid wastes and air emissions from particulates, H_2S, NO_x, SO_x. These have been covered in detail including means to reduce their emissions and treatment methods.

6.1 Introduction

Paper is a thin sheet of compressed connected fibers used for writing upon, printing or packaging. There are many uses to which paper is put. Principal among these are:

i. As a communication tool. It is written and printed upon for the purposes of documentation for record keeping or simply for direct transmission of information.

© The Author(s), under exclusive license to Springer Nature Switzerland AG 2021   171
O.-W. Achaw and E. Danso-Boateng, *Chemical and Process Industries*,
https://doi.org/10.1007/978-3-030-79139-1_6

ii. To represent value as in the case of paper money, bank notes, cheque voucher, ticket and security paper.

iii. For storing information, i.e. for personal use, e.g. dairy, or for communication to someone, e.g. letter or for the benefit of mute people.

iv. For packaging, e.g. corrugated box, paper bag, envelope, wrapping tissue and wallpaper.

v. For cleaning, e.g. as a toilet paper, handkerchief, paper towel, facial tissue, car litter and pampers.

vi. For construction, e.g. papier-mâché, origami, quilling, paper honeycomb, in composite materials, paper engineering, construction paper and clothing.

vii. Various other uses of paper include emery paper, sandpaper, blotting paper, litmus paper, universal indicator, paper chromatography and capacitor dielectrics among others.

6.2   Types of Paper

Paper is often characterized according to its weight. In Europe, and other regions using the ISO 216 paper sizing system (ISO—International Standards Organization), the weight is expressed in grams per square meter (g/m^2 or usually just g) of the paper. The weight of printing paper is between 60 and 120 g. When the weight is more than 160 g, it is called a card. The weight of a ream, therefore, depends on the dimensions of the paper and its thickness. The thickness of paper is often measured by the caliper, and it is typically given in thousandths of an inch. The thickness of paper is usually between 0.07 mm and 0.18 mm. Standard sizes of paper are A0 (A zero), A1 (A one), A2 (A two), A3 (A three) and A4 (A four). A0 is the largest standard size paper. Two sheets of A1 placed upright side by side fit exactly into one sheet of A0 laid on its side. Two sheets of A2 fit into one sheet of A1, two sheets of A2 fit into one A3 and two sheets of A3 fit into one A4. In the United States, the weight assigned to a paper is the weight of a ream, 500 sheets, of varying 'basic sizes', before the paper is cut into size, it is sold to end-users. In the United States, printing paper is generally 20 lb, 24 lb or 32 lb at most. Paper may also be characterized according to its properties or the application to which it is put. The various types of paper are presented in Table 6.1.

6.3   Raw Materials for Making Paper

Paper is made by pressing together moist fibers and drying them into flexible sheets. Therefore, any material from which fibers can be obtained could also be used for making paper. In practice, this is not so as factors such as the abundance of fibers, strength of the fiber, flexibility of the fiber and the length of the fibers in the source material determine whether or not a raw material can be used to process paper. The

Table 6.1 The various types of paper

Type of paper	Use	Characteristics of paper
Bank paper	It is used for typewriting and correspondence	Thin strong writing paper of less than 50 g/m^2
Bond paper	It is used for letterheads, electronic printing, graphic work and other stationery	The quality is similar to bank paper. However, the weight greater than 50 g/m2. It is made from rag pulp
Book paper (publishing paper)	Used specifically for the publication of printed books	Traditionally, book papers are off white or low white papers (easier to read), are opaque to minimize the show through of text from one side of the page to the other and are (usually) made to tighter caliper or thickness specifications, particularly for case bound books. Typically, book papers are light weight papers 60–90 g and often specified by their caliper/substance ratios (volume basis)
Construction paper (sugar paper)	Used for projects and crafts	Construction paper or sugar paper is a type of coarse colored paper typically available in large sheets. The surface is characterized by rough texture, the presence of small particles and looks unfinished
Cotton paper	Due to its durability, it is used for printing and storing important documents such as dissertation or thesis, bank notes and legal documents. It is also sometimes referred to as cotton rag or ragged paper	Cotton paper is made from 100% cotton fibers. It is superior to paper made from wood pulp in strength and durability. It can last for several hundred years without fading, discoloring or deteriorating
Electronic paper or E-paper	It mimics the appearance of ordinary ink paper and used mainly in display technology	Reflects light like ordinary paper. It can hold text and images indefinitely without drawing electricity, while allowing the image to be changed later. It portrays other characteristics like ordinary paper, e.g. it can be bent and crumpled
Inkjet paper	Mostly used for printing in inkjet printers	It is paper with good dimensional stability, good surface strength and no curling. It is classified by its brightness, smoothness, and sometimes by its opacity
Photo paper	It is used specifically for reproduction of photographs	The best of these papers, with suitable pigment-based ink systems, can match or exceed the image quality and longevity of traditional materials used for printing color photographs. For printing monochrome photographs, traditional sliver-based

(continued)

Table 6.1 (continued)

Type of paper	Use	Characteristics of paper
		papers are widely felt to retain some advantage over inkjet prints
Kraft paper	It is used for paper grocery bags, multi-wall sacks, envelopes and other packaging	This type of paper is strong and relatively coarse. It is usually of brown color but can be bleached to produce white paper. It is produced from wood pulp by the Kraft process
Laid paper	It is used as a support for charcoal drawings	It has a ribbed texture imparted to it as part of the manufacturing process
Tyvek	Example applications include, medical packaging, envelopes, car covers, air and water intrusion barriers (house wrap) under house siding, labels, wristbands, mycology, and graphics	Tyvek is a brand of flash spun high-density polyethylene fibers, a synthetic material; the name is a registered trademark of the DuPont Company. It is a very strong material permeable by water vapor but not liquid water
Paper towel	It is a disposable product made of paper. They are often chosen to avoid the contamination of germs	This is high moisture absorbency, soft paper with properties similar to conventional towels used for drying hands, dusting and wiping windows
Wall paper	Wallpaper is a material that is used to cover and decorate the interior walls of homes, offices, and other buildings	The surface of this type of paper is either plain so it can be painted or has graphic patterns. Wallpaper printing techniques include surface printing, gravure printing, silk screen-printing, and rotary printing. 'Wallpaper' is also a term for computer wallpaper. Wallpapers are usually sold in rolls and are put onto a wall using wallpaper paste
Washi (Japanese paper)	It is used in traditional arts mostly in Japan where it originated from. It is also called Wagami	Washi is commonly made using fibers from the bark of the gampi tree, the mitsumata shrub (Edgeworthia papyrifera), or the paper mulberry, but also can be made using bamboo, hemp, rice and wheat. It is generally tougher than paper made from wood pulp
Wax paper	Used in cooking, for its non-stick properties. It is also used for wrapping food for storage, as it keeps water out or in. Another area of use is in the in arts and crafts industry	This is a moisture proof paper achieved by the application of wax. The practice of oiling parchment or paper in order to make it semi-translucent or moisture-proof goes back at least to medieval times
Coated paper	For printing purposes, e.g. Newspaper insert, advertising materials, catalog, security papers, magazines	The surface of this type of paper is coated to impart certain qualities to the paper, such as surface gloss, smoothness, ink absorbency and protection against ultraviolet radiation. The coating materials used are mostly inorganic compounds, e.g. Kaolinite
Wove paper	It is used mainly as a writing paper	It has a uniform surface and is neither ribbed nor watermarked

common raw materials used to make paper include, esparto-grass, straw, wood, flax (linen), hemp, jute and rags-cotton and linen.

The process of paper manufacture consists of two main steps, namely, pulp making and the conversion of pulp into paper. During the first step, i.e. pulp making, fibers are extracted from the raw materials. In the second stage, the pulp in stage one is converted into paper proper ready for the market. Linen and cotton rags having already undergone a process of manufacture, consist of almost pure fibers with the addition of fatty and coloring matters, which can be got rid of by simple boiling under a low-pressure steam with a weak alkaline solution. On the other hand, esparto, wood, straw, flax, hemp and jute as they are from the soil/wood contain all the intercellular matter in its original form, which should be dissolved by strong chemical treatment under a high temperature in other to free the fiber for subsequent use.

Wood-pulp has grown to be one of the most important fibers for paper-making purposes. Scandinavia, Germany, the United States and Canada are the countries that mainly use wood as a material for paper-making, owing to their possession of large forest areas. These countries are the source of wood-pulp used by many other countries. Trees of medium age are usually selected, varying from 70–80 years' growth and running from 8 to 12 in. in diameter. They are fell in winter and reach the mill in logs about 4 ft long. After being freed from the bark and the knots taken out by machinery, the logs are cut into small cubical chips about 2 in. in size by a revolving cutter. The chips are then bruised by being passed between two heavy iron rolls to allow a boiling solution to thoroughly penetrate them, and are conveyed to boilers over a screen of coarse wire-cloth, which separates out the fine sawdust as well as any dirt or sand. In the soda process, the wood is boiled in large revolving or upright stationary boilers to extract the fiber.

Most kinds of **straw** can be utilized for making into paper, the varieties generally used are **rye, oat, wheat and barley**. Oat and rye are the most important, as they give the largest yield in fiber. Germany and France are the two principal users of straw, which closely resembles esparto in its chemical constitution and is reduced to a pulp by a somewhat similar process. Fibers like jute, hemp and manila are chiefly used for the manufacture of coarse papers where strength is of more importance than appearance, such as wrapping papers, paper for telegraph forms, etc. The boiling processes for these are similar to those used for esparto and straw.

6.4 Pulp Production

As a first step in making paper, fibers as pure and as strong as possible are extracted from fiber-bearing raw materials (mostly plant materials) in the form of pulp. The processes leading to the manufacture of pulp depend on the raw material and the nature of the product required. Pulping is the process of extracting cellulose of fibers from plant and other fiber-bearing materials. All pulping processes have the singular goal to release the fibrous cellulose from its surroundings while keeping the hemicelluloses and celluloses intact, thereby increasing the yield of the fibers. The exact processes of pulp production depend on the nature of the raw material used and the application to which the eventual paper is to be put. The separation of the fibers from the fiber-containing matrix may be achieved by chemical means, mechanical means or by a combination of both.

6.4.1 Chemical Pulping

During chemical pulping, the lignin binding the fibers is dissolved (broken down) by the use of a chemical. In the process, the fibers are released for further treatment and use. The particular chemical used gives rise to its own procedures. The most widely used chemical procedure is Kraft or sulfate pulping. Others are sulfite pulping and the NSSC (neutral sulfite semi-chemical) pulping. Another chemical used in the past but no longer used is soda (the process cooked wood with strong solution of NaOH and Na_2CO_3).

6.4.2 Kraft Pulping

Also called the sulfate pulping, this process cooks the raw material in a liquor to which sodium sulfate (Na_2SO_4) has been added (hence the name 'sulfate process'). This separates the lignin and wood resins from the cellulose fiber pulp, which is then washed and, if necessary, bleached. Kraft pulping is the dominant method of pulp production in the world by virtue of its versatility and the high strength, long fiber, very low lignin content pulp it produces. The raw material most adopted to this process is coniferous wood (soft wood) even though the method can be used for all sorts of wood and non-wood materials. Most Kraft mills operate a closed loop system whereby 95–98% of the chemicals used in the process are recovered and reused. This means that the Kraft process is superior to other processes in terms of chemical usage per ton of wood-pulp produced. The Kraft process may be operated batchwise or in a continuous manner. Batch processes are easier to control. Continuous processes, however, are more economical and are amenable to the installation of pollution control facilities. Paper produced from Kraft pulp is very strong because of the long fibers it produces and the mild action of the chemicals on the pulp. Depending on the nature of the bleach, Kraft pulp comes in all manner of colors from dark color (for making sacks, wrappers and paper board) to light colors and white color pulps used for all manner of papers.

A flowchart of the Kraft process is shown in Fig. 6.1 for the processing of wood pulp. The raw materials, the logs, when they arrive at the plant are first cut into convenient sizes and subsequently debarked. They are then conveyed to chippers where they are reduced to chips of preselected size. From the chippers, the chips are screened (rotating or vibrating screens are used) to separate the saw dust and oversized chips from the product (chips of required size). The oversized chips are returned to the chipper to reduce them to the required size. The product is then fed

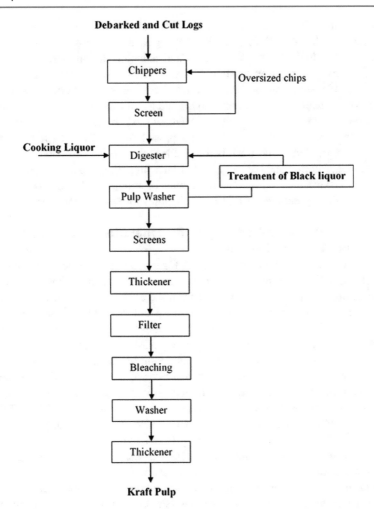

Fig. 6.1 A simplified flowchart of the Kraft pulping process for wood

to a digester. The digester is made up of two zones. In the first, zone steam at a pressure of about 100 kPa is used to volatilize non-condensable gases and turpentine out of the chips. Subsequently, they pass on to a high pressure zone where they are dosed in cooking liquor. The temperature and pressure of this zone are kept at 170 °C and 900 kPa, respectively. The chips stay in this zone for about 1.5 h to cook. During the cooking, the lignin that welds the useful cellulose and hemicellulose compounds and other organic compounds is hydrolyzed and dissolved releasing these compounds into the solution or the atmosphere. The organic acids (rosin acid) form sodium soaps, and mercaptan and organic sulfides are released into the environment, which are the sources of foul odor associated with pulp mills.

The cooking reaction is stopped by quickly introducing a flow of cold cooking liquor into the digester. Before leaving the digester, the chips are first washed to reduce the chemical content of the chips and reducing the pressure of the dosing zone. The process produces flashing steam, which is used to pre-steam the chips entering the digester. The product from the digester is known as brown stock. The brown stock carries with it adhering liquor and an enormous heat energy load. The heat energy is tapped before further treatment of the brown stock. From the digester, the brown stock is washed and passed over screens to remove knots, unreacted chips, slivers and trash before subjecting it to the subsequent operations of thickening, filtering and bleaching. After bleaching, the pulp is washed and ready for use to make paper. Alternatively, it may be rethickened and made into laps (sheets of pulp dry enough to fold into a bundle) for storage and shipment.

6.4.2.1 Treatment of Black Liquor (Regeneration and Recycle of Cooking Chemicals)

An attractive and important feature of the Kraft process is the recovery and recycling of spent liquor from the cooking process. This spent liquor is commonly known as black liquor and is extracted from the cooked pulp. The black liquor is treated to recover its chemical content for reuse and its organic content as a source of provision of heat energy. The spent liquor from the washer contains between 95 and 98% of the total chemicals fed to the digester. The chemicals present in the liquor are organic sulfur compounds, sodium sulfide, sodium carbonate and small quantities of sodium sulfite, salt and silica. Traces of lime, iron oxide, alumina and potash are also found in the liquor. The total solids content of the black liquor is generally, on average, about 20%. To recover these chemicals, the black liquor is first concentrated in evaporators, burnt in a furnace and finally limed.

In one such cooking liquor recovery process, the Thomlinson Kraft recovery process, the black liquor is first concentrated in multiple-effect evaporators to a concentration of about 35% solids. The resulting solution is sprayed into a Thomlinson furnace where it is burnt. In the process, sulfate is reduced to sulfide, steam is generated and a molten salt mixture (smelt) is produced. In the furnace, any remaining organic compounds are broken down, the carbon burnt away, and inorganic compounds are melted. The reduction of sulfate into sulfide takes place in the furnace according to the following reaction:

$$Na_2SO_4 + 2C \quad \rightarrow \quad Na_2S + 2CO_2 \tag{6.1}$$

The carbon (reducing agent) is supplied by the organics in the wood.

The smelt is allowed to fall and dissolve in a 'dissolving liquor' from a causticizing plant to give a green liquor. Insoluble impurities in the smelt settle out and any carbonate is causticized by the addition of slaked lime to the green liquor. The reaction proceeds quickly as follows:

$$Na_2CO_3(aq) + Ca(OH)_2(s) \quad \rightarrow \quad 2NaOH(aq) + CaCO_3(s) \tag{6.2}$$

The slurry is separated in settlers with the aid of Monel metal screens as the filtering medium. The filtrate is the white liquor used in the cooking of the fibers. It contains sodium hydroxide, sodium sulfide and small quantities of sodium carbonate, sodium sulfate, sodium sulfite and thiosulfate. These compounds are formed from the original sodium sulfate added to the digester. The calcium carbonate sludge, also called mud, is sent to a lime kiln to regenerate calcium oxide for reuse.

Important by-products of the Kraft process are tall oil and turpentine (a mixture of organic compounds called terpenes, e.g. α-pinene). It is used as an antiseptic in detergent products. The tall oil is a black sticky, viscous liquid composed mainly of resin and fatty acids. It is recovered from weak black liquor by means of centrifuges or by floatation from the concentrated liquor. The turpentine accompanies the relief gases from the digester from where it may be separated. About 11–40 L of turpentine may be won from every ton of pulp produced. Tall oil has a variety of uses in industry. It is used as a frothing agent in the flotation process for reclaiming low-grade metal-bearing ores (such as ores of copper, lead and zinc), and as a solvent or wetting agent in a variety of textile and synthetic fiber manufacturing processes. The distilled fatty acids of tall oil are used in soaps, detergents and disinfectants and as a base for lubricating greases, textile oils, cutting oils and metal polishes. They are also used as drying agents in paint. There are, however, a number of problems associated with the Kraft process, including:

   i. Strong odoriferous materials released into air (air pollutants) that are difficult to control.
   ii. Chemicals, mostly chlorine left in the bleach water are unfriendly to the environment.
   iii. The bleaching process may hurt the fibers.
   iv. The molten salt of the black liquor recovery process (is explosive and) can cause explosion when in contact with even small quantities of water.

6.4.3   Sulfite Pulping

The main chemicals used in this process are calcium bisulfate and sulfur dioxide. Magnesium bisulfate may and is increasingly used in place of calcium bisulfate. When magnesium bisulfate is used, the chemicals can be regenerated and recycled while this is not possible when calcium bisulfate is used. The process involves the cooking of an appropriately treated wood species in an aqueous solution containing calcium bisulfate (magnesium bisulfate) and an excess of sulfur dioxide. The main raw material used in this process is spruce wood, even though other wood species like hemlock and balsam are also used. The unit operations involved in the sulfite process are similar to the Kraft process. A simplified flowchart for the sulfite process (based on the use of magnesium bisulfite as the cooking chemical) is shown in Fig. 6.2.

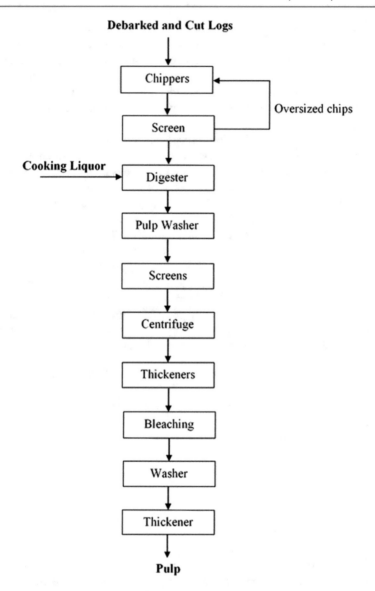

Fig. 6.2 A simplified flow chart of the sulfite pulping process for wood

As a first step in this process, the cooking liquor ($Ca(HSO_3)_2$, $Mg(HSO_3)_2$, $NH_4(HSO_3)_3$) is prepared. The liquor charged to the digester contains 4.5% total sulfur dioxide and about 3.5% free sulfur dioxide. The liquor is prepared by absorbing a gas of sulfur dioxide in water in the presence of an appropriate cooking chemical in a counter-current absorber. The following reactions proceed to produce the cooking liquor;

$$2SO_2 + H_2O + MgCO_3 \quad \rightarrow \quad Mg(HSO_3)_2 + 2CO_2 \qquad (6.3)$$

or

$$2SO_2 + Mg(OH)_2 \quad \rightarrow \quad Mg(HSO_3)_2 \qquad (6.4)$$

When calcium bisulfite or ammonium bisulfite is used, the corresponding reactions proceed as follows:

$$2SO_2 + H_2O + NH_3 \quad \rightarrow \quad NH_4(HSO_3)_3 + CO_2 \qquad (6.5)$$

$$2SO_2 + H_2O + CaCO_3 \quad \rightarrow \quad Ca(HSO_3)_2 + CO_2 \qquad (6.6)$$

The cooking liquor is charged to the digester together with appropriately prepared chips. The digester is either heated with direct steam or by heating the liquor outside of the digester in stainless steel tubes and subsequently circulating it through the chips with the help of pumps. The digester temperature is kept at 170–176 °C for between 6 and 12 h. The digester pressure varies from 480 to 1100 kPa. In the digester, two main reactions are believed to take place. The first is the sulfonation and dissolution of lignin with the bisulfite accompanied by a second reaction, which is the hydrolytic splitting of the cellulose-lignin complex. In the process, the hemicelluloses are hydrolyzed to simpler compounds and other extraneous wood components are acted upon.

From the digester, the reaction mixture is sent to a blow tank where the cooking liquor (a weak red liquor) is evaporated and burnt in a boiler during which magnesium oxide (MgO) and sulfur dioxide (SO$_2$) are formed. The MgO is slaked and send to the absorption tower together with the SO$_2$ to produce the cooking liquor. In the boiler, steam is generated that is used in the process. The residual pulp is washed in the tank with fresh water and passed on from the blow tank to a series of screens where knots and large lumps are removed. Additional foreign matter is removed in a centrifuge. Subsequently, the pulp is concentrated in thickeners and sent to a bleacher where it is bleached with chlorine dioxide. After bleaching, the pulp mass is neutralized by milk of lime. It is then once again washed and thickened and sent to a machine stock chest from where it is formed into laps. The dry-fiber content of the pulp at this stage is 35%. The laps are dried in steam-heated rolls dryer and bailed. The final pulp product has a fiber content of between 80 and 90%.

6.4.3.1 Preparation of the Sulfite Cooking Liquor

The process involves first slaking burnt lime that contains a high proportion of magnesia (Mg(OH)$_2$) with warm water to produce a suspension with a turbidity of 1°Be. The suspension is then treated with sulfur dioxide to produce the cooking liquor. The cooking liquor contains 4.5% total sulfur dioxide and about 3.5% free sulfur dioxide.

High-quality pulp is made from this process. The pulp is used in the manufacture of some of the finest papers, e.g. bond paper, writing paper and book-paper. The pulp is also used as an additive of non-wood products such as plastic and synthetic fibers. The pulp is easy to bleach but the resulting fibers have the disadvantage of being weak.

6.4.3.2 Treatment of the Waste Liquor

Only the magnesium-based cooking liquor can be treated to regenerate the cooking liquor. The process for regeneration of the liquor involves first sending the weak red liquor from the washer (after the blow tank) to multiple effect evaporators to be concentrated. The resulting heavy red liquor is first preheated and fed to a recovery furnace where it is burnt to produce magnesium oxide. Make-up magnesia is added to the product from the furnace and sent to an absorber where water and sulfur dioxide are added to produce the cooking liquor, which before being fed to the digester is first filtered and kept in a storage tank.

Calcium-based sulfite liquor is not amenable to recovery and reuse of either the Ca or S content. Accordingly, when used as a cooking liquor, the waste liquor must necessarily be discarded into the environment, a decision that does not please an increasingly environmentally conscious society. Calcium-based sulfite cooking liquor is, therefore, increasingly not used in processing pulp. Ammonia-based spent cooking liquor, like the magnesium-based liquor, can be regenerated, however, the ammonia itself cannot be recovered.

6.4.4 Mechanical Pulping

In mechanical pulping, the extraction of fiber from the wood is performed mostly by mechanical means without the use of any chemical. The pulp obtained by this process is of low quality in terms of color and strength. Because the pulp is not well reined, chemical decomposition of the non-cellulosic constituents sets in and results in the eventual brittleness and discoloration of the fiber. It is therefore used mostly for making cheaper grades of paper and board where permanency is not required. In practice, it is used in combination with pulp from other sources.

The mechanical pulping process uses mostly soft coniferous wood species to make pulp. The wood species commonly used are spruce and balsam. The wood log is first debarked and held at an acute angle against a rotating stone, which tears the fibers of the wood apart. The tilting of the log against the stone ensures that the fibers in the wood are not broken. The wood log is held against the stone by hydraulic pressure from hydraulic cylinders acting through pockets on the surface of the stone. Water is constantly poured onto the rotating stone to remove the heat due to friction between the stone and wood and also to carry the dislodged fiber away. The fibers are first carried to a stock sewer from where they are sent to be screened with a sliver screen. Finer fibers are collected through the screen into a

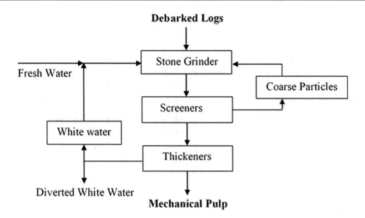

Fig. 6.3 Simplified flowchart of the mechanical pulping process

stock pit. Coarser ones trapped on the surface of the screen are collected and sent to be refined in a refiner and returned to the screen. From the stock pit, the fines are sent to thickeners to be concentrated into mechanical pulp. The pulp may be further treated by bleaching even though the brightness will not reach that of a chemical pulp. Bleaching agents commonly used for mechanical pulp are calcium oxide, oxygen, ozone, chlorine dioxide and hydrogen peroxide. Calcium and sodium peroxide may also be used even though these latter ones are not as popular as the earlier ones. The water overflow from the thickener has between 15 and 20% of the total fiber from the grinding stone. To retrieve this quantity of fiber, the thickener overflow is recycled to the grinder. With time, the temperature of water in the process rises and therefore, some of the process water is diverted out of the system while freshwater is introduced to the system. The diverted water contains some fiber, which is strained out of the water before discarding the water. A flowchart of the mechanical pulping process is shown in Fig. 6.3.

6.4.4.1 Thermomechanical Pulping

A variation of the mechanical pulping process is the thermomechanical pulping in which the wood is heated in steam before stripping out the fiber from the wood. This process takes advantage of the fact that at elevated temperatures, the lignin of the wood is softened and hence makes mechanical pulping easier. The steaming is done at a pressure of around 975 kPa and the refining is carried out at 170 °C. Under these conditions, the resultant fiber is coated with lignin. Even though this type of fiber is very poor for making paper, it is highly suitable for making fiberboard.

6.4.5   Semi-chemical or NSSC Pulping

In this pulping process, minimal amount of chemicals is used to treat the wood before separating the fibers. The chemicals are used to cook the wood in a digester to make them mild to facilitate mechanical separation of the fibers. Among the chemicals used are sodium sulfite buffered with sodium carbonate and Kraft green liquor (these chemicals along with elevated temperatures are used to soften the wood). The yield of pulp for this process is very high, near 65–80%. This makes for better use of the wood. However, the pulp quality and its bleachability are poor. The pulp from this process is used in making linerboard and corrugating paper.

6.4.6   Bleaching

Bleaching is an important process (stage or unit operation) in pulp manufacture. Lignin in pulp and other organic chemical impart color to the pulp which, often, must be removed to give value to the pulp. Bleaching decolorizes the pulp and gives it a whitish color. During bleaching, solutions of inorganic chemicals are applied to the pulp to change and improve its color. The type of bleach used depends on the type of pulp and the use to which the pulp and for that matter paper from the pulp is to be put. The most commonly used bleach in the pulp and paper industry are hydrogen peroxide, chlorine, ozone, sodium bisulfites, oxygen, chlorine oxide and hypochlorites. The last three are used in technologies that avoid the use of elemental chlorine in the bleaching process so as to protect the environment. The bleaching processes involving individual bleaches are described in the sections below.

6.4.6.1   Hydrogen Peroxide Brightening

Hydrogen peroxide is used mostly in the bleaching of mechanical pulp because of its high lignin content. Hydrogen peroxide alters the chemical structure of the lignin, lightening its color but leaving it present in the pulp. Other lignin removing chemicals such as chlorine are not used for mechanical pulps because this would result in a large reduction in pulp yield. Hydrogen peroxide is environmentally benign; however, the process does require quite high levels of energy and is relatively expensive. Other chemicals used to produce brighter mechanical pulps include hypochlorites and sodium bisulfite.

6.4.6.2   Chlorine Bleaching

Chemical pulp is routinely delignified to remove the 5–10% of lignin remaining after the pulping process. In the past, chlorine gas was used for this, followed by several stages of treatment with chlorine dioxide or hypochlorite to further whiten the pulp. About 50–80 kg of elemental chlorine are required to bleach 1.0 ton of pulp. Approximately, 10% of this chlorine combines with organic molecules to form a range of organochlorines that pass off into the process effluent. Among these are dioxins and furans, which have serious environmental consequences. As a result, increasingly, totally chlorine free (TCF) and elemental chlorine free (ECF) processes are used in beaching chemical pulp.

6.4.6.3   Totally Chlorine Free (TCF) Process

In TCF bleaching, oxygen is first used to delignify the pulp followed by peroxide brightening in a series of treatment stages. Other agents such as enzymes are sometimes used to enhance the bleaching process. 'Chelating' agents such as (EDTA) are also sometimes added to bind the metal ions contained in the pulp and prevent them from decomposing the hydrogen peroxide. TCF technology is improving, but as yet paper whiteness achieved by this method is only ISO 70–86% ('full brightness' is ISO 90%+). The addition of an ozone bleaching stage to the sequence can improve this to ISO 85–90%. However, the use of ozone is problematic as it readily attacks cellulose and the level of process control required contributes significantly to the cost.

6.4.6.4   Elemental Chlorine Free (ECF) Bleaching

ECF bleaching technology uses oxygen delignification followed in later stages by chlorine dioxide treatment, in combination with other chemical agents, in a sequence of processes to achieve 'full brightness'. In practice, the treatment begins with the application of chlorine dioxide followed by caustic soda, oxygen and finally hydrogen peroxide. Both ECF and TCF processes are likely to yield less pulp per ton of wood fiber than conventionally bleached pulp, because the increased number of bleach/wash stages results in greater fiber loss. TCF, which has the most protracted bleaching cycle, has the lowest yield. Market demand for TCF pulp is almost exclusively European (primarily German) and it is produced predominantly by the Nordic pulp and paper industry to give them a market advantage in Europe over North American pulp. The North American industry elected to install ECF technology, arguing that not only was it cheaper but that the environmental benefits of TCF over ECF were questionable.

6.5   Manufacture of Paper

In converting pulp to paper, the object is to turn the otherwise plump pulp into flat sheets with adequate strength, feel and proper surface to serve the purposes of a paper. To achieve this, three operations are normally carried out in succession on the pulp. These are beating, conversion of pulp to paper, and finishing. While in some plants, the first two processes are carried out differently, modern plants combine the two processes into one operation. Or rather, modern machines are capable of carrying out the two processes in one operation. These processes are carried out to give the paper a proper surface, opacity, strength and feel characteristics that make the paper useful for its purposes.

6.5.1 Beating

Beating is a pounding and squeezing process carried out in a tank. Beating the pulp makes the eventual paper stronger, more uniform, denser, opaquer and less porous. It also has the effect of increasing the bond between fibers of the pulp. Machines that carry out the beating process are called *beaters*. The beater is generally a tank with rounded ends whose essential features include a partition just below its middle, stationary knives and bars and revolving bars which are all positioned inside the tank. During operation, the pulp is circulated and forced between bars on a revolving roll and stationary bars. In this manner, the desired effect is achieved on the pulp. Various filler materials can be added to the pulp during beating to give the paper the desired characteristics and qualities such as opacity of the paper product. Examples of such additives include chalks, clays, or chemicals such as titanium oxide. Sizing may also be added during beating. Sizing modulates the behavior of ink on paper. Without any sizing at all, a paper will be too absorbent for most uses except as a desk blotter. Starch, when present as a sizing material during the beating of pulp, will make the eventual paper product resistant to water-based ink. The choice of sizing depends on the eventual use of the paper. For example, when paper is meant to be used to receive a printed design, it will require that a particular formula of sizing be applied that will facilitate the printing of the design on the paper. Examples of sizing materials are starch, rosin soap, wax emulsions and gums.

6.5.2 Conversion of Pulp to Paper

This process converts the bulky and wet pulp into dried sheets called paper. This occurs when the pulp from the beater is squeezed through a series of rollers whilst draining the water from it by suction. This process is carried out in an automated device called *Fourdrinier machine*. The process follows the following sequence of steps; (i) feeding the pulp onto a belt of fine mesh screen and conveying to the machine, (ii) where necessary, a dandy moves across the sheet of pulp to press a design into it, (iii) the paper is sent to be pressed in between rollers of wool felt in the press section of the machine and (iv) finally, the paper is passed over a series of steam-heated cylinders where the remaining water is removed.

6.5.3 Finishing

This operation enables paper manufacturers to give characteristic properties to the paper. To achieve this, the dried paper is wound onto large reels, where it will be further processed depending on its ultimate use. The paper is smoothed and compacted further by passing it through metal rollers called *calendars where a soft and dull or hard and shiny finishing can be imparted to the paper*. The paper may be further imparted with other finishing, including; (i) passing through a vat of sizing,

(ii) coating by brushing or using rollers, (iii) running the paper through extremely smooth calendar rollers in a process called supercalendered and (iv) finally cutting the paper to the desired size. The most commonly used coating material is clay. Coating supplements the earlier processes of sizing and filling by adding chemicals and pigments to the surface of the paper.

6.5.4   Paper Making Additives

Additives used in the manufacture of paper include dyes, fillers, sizings and coatings. Dyes are used to give color to the paper. Fillers are used to give smoother surface, a more brilliant whiteness, improved smoothness and printability and improved opacity to the paper. Fillers are added to the pulp during the beating process. Examples of fillers are chalk, clay or titanium oxide. Sizing affects the way paper reacts with ink. They are added to paper to improve resistance to penetration by ink. Rosin soap, gum, wax emulsion and starch are examples of sizing. Sizing may be added at the beating stage, in the Fourdrinier machine or applied to the surface of the finished paper. Coatings are applied to specialty paper to impart special properties such as printability and resistance to fluids. Coatings include fine clay, wax and plastic materials. They are applied to the finished paper.

6.6   Environmental Concerns and Pollution from Pulp and Paper Mills

Environmental problems arising out of the paper and pulp industries are two pronged, namely:

(i) Concerns arising out of use of trees and other vegetation and attendant depletion of forest covers.
(ii) The negative impact of the chemicals (air borne, water borne and solid chemicals) used in paper manufacture, including cooking reagents, dyes, inks, bleach and sizing agents on the environment.

Industries have sought to correct concerns of depleting forest cover by planting as many new trees as they cut down. Even though these new growth trees will take many years to grow to effectively replace the bigger and old trees that were harvested, the effort is worthwhile. Another approach aimed at dealing with this problem has been the increasing use of recycled paper in the production of pulp. The effort has proven effective in mitigating the need for destruction of woodlands. On the other hand, the industry has dealt with the potential environmental impact of chemicals by adopting technologies that ensure that these chemicals do not, at all, reach the environment or minimal quantities reach the environment. However, when the environment is eventually polluted by these chemicals, measures are normally taken to clean up the chemicals from the environment. Yet another approach to deal with the problem of environmental pollution due to the industry is strict compliance with governmental regulations, which sets guidelines as to the amount and quality of effluents allowable into the environment. The effects of liquid effluents, air emissions and solid discharges from the pulp and paper industries into the environment and how they impact the environment are discussed below.

6.6.1   Liquid Effluent Discharges

These fall into the following categories:

i. **General organic pollution and suspended solids**

The most common organic pollutants in effluents are lost cellulose fiber, carbohydrate, starch and hemi-cellulose (or the organic acids resulting from the breakdown of these compounds). The levels of these pollutants are measured in terms of biological oxygen demand (BOD) or chemical oxygen demand (COD). COD discharge can range from 25 to 125 kg per ton of pulp. This demand for oxygen depletes that available to fauna and flora, thus damaging wildlife near to, and downstream from, effluent discharges.

Suspended solid can occur in quantities such that it can result in water opacity. Another effect of suspended solids is the blanketing of river or lake beds. When the blanketing is severe, anaerobic decomposition may occur under the blanket resulting in the release of hydrogen sulfide into the aquatic medium. These problems are reasonably localized. However, organic solids can also absorb many of the toxins present in mill effluents, such as resin, fatty acids and heavy metals leading to bioaccumulation and movement of these materials through the food chain. This occurrence can affect large tracts of the environment over a long period of time.

ii. **Acidic compounds**

These are predominantly, natural resin acids that occur in high concentrations in softwood pulp, and are most concentrated in mechanical and (C)TMP pulp effluents (96–98 mg/l). They can be chlorinated in bleached Kraft pulp effluent. Although, both resin acids and chlorinated resin acids are moderately toxic and persistent, generally, acidic compounds are of relatively minor concern in these concentrations as they are readily biodegradable and do not bioaccumulate. Also, they are known as abietic and dehydroabietic acids.

iii. **General organochlorine products**

As mentioned earlier, 5–8 kg of elemental chlorine per ton of bleached chemical pulp combines with wood products to form organochlorines. About 300 such organochlorine compounds have been identified, although the effects of all of them are not known. It is likely that several hundred more remain unidentified. No direct correlation has been found between the observed toxicity of mill effluents and the quantity of organochlorines present. The well-known organochlorines include chlorophenols, dioxins (PCDD), furans (PCDF), chloroform, chloro-acetones, chloro-aldehydes and chloro-acetic acids; these are discussed below.

Chlorophenolics

The chlorophenolic group contains phenols, guaiacols and catechols. They are formed in chlorine-bleached chemical pulping processes. They are most prevalent in softwood pulp effluents. Not only are they themselves toxic, persistent and bioaccumulative, but they can transform into other compounds which are even more so. As a group, chlorophenolics and their derivative products are probably the most hazardous chemical groups in pulp and paper mill effluents, being present in higher concentrations than more toxic compounds such as dioxins. Bio-transformation products include anisoles and verathroles. The most common chlorophenolics are the extremely toxic and persistent trichlorophenol and pentachlorophenol. Substituting chlorine dioxide for elemental chlorine in some bleaching process stages significantly increases chlorophenolic production.

Dioxins (PCDD) and Furans (PCDF)

Dioxins are extremely toxic, persistent and carcinogenic. Furans are chemically similar but are an order of magnitude less toxic and less persistent than dioxins. The amount of these chemicals found in effluents does not seem to depend on the type of raw material used in the production of pulp. Generally, the dioxins and furans are less prevalent in the effluent than in the pulp. This has led to concern about dioxin levels in finished paper products and wastewater treatment sludge disposed via landfill or incineration, as well as in liquid mill effluents. Dioxins can also be found in the flue gases from the mill. These chemicals are known to have wide-ranging effects on fish and mammals but any link to effects on humans is proving hard to establish. They are, however, suspected of causing miscarriages, birth defects, liver damage, skin complaints and behavioral and neurological problems. Bio-concentration through the food chain, especially via fish, is a major concern.

Chloroform and other neutral chlorinated compounds

These groups include chloroform, chloro-acetones, chloro-aldehydes and chloro-acetic acids, which are formed during the bleaching process. They occur in lower concentrations than chlorophenolics. Compounds in this group are generally non-persistent and non-bioaccumulative. However, some are moderately toxic, mutagenic and/or suspected carcinogens. The major concern is the likely effect of human exposure to chloroform via drinking water and air. In the United States, the Environmental Protection Agency (US-EPA) has identified pulp bleaching as the single largest source of atmospheric chloroform.

iv. **Other environmental issues relating to effluent discharge**

Technological advances in paper-making processes, waste treatment and environmental monitoring have led to several 'new' environmental problems associated with the manufacturing process becoming apparent. For example, physiological changes in fish reproductive organs and hormone production consistent with oestrogenic pollutants have been observed. These effects do not appear to be related to the pollutants mentioned earlier above. As yet, the cause remains largely unidentified, but it is thought compounds present in the wood itself, which may be chemically modified by the Kraft process, are responsible. The use of chlorine dioxide, the basis of ECF bleaching, has led to problems associated with chlorate production. Chlorate can be harmful to waterborne algae. Chelating agents used in TCF technology bind trace metal ions in the pulp to prevent them degrading the peroxide. This increases the metal load in effluents and is an issue of concern. However, the environmental impact of the process is so far largely unknown.

6.6.2   Air Emissions

Air emissions from chemical pulp mills are primarily made up of particulates, hydrogen sulfide, oxides of sulfur and oxides of nitrogen. Micro-pollutants include chloroform, dioxins and furans, other organochlorines and other volatile organics. As with liquid effluent discharges, the levels of emissions are highly dependent upon the type of process technology employed and individual mill practice. Another important factor is the fuel type and quality. While older mills caused severe air pollution, mitigating technology now exists to eliminate most harmful gas and particulate emissions. Whether this technology is utilized depends on local factors such as legislation, company and mill policy and proximity to populated areas. The major pollutants and their effects are summarized in Table 6.2.

The contribution of the paper industry to global warming has been an ongoing debate for several years, with some suggesting that the absorption of carbon dioxide by plantation of more forestry offsets the emissions of greenhouse gases caused during the production, transportation and disposal of pulp and paper products. A recent study by the International Institute for Environment and Development (IIED) dismisses this argument, concluding that the paper cycle results in the net addition of some 450 million CO_2 equivalent units per year.

6.6.3   Solid Waste

Solid waste from paper manufacture ranges from 10 to 250 kg/ton (dry equivalent). They are disposed of by sending to landfill or are incinerated. Other experimental disposal techniques include using the waste as a soil improver but, as with all

Table 6.2 Air emissions from the pulp and paper industry

Pollutant	Effects	Source	Comments
Carbon dioxide	Greenhouse gas	Fuel combustion	See 'air emissions'
Hydrogen sulfide	Rotten egg smell	Kraft process	Very low odor threshold (5–10 ppb) makes it very difficult to remove smell completely. Kraft process is banned in Germany because of this
Sulfur dioxide	Acid rain	Fuel combustion and pulping process. (Kraft 1–3 kg SO_2/ton), sulfite 5 kg SO_2/ton)	Contribution to acid rain problem could be significantly reduced by switching from fossil to wood waste fuels
Volatile organics	Have toxic effects and react to form ozone	Various	
Chloroform	Toxic, possible carcinogen	Chlorine bleaching	See 'liquid effluent discharges'
Other organo-chlorines	Some highly toxic	Chlorine bleaching	See 'liquid effluent discharges'

ppb = parts per billion

disposal options; there are some concerns about possible dioxin and heavy metal contamination. Another approach that is gaining scientific interests in recent times is converting the solid wastes or waste sludges into a 'biological' char by using a process called 'hydrothermal carbonization' (HTC). This process involves heating the waste in a suspension of water under pressure at temperatures between 160 °C and 250 °C for several minutes or hours, converting the waste into carbonaceous solids, called 'hydrochar'. However, the HTC process also generates a liquid by-product (waste) containing high concentration of organics (COD > 2000 mg/l) and will require further treatment if discharging to the environment is the option. Also, a high-pressure resistance system is required, leading to an increase of the investment and operating costs of the process on an industrial scale. Furthermore, a continuous-scale operation requires feeding against pressure, hence, specialized pumps with back-pressure control are needed to avoid blockage and to ensure continuous operation.

6.7   Effluent Treatment in Pulp and Paper Mills

6.7.1   Primary Treatment

During primary treatment, suspended solids are removed from effluents by mechanical means such as settlement. It may be the only wastewater treatment employed in many mills worldwide; however, more often emission standards require secondary and possibly tertiary treatment. The effectiveness of these subsequent treatments depends upon efficient primary treatment, which itself reduces BOD and halogenated organic compounds (AOX).

6.7.2   Secondary Treatment

Micro-organisms are used to accelerate the natural decomposition of organic waste during secondary treatment. Two commonly used methods for secondary treatments are *aerated stabilization* and *activated sludge treatment.* These two methods are all aerobic treatments. The efficiency of these two systems varies widely, depending on climate, influent quality, pulp type, fiber source and mill practice. In ideal conditions, activated sludge performs better at reducing BOD and removing suspended solids. Disadvantages of both methods of aerobic treatment include high energy consumption, production of sludge waste and generation of greenhouse gases.

6.7.3   Tertiary Treatment

Tertiary or non-biological chemical treatments are expensive and only applied in a few mills. Aluminum oxide, ferric oxide and polyelectrolytes assist coagulation of waste in the effluents, which are then sand filtered.

Further Readings

Bajpai, P. (2015a). *Pulp and paper industry* (1st ed.). Elsevier.
Bajpai, P. (2015b). *Green chemistry and sustainability in paper and pulp industry*. Springer.
Bajpai, P. (2015c). *Pulp and paper industry: chemicals*. Elsevier.
Clarke, J. J. (2019). The manufacture of pulp and paper, vol 1: A textbook of modern pulp and paper mill practice. London: Forgotten Books.
Diesen M (2007) Economics of the pulp and paper industry (2nd ed.). Helsinki: Finnish Paper Engineers Association.
Grace, T. M., Malcolm, E., Kocurek, M. J. (eds.).(1989). Pulp and paper manufacture vol 5: Alkaline pulping (3rd ed.). Atlanta: TAPPI Press.
Joint Textbook Committee on Paper Industry. (1929). *The manufacture of pulp and paper: Papermaking machines; handmade papers; paper finishing; coated papers*. McGraw-Hill.
NIIR Board of Consultants and Engineers (2003) The complete technology book on pulp & paper industries. Asia Pacific Business Press Inc, New Delhi.

Smook, C. A. (2001). *Handbook of pulp & paper terminology: a guide to industrial and technological usage* (2nd ed.). Angus Wilde Publications Inc.

Smook, C. A. (2016). *Handbook for pulp & paper technologists (The Smook book)* (4th ed.). TAPPI Press.

Witham, G. S. (2018). Modern pulp and paper making: A practical treatise (2nd ed.). Franklin Classics Trade Press.

Manufacture of Crude Palm Oil and Refined Palm Oil

7

Abstract

Palm oil is currently the most widely produced and consumed vegetable oil in the world and has many uses as a raw material for food and non-food industries. Palm oil contains several phytonutrients such as tocols, carotenoids, phytosterols, squalene, lecithin, co-enzyme Q10, and polyphenols as well as minerals in measurable quantities namely calcium, potassium, iron, zinc and magnesium. The market of crude palm oil is determined by many indicators such as free fatty acid (FFA) content, iodine value, density, melting point, cloud point, flash point, etc. Crude palm oil (CPO) is obtained from the ripped fruit of the oil palm plant through five basic operations, namely, sterilization, threshing/striping, digestion, pressing and clarification. To produce refined palm oil (RPO), the objectionable impurities in the CPO are removed to obtain the required levels using two main methods: chemical or alkali refining and physical refining methods. Either of these is followed by fractionation as a final purification process to separate the oil into stearin and olein components. The composition and characterization of palm oil and the manufacturing of CPO and RPO have been covered comprehensively in this chapter. The impact of the palm oil industry on the environment has also been presented.

7.1 Introduction

Palm oil is an edible vegetable oil extracted from the orange-red mesocarp of the fruits of the oil palm. It has the color of a rich, deep red, which is said to be derived from its rich carotenoid contents, which are known pigments often found in plants and animals. In the unprocessed form, the oil is reddish brown in color, and it has a semi-solid consistency at ambient temperature. Palm oil should not be confused with palm kernel oil, which is another oil product from the oil palm fruit. While palm oil is obtained from the mesocarp of the oil palm fruit, palm kernel oil is obtained from the kernel of the oil palm fruit.

Palm oil is an important raw material, especially, for the food industry. However, it also finds extensive use in the non-food industries. It is probably the most produced and consumed vegetable oil in the world today. Most palm oil is currently produced in South East Asia, even though the oil palm is originally an African crop. The two largest producers are Malaysia and Indonesia, who together account for about 85% of the world's palm oil production. Indonesia is the world's largest producer of palm oil, followed by Malaysia. However, Malaysia exports more palm oil than any other country in the world. Other countries that produce palm oil in notable quantities are Thailand, Columbia and Nigeria. China, the European Union, Pakistan, United States and India are the major export destinations of the product. Table 7.1 lists the major palm oil-producing countries and their production capacities in the period 2015–2019.

Ghana was ranked as the 13th largest producer of crude palm oil in the world in 2018, producing 375,000 metric tonnes of palm oil on a cultivated area of over 360,00 ha. Production of oil in Ghana peaked in 2011 when it produced about 448,000 metric tonnes of palm oil. However, this figure has been reduced to the current figure of 375,000 metric tonnes. The major players in the palm oil industry in Ghana are listed in Table 7.2.

As far back as 1820, Ghana was manufacturing and exporting Crude Palm Oil (CPO). By 1850, palm oil had become the principal export of the then Gold Coast. Indeed, by 1880s, palm oil accounted for 75% of the country's export revenue. The industry at the time was driven by oil palm plantations established by Europeans including the Dutch, the German and the British near the coast during the eighteenth and nineteenth centuries. Among the plantations established at the time were those at Butre and Sese in Western Region, and Winneba in Central Region. The plantation system, however, failed to gain a significant hold, partly because of the internal political insecurity engendered by inter-tribal warfare and by rivalry among the European powers seeking territorial hegemony. In addition, the negative attitude towards the system by the British Crown, which, from 1850 onwards, gained the upper hand in the European struggle to colonize Ghana, also contributed to the failure of the plantation system in pre-colonial era. Things, however, changed after

Table 7.1 Global production of palm oil (adapted from USDA, 2020)

Country	Annual production of palm in 1000 MT				
	2015	2016	2017	2018	2019
Indonesia	32,000	36,000	39,500	41,500	43,000
Malaysia	17,700	18,858	19,683	20,800	21,00
Thailand	1,804	2,500	2,780	2,900	3,000
Columbia	1,268	1,099	1,633	1,625	1,680
Nigeria	955	990	1,025	1,015	1,015
Ghana	375	375	375	375	375
Others	4,293	4,293	5,365	6,974	5,624

Table 7.2 Major producers of palm oil in Ghana (adapted from MASDAR Report, 2011)

Industry	Location	Milling capacity (ton/hr)
Ghana Oil Palm Dev. Corp Ltd (GOPDC)	Kwae, Kade	60.0
Twifo Oil Palm Plantation (TOPP) Ltd	Twifo Prtaso	30.0
Norpalm Ghana Ltd	Prestea	30.0
Benso Oil Palm Plantation (BOPP) Ltd	Benso	27.0
Juaben Oil Mills Ltd	Juaben	15.0
National Oil Palm Plantation	Ayiem	10.0
Golden Star Oil Palm	Bogoso	–
Small Scale and Private Holders	Country wide	–

1957, during the post-independence period, when there was a policy change involving greater emphasis on the plantation system centered on oil palm and rubber. The policy change, up to the time of the 1966 military coup d 'etat, encouraged state-owned and state-operated plantations. As a result, the palm oil industry has once again emerged as a serious national industry, which hold promise as an export industry that can rake in millions, if not billions, of dollars for the country and offer employment to several thousands of people. In 2018, according to the USA Department of Agriculture, Ghana exported 160,000.00 metric tons of crude palm oil. The major destination of Ghana palm oil is Senegal. Other countries that import Ghana's palm oil are Burkina Faso, Niger, UK, Spain, Netherlands and United States. In 2018, Ghana imported 300,000 tons of palm oil, mostly, from Malaysia. Ghana is unable to meet is annual demand of oil and fats of about 680,000 metric tons from local production alone. The shortfall is met mostly with imports from South East Asia, primarily, Malaysia. In order to improve palm oil production in the country, the government of Ghana has initiated in recent times, specific programs for the growth of the oil palm sub-sector of the economy. The oil palm industry, and particularly following the recent interventions of the government, has offered a lot of employment for rural people, playing the role as a pillar of a number of rural economies in Central, Eastern, Western and Ashanti regions. The industry also supports other industries in the country such as food processing, soaps and detergents production, the body care industry and the activated carbon industry.

7.2   Composition of Palm Oil

The physical and chemical properties of palm oil, like other vegetable oils, are essentially determined by the nature of the constituent fatty acids, which make up approximately 95% of the glyceride molecule. Palm oil is made up mostly of glyceridic materials even though some nonglyceridic materials in small or trace quantities are also present in the oil. It is this chemical composition that defines the chemical and the physical characteristics of palm oil, which in turn determines the suitability of the oil in various processes and applications. The major components of palm oil are the three acylglycerols (TAGs). Over 95% of palm oil consists of mixtures of TAGs, i.e. glycerol molecules, each esterified with three fatty acids. The major fatty acids in palm oil are myristic, palmitic, stearic, oleic and linoleic. Most of the fatty acids of palm oil are present as TAGs. The different placement of fatty acids and fatty acid types on the glycerol molecule produces a number of different TAGs. The other group of constituents of palm oil called minor constituents can further be divided into two groups. The first group consists of fatty acid derivatives, such as partial glycerides, monoglycerides (MGs), diglycerides (DGs), phosphatides, esters and sterols. Included in this group are the phospholipids and glycolipids, which are the polar lipids of palm oil. The components in the second group are hydrocarbons, aliphatic alcohols, free sterols, tocopherols, pigments and trace metals. The major and minor constituents of palm oil are listed in Tables 7.3 and 7.4.

Besides the constituents listed in Tables 7.3 and 7.4, other components like water, and minerals like calcium, potassium, iron, zinc and magnesium are also present in palm oil in measurable quantities.

Palm oil quality and price are dependent on the free fatty acids (FFA) content. High content of free fatty acids in palm oil affects the quality of palm oil and leads to various health and environmental issues. The fatty acid composition of palm oil may vary with the country of origin, however, oils from the same country are reasonably consistent, particularly where the oil production is plantation-based and maximum benefit is obtained from control of ripeness at time of harvesting, and genetic uniformity of the fruit. The composition of the fatty acid content of palm oil is shown in Table 7.5.

Table 7.3 Composition of palm oil

Component	Percentage (wt.)
Triglycerides (TAGs)	≥ 90.0
Diglycirides (DAGs)	$\sim 2.0-7.0$
Monoglycerides (MADs)	≤ 1.0
Free Fatty Acids (FFA)	$\sim 3.0-5.0$
Phytonutrients	~ 1.0

Table 7.4 Minor components (phytonutrients) in palm oil

Component	Concentration (ppm)
Tocols	600.0–1000.0
Carotenoids	500.0–1000.0
Phytosterols	300.0–620.0
Squalene	250.0–800.0
Lecithin	20.0–100.0
Co-enzyme Q10/Ubiquinones	10.0–80.0
Polyphenols	40.0–70.0

Table 7.5 Fatty acid (FFA) composition of palm oil

Fatty acid	Systematic name	Symbols	Percentage (wt.)
Saturated acids			
Lauric	n-Dodecanoic	C12:0	<1.0
Myristic	n-Tetradecanoic	C14:0	0.8–6.0
Palmitic	n-Hexadecanoic	C16:0	32.0–47.0
Stearic	n-Octadecanioc	C18:0	1.0–6.0
Arachidic	n-Eicosanoic	C20:0	<1.0
Mono-unsaturated acid			
Palmitoleic	n-Hexadec-9-enoic	C16:1	<1.0
Oleic	n-Octadec-9-enoic	C18:1	40.0–52.0
Gadoleic	n-Eicos-9-enoic	C20:1	0.15–1.52
Poly-unsaturated acid			
Linoleic	n-Octadeca-9, 12-dienoic	C18:2	5.0–10.2

7.3 Characteristics of Palm Oil

In addition to the FFA content, other quality indicators such as the iodine value, the density and melting point among others all determine the market value of crude palm oil. The apparent density is used as a purity indicator of the oil. It is also an important commercial parameter used for the conversion between the weight and volume of the oil. The solid fat content of oil is a measure (in percentage terms) of the amount of solid fat present in the oil at any one temperature. The solid present in the oil at any one temperature is due to the process of crystallization occurring in the oil because of its chemical properties. The different triglyceride molecules of the oil have different chemical characteristics. The different molecules manifest their physical state at different temperatures and thus imparting characteristic crystallization and melting behavior to the oil. The physical and chemical properties of palm oil are listed in Tables 7.6 and 7.7.

i. Free Fatty Acid/Acid Value

The free fatty acid is a measure of the amount of free fatty acid present in fats/oils. Some of the deteriorations that take place during storage of either the raw materials from which the fat is obtained, or in the oil/fat itself after isolation, results in hydrolysis of triglycerides to yield free fatty acids.

ii. Peroxide Value

Peroxide value (PV) gives the milliequivalents of peroxide oxygen combined in a kilogram of oil and able, under testing, to liberate iodine from potassium iodide. The iodine liberated is next estimated using a standard sodium thiosulfate solution.

Table 7.6 Physical properties of palm oil

Property	Value
Specific gravity (20 °C)	0.9119
Viscosity	
Melting point (°C)	30.8–37.6
Boiling point	
Flash point (°C)	314
Fire point (°C)	341
Cloud point	
Refractive index (η, at 25 °C)	1.4550–1.4560
Smoke point (°C)	223
Moisture & impurities (wt%)	0.26–0.33

Source 1. Ullmann's encyclopedia of industrial chemistry, Vol A 10, Fats and oils, VCH, Weinheim 1995; 2. Bailey's industrial oil & fat products, 6th edition 2005, Wiley-Intersience New York; 3. Physical and chemical characteristics of oils, fats, and waxes, Champaign, Illinois, AOCS Press, 2006; 4. Manuel des corps gras, AFCEG, Paris 1992; 5. World Scientific News 88(2) (2017) 152–167

Table 7.7 Chemical properties of palm oil

Property	Value
Iodine value (g/100 g)	50.6–55.0
Saponification value (mg KOH/g Oil)	190–202
Specific heat (J/g °C)	2,400
Unsaponifiable (g/kg)	7.6–7.8
Free fatty acid (FFA) (mg KOH/g Oil)	273–289
Peroxide value (meq/kg)	7.90–8.80
Acid value (mg KOH/g Oil)	28.3
Ester value (mg KOH/g Oil)	168.77

Source 1. Ullmann's encyclopedia of industrial chemistry, Vol A 10, Fats and oils, VCH, Weinheim 1995; 2. Bailey's industrial oil & fat products, 6th edition 2005, Wiley-Interscience New York; 3. Physical and chemical characteristics of oils, fats, and waxes, Champaign, Illinois, AOCS Press, 2006; 4. Manuel des corps gras, AFCEG, Paris 1992; 5. World Scientific News 88(2) (2017) 152–167

The magnitude of the result is an indication of the extent of the first stage oxidation in the present oil. The peroxide value (PV) is a quantity that measures the oxygen content of the oil. Fresh oils have lower values of peroxide value often below 10 meq/kg. When the peroxide value goes up between 20 and 40 meq/kg, the oil starts to develop a rancid taste.

iii. **Saponification Value**

Saponification value of fat or oil can be defined as the number of milligrams of potassium hydroxide required to neutralize the fatty acids resulting from the complete hydrolysis of 1.0 g of the sample. Saponification values are usually large when compared with the acid value of most edible oils. The saponification value is inversely proportional to the mean molecular weight of the fatty acids in the glycerides present.

iv. **Unsaponifiable Matter**

The term "unsaponifiable matter" refers to those substances in oil and fats that are soluble in ordinary fat solvents, but which are not saponifiable by alkali hydroxides.

v. **Iodine Value**

The degree of unsaturation in vegetable oil is referred to as the iodine value of the oil. For fats and oils, it is the weight of iodine absorbed per 100 parts of the sample. The glyceride of the unsaturated fatty acids present react with a definite amount of the halogen and the iodine value is, therefore, a measure of the extent of unsaturation. Iodine value is constant for an oil or fat, but the exact figure obtained depends on the techniques employed.

vi. **Ester Value**

The Ester value is the quantity of potassium hydroxide measured in mg required to saponify the esters in 1.0 g of the oil or fat. If the saponification value and the acid value have been determined, the difference between these two represents the Ester value.

7.4   Uses of Palm Oil

Palm oil has a wide range of applications, about 80% are used for food applications while the rest is feedstock for many non-food applications. Among the food uses, refined, bleached and deodorized (RBD) olein is used mainly as cooking and frying oils, shortenings and margarine while RBD stearin is used to produce shortenings and margarine. RBD palm oil (i.e. unfractionated palm oil) is used for producing margarine, shortenings, frying fats and ice cream. Several blends have been developed to produce solid fats with a zero content of trans-fatty acids. In the production of ice cream, milk fats are replaced by a combination of palm oil and palm kernel oil. A mixture consisting of palm oil, palm kernel oil and other fats are used to replace milk fat during the production of non-diary creamers or whiteners. Palm oil is also an ingredient for production of specialty fats, which include Cocoa Butter Equivalents (CBE), Cocoa Butter Substitutes (CBS) and general-purpose coating fats. In the production of chocolate confectionaries, these specialty fats, namely, CBE and CBS are widely used. Crude palm oil (CPO) is also used as a diesel fuel substitute, in drilling mud, and in soaps and epoxidized palm oil products (EPOP), polyols, polyurethanes and polyacrylate. Other uses of CPO include as a fuel for cars with suitable modified engines, and as a non-toxic alternative to diesel as a base for drilling mud.

Palm oil is an important input in industrial production of non-dairy creams, ice-cream powder, salad dressing, fat spread and chocolate. Its substitute, stearin, is traded as a substitute in the formulation of soaps, detergents, body care products, margarine and baking fat. Vitamins A, D and E, which are very important substances in the pharmaceutical industry can all found in palm oil. Red palm oil has been found to be nutritionally healthier than refined palm oil. This is a result of the unique substances found in red palm oil. These compounds are beta carotenes, ubiquinone, squalene, vitamin A and vitamin E. Finally, the oil is used as a biofuel.

7.5 The Manufacture of Crude Palm Oil (CPO) and Refined Palm Oil

In Ghana, crude oil extraction and refinery are not necessarily integrated processes. There is a historical reason for this. The earlier plants produced crude palm oil mostly for the local market and therefore did not see much need to refine it. Lately, competition in the market and the need to add value to facilitate export have compelled most plants to add refinery units to their crude oil existing plants (e.g. Juaben Oil Mills). In discussing palm oil manufacture, therefore, this chapter will look at the CPO manufacture and the CPO refinery processes separately.

7.5.1 Crude Palm Oil (CPO) Extraction Processes

The raw material for the production of crude palm oil is the ripped fruit of the oil palm plant. The physical composition of the palm fruit bunch is provided in Table 7.8 and illustrated in Fig. 7.1.

Table 7.8 Physical composition of palm fruit bunch

Component	Composition
Bunch	23.0–27.0 kg
Fruit/bunch	60.0–65.0%
Oil/bunch	21.0–23.0%
Kernel/bunch	5.0–7.0%
Mesocarp/bunch	44.0–46.0%
Mesocarp/fruit	71.0–76.0%
Kernel/fruit	22.0–21.0%

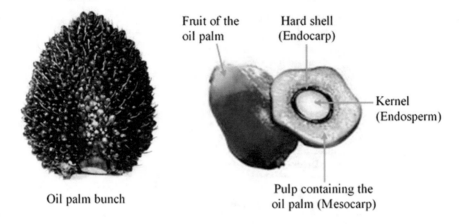

Fig. 7.1 Oil palm bunch and its fruit

There are five basic operations involved in the extraction of CPO from the fruit-containing bunch. These are sterilization, threshing (also called striping), digestion, pressing and clarification. These operations, for the most part, follow sequentially as listed. Each operation is unique and plays an important role in the extraction process, affecting yield and quality depending on how it is carried out. The overall yield of the process depends on the specific technology employed, the processing conditions and the quality of fruits employed. The main product of the process is the crude palm oil (CPO); however, palm kernel from which palm kernel oil is processed is an important by-product of the process. The depleted bunch from the thresher and the fiber from the blower are also important by-products, which may be used as fuel or as fertilizers. A generalized flowchart of the extraction process is shown in Fig. 7.2.

7.5.2 Description of Unit Operations

The first of the unit operation in the CPO extraction process is sterilization. During sterilization, the fresh fruit bunch is directly contacted with pressurized steam at 3 bars (T = 133.5 °C) for a period between 1 and 1.5 h for fresh fruits and between 30 and 45 min for ripped fruits. The bunches are first loaded into a cage and carried using cranes onto budgies (rails) and then pushed into the sterilizer. A loaded cage can weigh 1.3 tons or more. Sterilization serves a number of purposes as follows:

i. It destroys oil-splitting enzymes and arrests hydrolysis and autoxidation.
ii. It weakens the stem of the palm fruit and facilitates its removal from the bunch during the subsequent threshing operation.
iii. It weakens the pulp structure, softening it and making it easier to detach the fibrous material (mesocarp) and its contents from the kernel during the

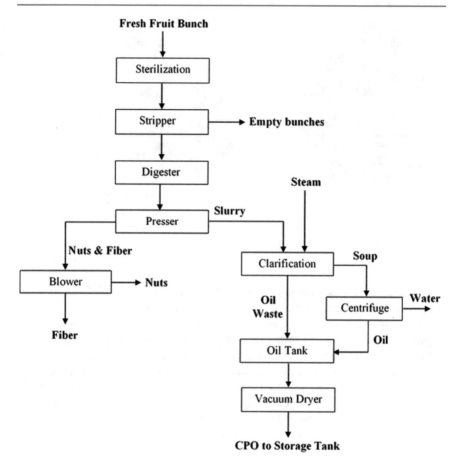

Fig. 7.2 A simplified flowchart of the CPO extraction process

    digestion process. Also, the high heat disrupts the oil-containing cells in the mesocarp and permits oil to be released more readily.

iv. The moisture of the steam acts chemically to break down gums and resins so that they can be removed during clarification. When present in the oil, the gums and resins cause the foaming of the oil during frying. Starches present in the fruit are also hydrolyzed and removed similarly.

v. The heat of the high-pressure steam causes the moisture in the nut to expand. When the pressure is reduced, the nut contracts leading to the detachment of the kernel from the shell wall. This facilitates the later nut-cracking process during which the kernel is separated from the shell.

    After sterilization, the cages are removed from the sterilizer and transferred to a drum stripper using the crane. The drum stripper has a rotating drum equipped with rotary beater bars that detach the fruit from the bunch. The process of detaching the fruits from the bunch is called threshing. Following threshing, the released fruits are transferred to the digester using a bucket conveyer. The palm oil in the fruit is released through the rupture of the oil-bearing cells in the mesocarp during digestion. The digester is a steam jacketed cylindrical vessel fitted with a central rotating shaft carrying a number of beater arms. Through the action of the rotating arms, the fruit is pounded to rupture the cells. The high temperature process also results in the reduction of the viscosity of the oil which helps with its flow. Contamination from iron is greatest during this process where a very high rate of metal wear is encountered. Iron contamination makes the oil more liable to oxidation and becoming rancid. The product of the digestion is a mash—a mixture of oil, water and fiber. The oil in the mash can be retrieved in one of two ways. In the 'dry' method, the mash is sent to a press where the oil is pressed out of the mash by application of mechanical pressure. In the 'wet' method the oil is leached out of the mash by use of hot water. Different designs of the presses are extant. They may be a screw press (spindle press) or a hydraulic press. All the presses operate on the same principle, namely, the application of pressure to squeeze the oil out of the mash.

The screw presses consist of a cylindrical perforated cage through which runs a closely fitting screw. Digested fruit is continuously conveyed through the cage towards an outlet restricted by a cone, which creates the pressure to expel the oil through the cage perforations. Oil-bearing cells that are not ruptured in the digester are ruptured in the screw press as a result of turbulence and kneading action exerted on the fruit mass in the cage to release more oil. Thus, this press acts as an additional digester and is very efficient in oil extraction. Hydraulic presses exert pressure on the mash through a fluid pressure. They are faster than the screw presses and can also attain higher pressures. However, there is the danger of the hydraulic fluid leaking into the oil to contaminate it. Again, the hydraulic fluid can absorb moisture from the air and lose its effectiveness. Also, the plungers wear out and need frequent replacement. Spindle press screw threads are made from hard steel and held by softer steel nuts so that the nuts wear out faster than the screw and are easier to replace. During operation of the press, the pressure should be increased gradually to allow time for the oil to escape. If the depth of material is too great, oil will be trapped in the center. Heavy plates are sometimes inserted into the raw material in order to avoid the occurrence of this phenomenon. During pressing, metal wear occurs and this becomes the source of iron contamination of the oil. The rate of wear depends on the type of press, method of pressing and nut-to-fiber ratio. High pressing pressures are known to have an adverse effect on the bleachability and oxidative conservation of the extracted oil.

From the press, the soup (mixture of oil, water and other impurities) is sent by a conveyor to a clarifier to be clarified. The kernels and the fiber, however, are sent to a blower where they are separated. In most mills, the fiber is sent to a boiler where it is used as a fuel. The kernel, however, is further processed into palm kernel oil. The main objective of the clarification operation is to separate the oil from its entrained impurities. The fluid coming out of the press is very thick and viscous mixture of oil, water, cell debris, fibrous material and non-oil solids. The oil is separated from the other foreign materials through a series of four (4) operations. First, hot water is added to the mixture in a ratio of 3:1 in a tank to thin it. The water added serves as a barrier allowing the heavy solids to settle down the tank whilst the oil settles on top of the water. Second, the diluted mixture is passed through a vibrating screen to remove coarse fibers. Third, the screened mixture is boiled in a tank for about 1–2 h. The boiling facilitates the breakage of oil/water emulsion and hence the separation of the oil from the water. Finally, the mixture is pumped into a continuous stirred tank and allowed to settle by gravity. In the process, the oil being lighter settles at the top of the water and is decanted into a tank. In the reception tank, the clarified oil is heated with pressurized steam to dry the oil. Before sending to storage, the water content is further reduced under vacuum to about 0.15 and 0.25% to prevent increasing free fatty acid (FFA) through autoxidation hydrolysis of the oil. The soup from the settling tank is first sent to a sludge tank and then to a strainer where its impurities are removed with an inner brush. From the strainer, the soup is sent to a centrifuge where the residual oil is extracted, and the wastewater drained into a sludge pit where it is treated before disposing of it. One use for the treated sludge is to use it to kill weeds. Unless immediately dispatched for use, the oil is stored. The oil is stored at a temperature of about 50 °C using low pressure steam in heating coils to prevent it from going rancid through oxidation and to prevent solidification and fractionation. To prevent iron oxidation, the storage tank is lined in a suitable protective coating. The oil at this stage is called the crude palm oil (CPO).

7.6 Processing of Refined Palm Oil (RPO)

The aim of refining is to convert the CPO to quality edible oil by removal of objectionable impurities to the desired levels in the most efficient manner. CPO consists of glycerides and small and variable portions of non-glyceride component as well. The non-glycerides are of two main types namely: (i) *oil insoluble components*, e.g. fruit fibers, nut shells and free moisture and (ii) *oil soluble components,* e.g. free fatty acids, phospholipids, trace metals, carotenoids, tocopherols, tocotrienols, oxidation products and sterols. These are more difficult to remove than the oil insoluble impurities.

Not all the non-glycerides are undesirable. The tocopherols and the tocotrienols are useful as antioxidants in the oil and as nutritional supplements. The carotenoids are precursors of vitamin A and are therefore useful and desirable in oil. The other impurities are generally a nuisance in the oil and detrimental to the flavor, odor, color and keepability of the oil. It is therefore the removal of the objectionable non-glycerides to harmless levels that constitute refining of the oil. There are two main methods for refining CPO (the methods are applicable to PKO refining as well). These are the chemical or alkali refining and the physical refining methods. There are very little, if any, differences between the qualities of refined oils produced by these methods. However, the chemical method is more expensive and uses an alkali to neutralize most of the fatty acids, which are removed as soap stock. It is used mostly in older plants. The physical process removes the fatty acids in the oil by subjecting the oil to steam distillation under high temperature and vacuum. Clearly, the physical method does not require an effluent plant for the soap stock. The advantages of the physical processes are discussed in detail later in Sect. 7.6.2.

7.6.1 The Chemical Refining Method

A simplified flowchart for the method is shown below in Fig. 7.3. Before pumping the crude oil to the degummer, the oil in the crude oil tank is heated to about 45 °C to ease pumping and kept homogenized to provide a consistent final product. In the gum conditioning tank, oil is first heated to about 80 °C and subsequently treated with food grade ortho-phosphoric acid and mixed for 15 min for the gums to precipitate so that they are easily removable at the next stage. The acid-treated oil is subsequently dosed with caustic soda. The quantity of the caustic depends on the fatty acid content of the oil. Upon mixing, the alkali reacts with the acid-forming precipitated soaps, which are removed either through a centrifuge or washing. The light phase is the neutralized oil containing traces of soap while the heavy phase is the soap stock, insoluble materials, gums, free alkali and minute quantities of neutralized oil. The neutralized oil is then washed with water and passed through a centrifuge to remove the remnant soap in the oil. The resultant oil has a soap content of about 80 ppm. This residual soap is removed at the bleaching stage. The water-washed oil is sent to a vacuum dryer to dry and the product, a semi-refined oil, is termed *Neutral Palm Oil* (NPO). At this point, the oil still contains undesirable impurities, odors and color pigments that need to be removed. To achieve this, the NPO is first bleached using an activated earth.

Bleaching performs a number of functions, besides removing the impurities, odor and color pigments, it also adsorbs soap traces, pro-oxidant metal ions and decomposes peroxides. Bleaching is carried out under vacuum at a temperature of 100 °C and in a reaction time of 30 min. The amount of activated earth used is between 0.5 and 1.0% of the weight of the oil depending on the type and quantity of the starting oil to be treated. The slurry of oil and earth is next filtered twice to make sure that the activated earth is completely removed from the oil. The presence of the activated earth in the filtered oil would otherwise foul and reduce the oxidative stability of the product oil. It also acts as a catalyst for dimerization and polymerization activities. The oil at this stage is called *Neutralized Bleached Oil* (NBO). After filtering, the NBO is sent to the deodorizer where the residual free fatty acid content and the color of the oil are further reduced, and more importantly a stable product with the requisite flavor is produced. The deodorization process deaerates the oil, heats up the oil, steam strips the oil and cools the oil before it leaves the system. The heating occurs at a high temperature of between 220 and 240 °C and under a vacuum of 2–5 mbar. The process results in the thermal destruction of carotenoid pigments (color pigments). The direct steam stripping ensures the removal of the free fatty acids, aldehydes, ketones, which are responsible for the

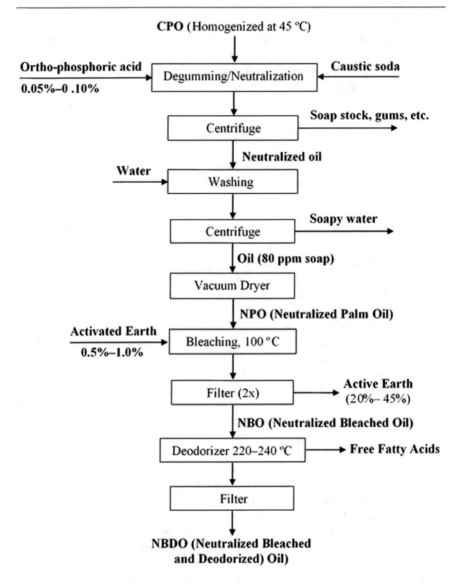

Fig. 7.3 A flowchart of the chemical refinery process

unacceptable odor of the oil. From the deodorizer, the oil passes through a polishing filter before it is sent to a storage tank. At this stage, the oil is termed *Neutralized Bleached and Deodorized Oil* (NBD oil).

The quality indicators of NBD oil and refined bleached and deodorized oil are provided in Table 7.9.

Table 7.9 Quality indicators for NBD/RBD oil	Indicator	Component
	FFA (as palmitic)	0.1% max
	Moisture & Insolubles (M & I)	0.1% max
	Melting Point	33.0–39.0 °C
	Color (5¼"Lovibond cell)	3.0 or 6.0 red max

7.6.2   Physical Refining

This is the more commonly used method because of its high efficiency, less losses (R.F. < 1.3), less operating cost, less capital input and less effluent to handle. The following essential points mark the point of departure of the physical method from the chemical method;

i. No chemical is used.
ii. Higher dosages of activated earth are used (1% of the oil weight).
iii. The filtered bleached oil is termed *Degummed Bleached Oil* (DBO).
iv. Heavier duty deodorization (250–270 °C).
v. The fatty acids distilled-off are condensed and collected and termed as *Fatty Acid Distillate*.
vi. The oil that leaves the deodorizer is termed RBD oil.

7.6.3   Fractionation of Palm Oil

This is the last of the purification processes to which CPO is subjected. Fractionating separates the refined oil into 'harder' (stearin) and 'more liquid' (olein) components. It is undertaken primarily to increase the market value of the oil. Fractionation is based on the principle that the triglycerides in the oil have different melting points and therefore at a certain temperature the lower melting point will stay in liquid while the higher melting point temperature component will crystallize into solid separating the oil into liquid (olein) and solid (stearin) fractions. The fractions are subsequently separated by filtration. Fractionation can be accomplished by three methods, namely, dry, detergent and solvent fractionation. A flowchart of the fractionation process is shown Fig. 7.4.

7.6.3.1   Dry Fractionation

The oil is first homogenized at 70 °C before the start of crystallization to destroy any crystals present and to induce crystallization in a controlled manner. Thereafter, the oil is cooled (cooling is done with chilled water in a jacket or in cooling coils) while agitating. When the oil reaches about 22 °C, crystallization begins, and the cooling is stopped. At this point, the oil turns into a semi-solid mass, termed *slurry*, and contains stearin crystals in liquid olein. The slurry is then pumped to a filter in a

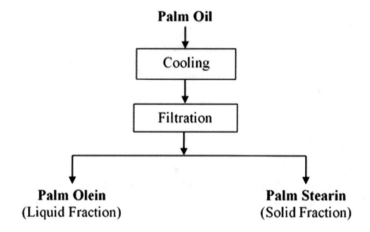

Fig. 7.4 Simplified flowchart of the palm oil fractionation process

Table 7.10 Quality indicators of the fractionated components

Indicator	Component	
	Olein	Stearin
Cloud point (°C)	8.0–10.0	–
Iodine value (Wijs)	56.0–59.0	42.0–46.0
Melting point (°C)	22.0–24.0	48.0–52.0

controlled manner to separate the two fractions. The filters commonly used are drum rotary filters or membrane recessed plate filters. Olein is the main product of the fractionation process. The type of filter used determines the yield of the olein. Olein yield of 65–68% is normal for filtration using vacuum suction filters. When membrane filters are used, olein yield could be as high as 75–78%. The quality of the olein is unchanged by the type of filter used but the stearin obtained is harder if the membrane filter is used. The quality indicators of fractionation refined oil are presented in Table 7.10.

Palm stearin from membrane filters has iodine value (Wijs) in the range 33.0–37.0 and melting point between 52 and 64.0 °C.

7.7 Environmental Impact of Palm Oil Production

The impact of the palm oil industry on the environment includes deforestation, loss of biodiversity and increase in greenhouse gas emissions. Also, the effluent from oil mills if not treated before allowing into the environment adversely affects the flora and fauna as the effluent is acidic. During milling activities, the following palm oil mill wastes are generated: palm oil mill effluent (POME), empty fruit bunches

(EFB), palm fiber and palm kernel; which may still be put to a good use. The EFB, palm fiber and palm kernel (solid wastes) can be put to economically useful purposes such as fuel material and mulch in agriculture–despite their high carbon to nitrogen (C/N) ratio.

During extraction of CPO from the fresh fruits, about 50% of the large quantity of water used results in POME. It is estimated that each ton of fresh fruit bunches at the extraction mill generates about 0.22 tons of empty bunches, and between 800 and 900 L of effluents and other by-products. Also, 1 ton of CPO produce between 5.0 and 7.5 tons of water ends up as POME. The raw partially treated POME contains extremely high content of degradable organic matter, which is partly due to the presence of unrecovered palm oil. The average biological oxygen demand (BOD) values of raw POME is about 25,000 mg/L, which is about 100 times more polluting than domestic sewage sludge. The effluent must be treated before discharge to avoid serious environmental problems. Liquid waste treatment involves anaerobic fermentation followed by aerobic fermentation in large ponds until the effluent quality is suitable for discharge. The treated effluent can serve as manure on farms and as a source of water for irrigation. The sludge accumulating in the fermentation ponds is regularly removed and fed to the land.

Further Readings

Afoakwa, E. O., & Sakyi-Dawson E. (2013). *Palm oil processing technology* (1st ed.). Germany: Scholars Press Inc.

Carrere, R. (2010). Oil palm in Africa: Past, present and future scenarios. World Rainforest Movement Series on Tree plantation No. 5. World Rainforest Movement.

Chakrabarty, M. M. (2003). *Chemistry and technology of oils and fats* (1st ed.). New Delhi: Allied Publishers Ltd.

Chemat, S. (Ed.). (2017). *Edible oils: Extraction, processing and applications* (1st ed.). Boca Raton: CRC Press.

Corley, R. H. V., & Tinker, P. B. H. (2015). *The palm oil* (5th ed.). Hoboken: Wiley-Blackwell.

Gupta, M. (2017). *Practical guide to vegetable oil processing* (2nd ed.). London: Academic Press & AOCS Press.

Hamm, W., Hamilton, R. J., Gijs Calliaiuw, G. (Eds) (2013). *Edible oil processing* (2nd ed). Hoboken: Wiley-Blackwell.

Hui, Y. H. (Ed) (1996). *Baileys's industrial oil and fat products: Edible oil and fat products: Processing technology* (Vol. 4, 5th ed.). Hoboken: Wiley-Blackwell.

Lai, Q.-M., Tan, C.-P., & Akoh, C. C. (Eds) (2012) Palm oil: Production, processing, characterization, and uses. Academic Press and AOCS Press, 1st Edition

MASDAR. (2011). Ministry of food and agriculture. Master plan study on the oil palm industry in Ghana. Final report.

United States Department of Agriculture. (2020). Oilseeds: world markets and trade. FAS-USDA.

Wu, W., & Weng, X. (Ed.). (2019). *Edible oil: Compounds, production and applications* (1st ed.). New York: Nova Science Publishers Inc.

Industrial Gases

8

Abstract

Industrial gases are used in many industries such as in chemical synthesis, food preservation, metal works and in medicine in the form of gases, liquid or cryogenic liquid and sometimes as solids. There are two groups of industrial gases based on their occurrence and/or mode of synthesis, that is naturally occurring gases (i.e. atmospheric gases—e.g. nitrogen, hydrogen, carbon dioxide, oxygen, liquid natural gases etc.) and gases that are synthesized from other substances (acetylene, liquefied petroleum gas etc.). The atmospheric gases are produced by three methods, i.e. physical separation of air by cryogenic method, membrane separation and adsorption method, or by a combination of any of three methods depending on the desired purity level of the product and required production capacity. Separation by the cryogenic methods is based on the differences in the boiling points of the components. The adsorption methods combine the selective attraction of an adsorbent for the components of air with swings in pressure to separate air into its components. In the membrane methods, the mixture is separated into its components based on the relative differences in the sizes of the components. This chapter comprehensively discusses the methods of manufacturing the major industrial gases.

8.1 Introduction

Industrial gases are those gases that are used in their pure form by industry in commercial quantities. These gases find applications in such diverse areas as energy generation, chemical synthesis, food preservation, metal works and medicine. In industry, they are used as gases, liquids or cryogenic liquids, and occasionally as solids as in the case of carbon dioxide (i.e. dry ice). They can be generally grouped into three main categories as oxidizers, inert gases and flammable gases. Table 8.1 shows the list of some industrial gases and their areas of application.

Table 8.1 Examples of major industrial gases, their methods of manufacture and their uses

Gas	Category	Area of application	Method of manufacture
Oxygen	Oxidizer	– Supports life and combustion; in medicine for resuscitation and in anesthesia; in high altitude and deep-sea diving; in welding and metal cutting; in chemical synthesis; etc.	– Cryogenic air separation – Membrane separation – Adsorption (pressure swing adsorption or vacuum swing adsorption) – Electrolysis of water
Nitrous oxide	Oxidizer	– Anesthetics – Propellant – Oxidizing agent in blow torches – Chemical synthesis – Ingredient of rocket fuel	– It is generally made by heating ammonium nitrate
Nitrogen	Inert gas	– Inerting systems (gas) – Pressurizing systems (gas) – Purging systems – In food preservation (storage freezing) – Coolant for electronic equipment – Pulverizing rubber or plastic materials – Freezing liquids in pipelines during repairs – In medicine for preservation of biological media, e.g. blood	– Cryogenic air separation – Membrane separation – Adsorption (pressure swing adsorption or vacuum swing adsorption) – Decomposition of ammonia
Argon	Inert gas	– In lighting devices – As an inert gas during manufacturing, metal cutting and chemical reactions	– Cryogenic air separation – Membrane separation – Adsorption (pressure swing adsorption or vacuum swing adsorption)
Helium	Inert gas	– In balloons – Coolant in nuclear reactors; refrigerant; in cooling superconducting magnets for MRI – Welding – Lighting – Tracer for leaks	– Cryogenic fractionation – Liquefaction and stripping of natural gas
Carbon dioxide	Inert gas	– Urea manufacture; Oil recovery; Preservative (food and beverage industries); In fire extinguishers; Coolant (refrigerant); pH control of water; chemical synthesis	– By-product of steam methane reforming followed by; (i) membrane separation (95% purity) or (ii) PSA (99% purity) or (iii) by absorption of liquid solvents

(continued)

Table 8.1 (continued)

Gas	Category	Area of application	Method of manufacture
Acetylene	Inflammable	– Chemical synthesis – Welding – Metal cutting – Lighting	– Reaction of calcium carbide with water – Thermal or catalytic cracking of hydrocarbons
Methane	Inflammable	– Fuel – Organic synthesis	– Anaerobic decomposition of biomass
Liquefied natural gas (LNG)	Inflammable	– As fuel – Production of carbon black – Organic synthesis	– From natural gas purification and processing
Liquefied petroleum gas (LPG)	Inflammable	– Fuel – Aerosol propellant – Refrigerant – Organic synthesis	– Cracking of hydrocarbons – Stripping of natural gas
Hydrogen	Inflammable	– Fuel (e.g. para-hydrogen as rocket fuel) – Food processing – Fertilizer manufacture – Manufacture of polyurethane – Chemical synthesis – Annealing of metals – Welding – Under water torches – Under water cutting – Used to remove oxygen from other gas streams	– Electrolysis of water – Steam methane reforming – Off gas purification (PSA, membrane separation, catalytic purification and cryogenic separation) – Decomposition of ammonia

In terms of their occurrence and/or mode of synthesis, the industrial gases may be grouped into two, namely: (i) naturally occurring gases and (ii) gases that are synthesized from other materials. The naturally occurring gases, for example, components of atmospheric air (i.e. nitrogen, hydrogen, carbon dioxide, oxygen, argon, helium, xenon and krypton) and liquid natural gas (LNG) are produced by purification of source gases using cryogenic fractionation, membrane separation and adsorption processes or a combination of these processes. The others are manufactured through chemical reactions or by-products of chemical processes and purification via conventional processes. Yet others, such as hydrogen and carbon dioxide, even though they occur naturally, may also be produced through chemical synthesis.

In Ghana, the gases that are produced commercially are nitrogen, oxygen, argon, helium carbon dioxide, hydrogen, acetylene and liquefied petroleum gas (LPG). With the discovery of natural gas, liquefied natural gas (LNG) will increasingly be an important commercial gas in the country. The major manufacturers of these gases are Air L'Quide, Air Mate, Jokumans and Katsena Air Products, Ghana Gas Company and Tema Oil Refinery, among others. The sections that follow will focus on the locally manufactured industrial gases.

8.2 Production of Atmospheric Gases

Atmospheric gases are those gases that occur in the atmosphere as a component of air. They are gases under normal conditions. However, they are not necessarily stored or transported as such. These gases are produced mostly by physical separation of air by the cryogenic air separation, membrane separation or adsorption methods. Most methods use a combination of any number of the three processes. The choice of method of manufacture depends on the level of purity required of the product and the expected production capacity.

8.2.1 Production of Nitrogen, Oxygen and Argon by the Cryogenic Air Separation Method

The cryogenic method relies on the differences in the boiling points of the components of air to separate them. Five major unit operations are involved in this process. First, atmospheric air is drawn and filtered to rid it of dust and other particulates, followed by cooling to condense moisture out of the air. Next, other unwanted components of air are separated based on size. Now the clean air is cooled to liquefy it, and then finally distilled to separate the air components based on the relative differences in their boiling points. During the distillation, nitrogen, which has a lower boiling point and therefore more volatile than oxygen or argon, is separated first. It is important to ensure that during the process the oxygen is removed from the argon and nitrogen streams as these are used as blankets to prevent oxidation. Oxygen, if present in these gases will rather promote oxidation. However, the other constituents of air need not be removed because of their low concentration and inertness. In Table 8.2 the atmospheric gases and their properties are listed, which are used as the basis for their purification. The operations of liquefaction and distillation, which are the key process units used for the separation of the components of air are based on a number of established technical principles, namely:

Table 8.2 Atmospheric gases, their concentration in air and their normal boiling points

Gas	Dry concentration (% vol.)	Dry concentration (% wt.)	Boiling point (°C)
Argon	0.93	1.28	−185.9
Carbon dioxide	0.038	0.0590	−78.5
Helium	0.0005	0.00007	−268.9
Hydrogen	0.00005	∼0.0	−252.7
Krypton	0.00011	0.003	−153.3
Neon	0.0018	0.0012	−245.9
Nitrogen	78.1	75.47	−195.8
Oxygen	20.94	23.20	−182.9
Xenon	0.000008	0.00004	−108.33

i. When work is done on air, by compressing it, it becomes hotter.
ii. When compressed air is expanded through an opening or a valve, it becomes cooler.
iii. When air is expanded in a turbine it does work on the rotors and cools by approximately ten times as much as in simple expansion.
iv. When a mixture of liquids is in equilibrium with its vapor, the vapor above the liquid is richer in the more volatile component (i.e. more of the lower boiling liquid vaporizes).
v. The boiling point of a liquid is lower at lower pressure.

A flowchart of the cryogenic method for separating the components of air is shown in Fig. 8.1. The feed, atmospheric air, is drawn into the process with suction filters. The filters remove dust and other particulate matter from the air as a first step in the purification and separation process. Again, the dust and particulates if not removed disrupt the efficiency of the subsequent unit operations. After filtration, the air is compressed in a multi-stage compressor to a pressure of about 45 bars. The compression process causes the temperature of the air to rise; therefore, to lower the temperature of the air, inter-coolers are installed in-between the compression stages.

From the compressor, the air is passed through a series of coolers during which its temperature drops considerably. The cooling is achieved using cooled water and nitrogen from the distillation unit. Cooling causes moisture in the air to condense; therefore, it is sent to a moisture separator (a column of packed alumina) to be separated. Any remnant water in the air is condensed and finally removed when the air is passed through a chilling unit and another moisture separator. The air at this stage has a temperature of between 7 and 12 °C and a pressure of 45 bars.

The moisture-free air is next sent through a series of filters to further purify it. The first of the filters is a carbon filter which removes any carbon in its natural form from the chilled air. This is followed by a passage through a molecular sieve filter which removes carbon dioxide from the air. The molecular sieve in this instance is a cylindrical vessel containing packed beads of potassium hydroxide or magnesium hydroxide. When exhausted, the molecular sieve may be regenerated by heating and cooling with the outgoing nitrogen from the distillation unit. It is important to remove the carbon dioxide before the air enters the heat exchangers where it is liquefied. Otherwise, the carbon dioxide will form dry ice in the heat exchangers and choke the heat exchanger tubes. From the molecular sieve, the air is again filtered in an air filter to rid it of any remnant dust before it finally enters a cold box.

From the cold box, the air is sent to the air separation unit where it is separated into its various components. The air separator is made up of a series of heat exchangers, and a mechanical expansion valve through which the air is further cooled until it liquefies. The cooling in the heat exchanger is carried out using the nitrogen and oxygen from the distillation columns. The cooled air from the heat exchanger liquefies upon passing through an expansion valve from where it enters the fractionating column. The method used is referred to as 'double column' as it contains two separate distillation columns mounted one on top of the other and operated at two different pressures. In between the two columns there is a

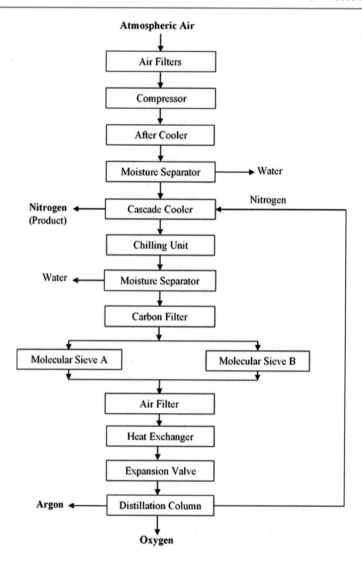

Fig. 8.1 Flowchart for the production of oxygen and nitrogen from air based on the cryogenic method

condenser, which acts as a reflux for the lower part and as a reboiler for the upper column. The cooled air is fed at the bottom of the high-pressure column where the lower boiling nitrogen boils and rises to the top as the product. The higher boiling crude oxygen which stays at the bottom (in the reboiler/condenser) is flashed into the lower pressure column. Vapor is withdrawn from the middle of the low-pressure column and fed to a crude argon column where it is condensed and taken as crude argon. The vapor escaping to the top of the column is pure nitrogen and is

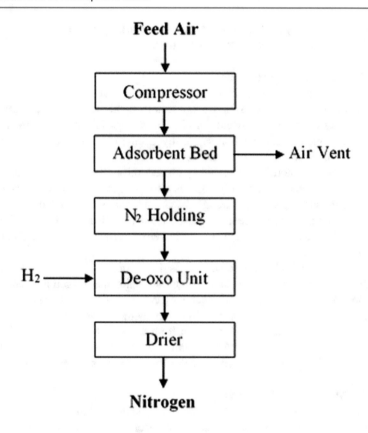

Fig. 8.2 A simplified flowchart for the manufacture of nitrogen by the PSA method

condensed for storage. Liquid oxygen remains at the bottom of the column. The gases are removed in separate paths in the heat exchangers where they are used for cooling the incoming air. The nitrogen produced by this method is 99.999% pure. Contaminants in the nitrogen are predominantly argon ad oxygen. The oxygen, on the other hand, is 99.5% pure with nitrogen, argon, water and methane as the principal contaminants. The argon is 99.999% pure with nitrogen, oxygen and water as the main contaminants. The main contaminants are in minute quantities, and in the case of nitrogen and argon they are inert, therefore do not affect the use to which the gases are put.

8.2.2   Non-Cryogenic Methods of Nitrogen, Oxygen and Carbon Dioxide Production

8.2.2.1   Adsorption Methods

There are two types of this method, namely pressure swing adsorption (PSA) and vacuum swing adsorption (VSA). These methods combine the selective attraction of an adsorbent for the components of air with swings in pressure to separate air into its components. The method is mostly used in the separation of oxygen, nitrogen and carbon dioxide from the air.

In the VSA method, ambient air is passed through a column of adsorbent material. The adsorbent selectively adsorbs nitrogen molecule on its surface while the oxygen molecule passes on through the column and is collected at the exit. The adsorbed nitrogen is subsequently released from the adsorbent using differential pressure from a vacuum. Once the column is free of nitrogen it is thus activated and can be re-used.

In the PSA method, compressed air is first dried and subsequently passed through a column of chemical adsorbents during which the oxygen molecule in the air is trapped while allowing the nitrogen molecule to pass through the column. The system can produce 90.0–99.5% pure nitrogen without the need for downstream purification. When necessary, the nitrogen from the adsorbent is first treated to get rid of any residual oxygen by reaction with hydrogen in a de-oxo unit and finally dried before sale or use. A flowchart of the PSA system is shown in Fig. 8.2.

8.2.2.2   Membrane Methods

Membranes separate a mixture into its components based on the relative differences in the sizes of the components. The membrane itself is a microporous structure that may consist of a cellulose acetate substructure and a thin layer of dense cellulose acetate on the upper surface. This upper layer is the active component of the membrane, which serves as the separating barrier. The membrane selectively allows smaller-sized molecular components to pass through the membrane substructure whilst refusing bigger molecules passage, thereby accomplishing the separation process. Therefore, by designing or selecting the appropriate molecular-sized membrane or a collection of membranes, it is possible to separate air into all its components.

8.3   Gases Produced by Chemical Synthesis

Besides those gases that occur naturally as witnessed in Sect. 8.1, most other industrial gases including hydrogen and carbon dioxide that are commonly found in the atmosphere are manufactured through chemical synthesis. The methods of manufacture of some of such gases are described in this section.

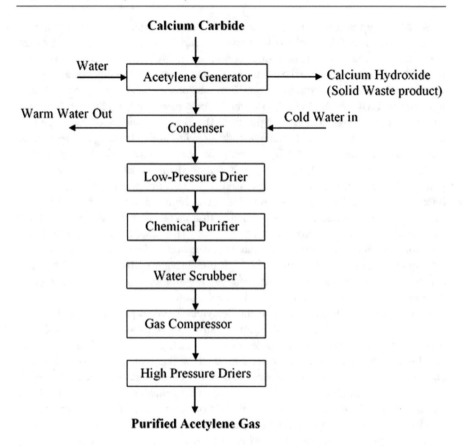

Fig. 8.3 A flowchart for the production of acetylene from calcium carbide and water

8.3.1   Acetylene

Acetylene is a product of the reaction between calcium carbide and water. This reaction, therefore, serves as the basis for the industrial manufacture of acetylene. The reaction proceeds as follows:

$$CaC_2 \quad + \quad 2H_2O \quad \rightarrow \quad C_2H_2 \quad + \quad Ca(OH)_2$$

| Calcium carbide | Water | Acetylene | Calcium hydroxide | (8.1) |

Following the reaction, the product gases (mainly acetylene and water vapor) are cooled to condense the water vapor and dried. Thereafter, it is purified, compressed, dried again and filled into storage tanks. A flowchart for the manufacture of acetylene is shown in Fig. 8.3.

As a first step in the process, a measured quantity of water is fed to the acetylene generator which is afterward fed with calcium carbide from a sealed hoper. In the generator, the reaction occurs spontaneously upon mixing and the gas produced passes to a condenser where the hot acetylene gas is cooled, and any accompanying water vapor is condensed out of the gas. From the condenser, the gas goes through a drier charged with calcium chloride which absorbs moisture from the gas. Next, the gas is passed through a chemical purifier. The purifier consists of two layers of regenerative chemicals, which remove additional impurities in the gas. The purifier may be regenerated for re-use by the passage of atmospheric air through it. Any remnant impurity in the acetylene gas is removed in the scrubber following the purifier. Thereafter, the gas is compressed and dried again using a high-pressure drier where oil and moisture are removed. The high-pressure drier consists of a battery of three vessels. The first is filled with a metal filter while the second and third are filled with anhydrous calcium chloride. The calcium chloride vessels are fully filled as any void in a high-pressure area is against safety rules. Finally, the gas leaving the drier is stored or filled into high-pressure gas cylinders/tanks. The main contaminants that usually accompany the gas are ammonia and air. The acetylene produced by this process is about 99.5% (vol.) pure.

Great care must be taken around the acetylene plant to prevent fires, as it is a highly flammable substance. Consequently, stringent controls are maintained over sources of ignition and incompatible materials such as moisture. In some cases, metal tools are made of bronze (rather than steel) so that they do not give off sparks during use; cigarettes are not allowed, and cell phones must be turned off near an acetylene plant. In particular, calcium carbide poses danger as it gives off a large volume of acetylene gas when wet, so it must be stored in the complete absence of moisture. Again, the large volume of acetylene gas generated during production, if not controlled could increase the pressure of the system and lead to an explosion. To obviate danger occurring at acetylene plants, therefore, the temperature, pressure, water and calcium carbide feeds are all automatically controlled. The danger areas in an acetylene plant include:

  i. Acetylene is highly inflammable and must be stored and/or transported away from sources of ignition.
 ii. Calcium carbide when in contact with water reacts violently giving off large volumes of gas leading to increased pressure and eventually explosion in enclosed places.
iii. The reaction of calcium carbide with water is exothermic. The heat generated if not controlled could lead to an explosion.
 iv. The reactor/generator cooling water feed if not properly controlled could lead to overheating in the reactor.

To avoid any of the above happenings, the carbide is stored such that it does not get into contact with moisture and is fed using automatic feeders into the generator/reactor. Again, the acetylene loading/filling into tanks is automatically done, in such conditions, so as to avoid all sources of fire or ignition/spark. The water fed to the generator to absorb the exothermic heat of the reaction and hence to check the reactor temperature is equally automatically regulated. These and other precautional measures ensure that the acetylene production process occurs under safe conditions.

Slaked lime $Ca(OH)_2$ is the major by-product of the acetylene production process. Since slake lime is an important raw material in water treatment processes where it is used to control the pH of potable water; the majority of the slake lime from the plant is therefore dewatered in a series of settling ponds and supplied to water treatment companies. Otherwise, it is dumped as waste.

8.3.2   Production of Hydrogen

Several methods exist for the production of hydrogen. Notable among these are:

i. Electrolysis of water;
ii. Steam reforming of light hydrocarbons (e.g. methane);
iii. Dissociation of ammonia; and
iv. Off-gas purification of product streams from chemical and petroleum processing.

In the last three methods, hydrogen is produced as a result of a chemical reaction, often in the presence of other products or chemicals. Therefore, to get pure hydrogen, the reaction mixture must be taken through a number of separation steps (mostly physical but occasionally chemical) during which the unwanted components of the reaction mixture are gotten rid of to get pure hydrogen.

The first method, the electrolysis method, uses electrical energy to decompose water into hydrogen and oxygen. There are three main steps involved in this method. Initially, the water to be decomposed is purified. Following the purification, the electrolysis is carried out, and finally, the products of the electrolysis are purified. A flowchart of the electrolysis process is shown in Fig. 8.4.

As a first step during the production of hydrogen via electrolysis, raw water is purified by distillation then subsequently passed through a series of ion exchangers to rid it of mineral contaminants. The purified water is next sent to an electrolytic cell where it is decomposed into hydrogen and oxygen. The cell is a steel tank containing potassium hydroxide, which acts as an electrolyte. The electrodes of the cell are made from nickel-coated steel plates. At the cathode, hydrogen is formed while oxygen is formed at the anode. The electrode reactions proceed as follows:

Cathodic reaction

$$2H_2O(l) \ + \ 2e \ \rightarrow \ H_2(g) \ + \ 2OH^- \tag{8.2}$$

Fig. 8.4 A flow diagram for
the production of hydrogen by
electrolysis

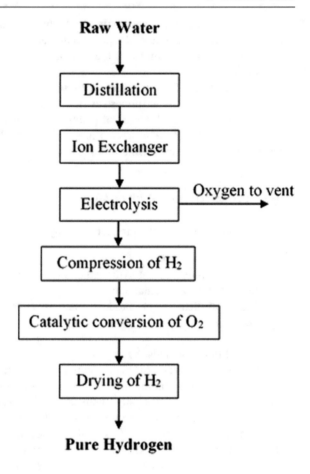

Raw Water

Distillation

Ion Exchanger

Electrolysis — Oxygen to vent

Compression of H$_2$

Catalytic conversion of O$_2$

Drying of H$_2$

Pure Hydrogen

Anodic reaction

$$2H_2O(l) \quad \rightarrow \quad O_2(g) + 4H^+(aq) + 4e \qquad (8.3)$$

Overall reaction

$$2H_2O(l) \quad \rightarrow \quad 2H_2(g) + O_2(g) \qquad (8.4)$$

The hydrogen and the oxygen are collected separately through different pas-
sages. While the oxygen is vented off into the atmosphere, the hydrogen is further
treated to purify it and finally collected for storage or use. The hydrogen from the
electrolytic cell often contains some amount of oxygen which is gotten rid of by
first heating the cathodic product to 90 °C by compression and then by application
of heat energy. The compression heats the gas to a temperature of 70 °C at a
pressure of 172 bar. The application of external heat raises the temperature of the

hydrogen finally to 90 °C. The heated hydrogen is passed over a palladium/alumina catalyst where any contaminating oxygen is reduced into water vapor. The wet hydrogen from the catalytic column is next dried over a mixture of alumina and silica gel column. It is finally filtered, compressed again and stored in cylinders and tanks. The potassium hydroxide electrolyte does not get consumed in the process and it is therefore rarely replenished. It only acts as a medium for the flow of electricity through the cell. It is the water that gets consumed in the process and must therefore be continually replenished.

8.4 Liquefied Petroleum Gas (LPG)

LPG is mostly used as fuel both domestically and industrially in Ghana. It is also used as fuel in the transportation sector to power vehicles. In some jurisdictions, it is used as a feedstock in chemical synthesis. LPG is a gas under normal conditions but can be liquefied by cooling and/or by compression. It is often a mixture of gases whose main components are propane (C_3H_8), propene (C_3H_6), isobutane (C_4H_{10}), n-butane (C_4H_{10}), and butene (C_4H_8). The mix can be predominantly propane or butane. In temperate lands, the mix is predominantly propane during winter and predominantly butane during summer. The butenes and propenes are present only in small quantities. Typically, LPG is 60% (vol.) propane and 40% (vol.) butane. It burns cleanly with no soot and very few sulfur emissions, hence, posing no ground or water pollution hazards. It has a typical calorific value of 46.1 MJ/kg. Comparatively, fuel–oil has a calorific value of 42.5 MJ/kg and the corresponding value for premium grade petrol (gasoline) is 43.5 MJ/kg. However, its energy density per unit volume of 26 MJ/l is lower than either that of petrol or fuel–oil. Again, LPG has a higher calorific value (94 M J/m^3 or 26.1 kWh/m^3) than natural gas (methane) (38 MJ/m^3 or 10.6 kWh/m^3). This means that LPG cannot be directly substituted for natural gas. In order to allow the use of the same burner controls and to provide for similar combustion characteristics, LPG can be mixed with air to produce a synthetic natural gas (SNG) that can be then substituted.

LPG is derived from fossil fuel sources from which there are, generally, two main production routes. LPG may be obtained from refineries where it is a by-product of the refinery process—high molecular weight hydrocarbons when cracked give propane and butane in addition to other paraffin. These are subsequently processed to extract the butane and propane as LPG. It is also processed from natural gas where it accompanies oil and gas streams as they emerge from the ground (gas/oil fields).

8.4.1   Process Description for the Production of LPG from Refinery By-Products

When crude oil is refined, a non-volatile residue made up principally of high molecular weight hydrocarbons (heavy crude oil) remains. The residue is often rendered more valuable by converting it into the more valuable lighter hydrocarbons. The conversion is performed by cracking the heavy crude in a catalytic reactor. The conversion could be carried out by thermal cracking where heat energy is applied to break down the heavy molecular weight hydrocarbons. One of the main products of the cracking process is the LPG. The most used catalytic cracking reactor for this purpose is the fluid catalytic cracking (FCC) reactor (or the residue fluid catalytic cracking (RFCC) reactor). The reactor uses a silica/alumina catalyst at sufficiently high temperature to break down the high molecular weight hydrocarbons of the heavy oil into the light molecular weight products. The product from the RFCC is a mixture of gasoline, LPG (butane and propane fractions) and non-condensable lean gases, which are separated into the various components in a gas concentration unit. The composition of the gas mixture from the RFCC is shown in Table 8.3.

The LPG from this unit is subsequently treated to remove sulfur (mostly in the form of hydrogen sulfide and mercaptans—organic sulfur compounds) and other contaminants before use or sale. A process flowchart for the production of LPG as a refinery by-product is shown in Fig. 8.5.

Economic considerations dictate that the removal of sulfur compounds from LPG should be performed in a three-step process. The first two steps, namely, the

Table 8.3 Components of the product stream from the RFCC

Component	Formula	Boiling point (°C)
Gasoline	C_5 + Hydrocarbons	38–221
Normal Butane	$n\text{-}C_4H_{10}$	−0.6
Isobutane	$i\text{-}C_4H_{10}$	−12
Butylenes	C_4H_8	−7 to + 4
Propane	C_3H_8	−42
Propylene	C_3H_6	−48
Hydrogen Sulfide	H_2S	−61
Ethane	C_2H_6	−88
Ethylene	C_2H_4	−104
Methane	CH_4	−162
Hydrogen	H_2	−253
Nitrogen	N_2	−78
Carbon Monoxide	CO	−192
Carbon Dioxide	CO_2	−196

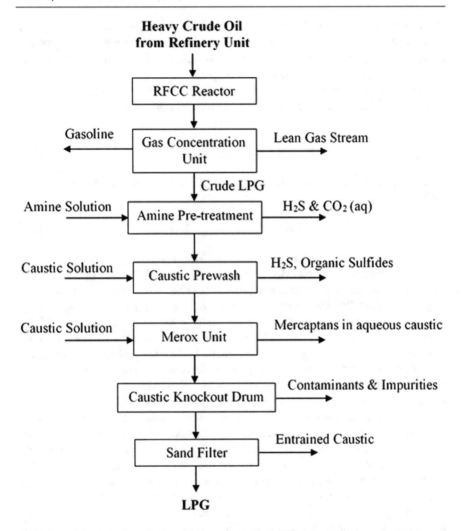

Fig. 8.5 A schematic flow diagram for the manufacture of LPG from heavy crude oil

amine treatment and caustic pre-wash are carried out to reduce the hydrogen sulfide content of the gas to 5 ppm or less before sending it for final treatment in the Merox unit. The amine treatment and the caustic pre-wash are necessary to prevent premature spending (weakening) of the Merox circulating caustic and hence maximize the life of the circulating caustic. Contamination of the circulating caustic with sulfide can prevent the extractor-rich caustic from being regenerated sufficiently, causing poor extraction efficiency. The amine treatment reduces the hydrogen sulfide content of the LPG to 50 ppm or less. The process takes place in an absorber where an aqueous solution of diethanolamine is utilized in a countercurrent process to extract the acid gas from the LPG. The caustic pre-wash following the amine treatment occurs in a batch-type vessel. The LPG enters from the bottom of the vessel and jets through a distributor with holes oriented downwards. The gas thus bubbles through the caustic already in the vessel. The hydrogen sulfide concentration in the LPG after the pre-wash is 5 ppm or less.

Two reactions take place during the caustic pre-wash. In the first instance, the hydrogen sulfide reacts with the caustic to form sodium sulfide according to the following reaction:

$$H_2S + NaOH \rightarrow Na_2S + 2H_2O \tag{8.5}$$

Subsequently, the sodium sulfide produced will further neutralize hydrogen sulfide as follows:

$$H_2S + Na_2S \rightarrow 2NaHS \tag{8.6}$$

From the pre-wash, the LPG is sent to the Merox unit where mercaptans in the gas are treated and removed. Merox is an acronym for **Mer**captan **Ox**idation. It is an efficient and economical catalytic process developed for the chemical treatment of petroleum distillates in general for the removal of sulfur present as mercaptan. There are two versions of the process, namely Merox extraction and Merox sweetening. In the former process, the mercaptans are completely removed from the fuel stream as a result of the catalytic treatment. In the latter process, the mercaptans are rather converted to other less objectionable forms. In both processes, special catalysts are used to accelerate the oxidation of the mercaptans. The two processes, namely, extraction and sweetening, can be carried out separately or as a combination depending on the application. The catalysts used are composed of iron group metal chelates. A flowchart for the Merox extraction process is shown in Fig. 8.6.

The mercaptan extraction is applied to both liquid and gaseous hydrocarbons. The extent of mercaptan extraction from the fuel medium depends on the solubility of the mercaptan in the alkaline solution. This in turn depends primarily upon the molecular weight of the mercaptan, the degree of branching of the mercaptan molecule, the caustic soda concentration and the temperature of the system. Increasing molecular weight and branching both decrease mercaptan solubility in the alkaline solution. The mercaptan extraction proceeds according to the following mechanism:

$$
\begin{array}{cccccc}
RSH & + & NaOH & \leftrightarrows & NaSR & + & H_2S \\
\text{Oil phase} & & \text{Aqueous} & & \text{Aqueous} & & \text{Gaseous} \quad (8.7) \\
\text{(Mercaptan)} & & \text{phase} & & \text{phase} & & \text{phase}
\end{array}
$$

The pre-wash LPG enters the extraction unit where the mercaptans are extracted into a caustic solution. The extractor is a multi-stage column through which the LPG and the caustic solution flow in a countercurrent manner. The untreated LPG enters near the bottom of the column and the regenerated caustic is pumped to the top. In the column, the two phases mix intimately and the mercaptans are

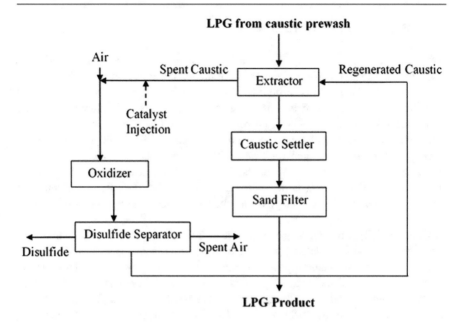

Fig. 8.6 A flowchart of the Merox process for the extraction of mercaptans from LPG

transferred into the caustic solution in which the Merox catalyst is dispersed. The treated LPG gas emerges from the top of the column and is sent to a caustic knockout drum (caustic settler) where any entrained caustic solution is separated and returned to the circulation system. Before storage or loading, the LPG is filtered through a sand filter to coalesce and get rid of any residual entrained caustic. If properly done, the LPG from the filter would be visibly clear and must contain less than 1 ppm sodium.

The mercaptan-loaded caustic is withdrawn from the bottom of the column and regenerated and returned to the extractor. The regeneration is performed by first heating the mercaptan-rich caustic and sending it to an oxidizer, where it is mixed with a controlled amount of air. The resulting oxidation converts the aqueous soluble mercaptides to oil-soluble disulfides, which are relatively insoluble in a caustic solution. The mixture of air, disulfides and caustic is pumped to a disulfide separator, where first, the oxygen-depleted air is vented off following which the regenerated caustic and the disulfide are separated in the main body of the separator. In the separator, the disulfides coalesce and are decanted as the upper oil phase. The lower regenerated caustic phase, which settles by gravity and now freed of substantially all disulfide, is then recycled to the extractor.

8.5   Processing of Natural Gas

Natural gas consists primarily of methane. Indeed, on the market or as a product, it is almost pure methane. Natural gas found at the wellhead is, however, by no means pure although it is composed primarily of methane. In nature, or when first brought from underground where it is most often mined, it exists in mixtures with other hydrocarbons, principally ethane, propane, butane and pentanes. Other substances found with raw natural gas are: water vapor, hydrogen sulfide (H_2S), carbon dioxide (CO_2), helium (He), nitrogen (N_2) and other compounds. It is the separation of methane from these other components of natural gas that constitutes natural gas processing. The processing is necessary as it ensures that the natural gas intended for use is clean and as pure as possible, making it clean burning and an environmentally sound energy choice.

Commercially, raw natural gas is produced from three types of wells, namely oil wells, gas wells and condensate wells. It is also found associated with coal beds and methane clathrates. Natural gas is created in marshes, bogs and landfills by methanogenic organisms. Natural gas that comes from oil wells is typically termed '*associated gas*'. In oil wells, the gas can exist separate from oil in the formation (in such instances it is termed '*free gas*') or dissolved in the crude oil (*dissolved gas*). When obtained or produced from gas wells or condensate wells, natural gas is called 'non-associated gas'. Gas wells mostly produce raw natural gas by themselves while condensate wells produce free natural gas alongside a semi-liquid hydrocarbon condensate. Natural gases from these latter sources may contain little or no crude oil.

Raw natural gas from the well is transported through pipes to natural gas processing plants where it is processed into highly pure natural gas, also called 'pipe quality' *dry natural gas*. Typically, the raw natural gas is taken through four (4) steps to get a market-quality product. The first step removes associated oil and condensate from the gas; the second step removes water; the third step separates natural gas liquids (NGLs); and the fourth step removes sulfur and carbon dioxide from the gas. A simplified flowchart for the processing of raw natural gas is shown in Fig. 8.7.

Raw natural gas from a well is often accompanied by sand and other solid materials. These impurities are removed in a scrubber before piping the gas to a natural gas processing plant. The scrubbing is done at or near the wellhead. When fluctuations in pressure and temperature are unfavorable, it could cause the gas to cool leading to the formation of natural gas hydrates from the gas. Hydrates are ice-like crystals found in the solid or semi-solid state. If the hydrates are allowed to form and accumulate, they could block the passage of the gas around valves and other obstacles in the pipeline. To avoid this happening, natural gas-fired heating units are installed at specific locations along the pipeline where there is the tendency of formation of hydrates.

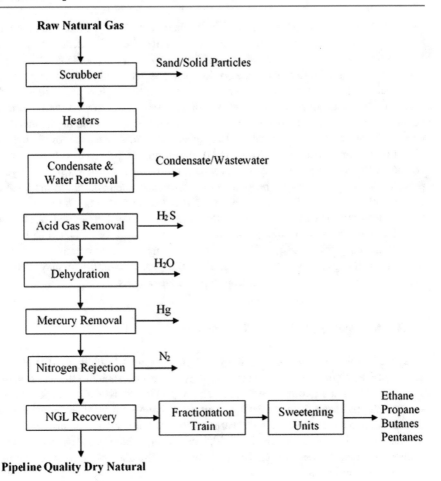

Fig. 8.7 A flowchart for the processing of natural gas

If the gas is from an oil well and for that matter it is accompanied by oil or dissolved in oil, then necessarily, the oil must be separated from the gas. If the gas was dissolved in the oil as a result of high well pressure, then upon production of the oil/gas, the attendant reduced pressure causes the gas to separate from the oil naturally. Where this is not possible, a conventional separator such as a simple closed tank may be used in which the oil and gas separate under gravity. There are instances where specialized equipment such as a low-temperature separator (LTX) may be necessary to separate the oil and condensate from the gas. The LTX uses pressure differentials to cool the wet natural gas and separate the gas from the oil and condensate.

Next, the gas is sent to a column absorber where it is treated with diethanolamine (DEA) to remove hydrogen sulfide from the gas. After the amine treatment, the gas goes to a dehydrating unit to remove water vapor in the gas. The dehydrating unit employs either a solid desiccant (adsorption) such as activated alumina or granular silica gel. In this case, the wet natural gas is passed through a tower filled with desiccant. The gas enters the tower at the top and as it flows toward the bottom, it contacts the desiccant which absorbs the water from the gas. Liquid desiccants may be used in place of solid desiccants. Examples of liquid desiccants are diethylene glycol (DEG) and triethylene glycol (TEG). A solution of the glycol is brought into contact with the wet gas in a contactor where the water is absorbed by the glycol particles. The glycol particles become heavier once absorbed and fall to the bottom of the contactor from where they are removed out of the system. The natural gas, having been stripped of most of its water content, is then transported out of the dehydrator and passed through a column of molecular sieve or activated carbon to remove mercury from the gas. Thereafter, it is sent into a cryogenic unit where nitrogen in the gas is separated by cooling the gas. The nitrogen may also be removed from the gas by using the appropriate adsorbents.

8.5.1 Natural Gas Liquids Separation

After the nitrogen cryogenic unit, the gas is now essentially methane (the natural gas proper) and other low molecular weight hydrocarbons such as ethane, propane, butanes and pentanes, commonly called natural gas liquids (NGLs). It is essential to separate the methane from the NGLs in order to have both cleaner, purer natural gas, as well as the valuable hydrocarbons that are the NGLs themselves. The separation of the NGLs occurs in the demethanizer, which employs one of two methods to effect the separation. These methods are the absorption method and the cryogenic separation method.

In the absorption process, a lean oil (i.e. an admixture with a liquid hydrocarbon) is contacted with the natural gas during which the bulk of the NGLs is absorbed into the oil leaving methane-rich gas (natural gas), which may be then stored or packaged for sale. The natural gas obtained in this manner contains portions of the other hydrocarbons, principally ethane and propane. For a cleaner natural gas, however, cryogenic methods are preferred. In this process, the natural gas from the nitrogen removal unit is cooled to a temperature of about −85 °C to liquefy all the components in the gas except methane. This process allows for the recovery of about 90–95% of the ethane and even higher percentages of the other hydrocarbons originally in the gas stream. The natural gas at this stage can be sold on the market.

To be more useful, the NGLs themselves must be separated into the base components. This is achieved in a train of fractionating units. A fractionation unit utilizes the difference in the boiling points of the components of NGLs to separate them. Each unit of the train separates a particular component after which it is so named. Thus, the lightest of the components, ethane, is first separated in the deethanizer followed by the separation of propane in the depropanizer. Finally, a debutanizer separates the butane from the residual higher molecular weight hydrocarbons like pentane. A separate fractionating unit, the deisobutanizer (or butane splitter), separates the isobutane from the normal butane.

The ethane, propane and butanes still contain significant hydrogen sulfide which must be cleaned before the gases can be sold and used. When the concentration of hydrogen sulfide in the gas is more than 5.7 mg/m^3 of natural gas, the gas is considered sour. The process of removing the hydrogen sulfide from the gas is termed '*sweetening*'. The sulfur compounds contained in sour gas can make it extremely harmful and therefore undesirable. It can cause death if breathed in and it is very corrosive to metal parts. Another incentive for removing the sulfur is that it can be sold or used for other processes. The cleansing of the hydrogen sulfide takes place in a sweetening unit where the Merox process, described earlier in Fig. 8.6 of Sect. 8.3, may be employed to rid the hydrocarbons of hydrogen sulfide. Alternatively, molecular sieves may be employed to clean the gases. Another method often used for this purpose is the Sulfrex method. The effluent gases from the Merox or Sulfrex units are virtually free of sulfur compounds and thus lose their sour gas status.

Further Readings

Downie, N. A. (1996). *Industrial gases* (2002nd ed.). Netherlands: Springer.

Gunardson, H. H. (1997). *Industrial gases in petrochemical processing: Chemical industries.* Boca Raton, USA: CRC Press.

Haring, H. W. (2007). *Industrial gases processing.* KGaA: Wiley-VCH Verlag GmbH & Co.

Haring, H. -W. (Ed) (2008). *Industrial gases processing digital.* Wiley-VCH Verlag GmbH.

Kerry, F. G. (2016). *Industrial gas handbook: Gas separation and purification* (1st ed.). Boca Raton, USA: CRC Press.

Kidnay, A. J., Parrish, K. W., & Macartney, D. G. (2020). *Fundamentals of natural gas processing* (3rd ed.). Boca Raton, USA: CRC Press.

Speight, J. (2007). *Natural gas: A basic handbook* (1st ed.). Houston, USA: Gulf Publishing Company.

Speight, J. (2019). *Handbook of industrial hydrocarbon processes.* Oxford, UK: Gulf Professional Publishing.

Crude Oil Refinery and Refinery Products

9

Abstract

Crude oil has remained the leading source of fuel or energy in the world. It is a flammable and toxic liquid that contains a complex mixture of hydrocarbons having different molecular weights and other liquid organics. There are varieties of crude oil each with an inimitable mix of molecules that gives its physical and chemical properties and market value. The market value of the crude oil is influenced by the geographical location of its production, the sulfur content and its specific gravity according to the American Petroleum Institute's (API) system. The crude oil can be separated into several hydrocarbons of useful commercial products by using a process called 'refining', which is carried in an oil refinery. The refinery process produced three groups of products: light fraction (e.g. gas such as liquid petroleum gas—LPG, gasoline, aviation fuel, etc.); middle distillates or medium fraction (e.g. kerosene, lubrication oil, light gas oil, diesel oil, etc.); and heavy fraction or bottom residue (e.g. asphalt and bitumen, waxes, etc.). The intermediate distillates can be altered using catalytic reforming to produce new compounds. The refinery plants generate a lot of waste which must be treated to meet the requirement of environmental agencies. In this chapter, the methods and technologies employed in the refinery of crude oil into its various product fractions as well as the types of waste generated at crude oil refineries and their methods of treatment are discussed. An overview of the downstream oil industry in Ghana is also presented.

9.1 Introduction

The world today derives most of its energy from crude oil which remains the world's leading fuel, accounting for 32.9% of total global energy consumption. Indeed, estimates indicate that until well into this century, over half to two-thirds of all commercial energy consumed will be met by oil and gas. Crude oil, therefore, is expected to continue to be the most important source of energy for the foreseeable future. That the world continues to rely on crude oil is also in part due to a number of advantages it has over other forms of commercial energy sources. In comparison to other energy sources, crude oil is relatively readily available; it is relatively easy to extract; it has high energy density; it is easy to transport and handle; and its use comparable has fewer environmental drawbacks.

In 2015, world oil production reached 4,461 Mt (94.2 million barrels/day), with the total world-proven crude oil reserves standing at 1,493 billion barrels. According to 2015 estimates, Russia is the highest producer of crude oil (10,250,000 barrels/day) followed by Saudi Arabia (10,050,000 barrels/day) and then the United States of America (9,415,000 barrels/day). Nigeria is the highest oil producer in Africa and number 12 in the world (2,317,000 barrels/day), while Ghana occupies the 11th position in African and the 42nd position globally with a production capacity of 102,400 barrels/day, according to 2015 estimates.

Crude oil, also commonly called petroleum, is a naturally occurring, toxic and flammable liquid, consisting of a complex mixture of hydrocarbons of varied molecular weights and other liquid organic compounds. It is formed as a result of the decomposition of sea organisms over a long period of time, sometimes millions of years. It is located in geological formations within the earth's crust. Crude oil is recovered through drilling beneath the earth's surface and causing the oil to flow to the earth's surface either naturally or by application of mechanical or other forces. In its natural-crude form, it has very limited uses but when refined and separated into its components it finds wide use in various applications. Products from crude oil find their use in such varied applications as fuels (petrol, diesel, kerosene, LPG, etc.), lubricating oil, bitumen, petroleum coke, paraffin wax, sulfur and chemical agents for making plastics, detergents, solvents, fibers, elastomers and pharmaceuticals.

The most common form of molecules in crude oil are hydrocarbons; however, atoms such as nitrogen and sulfur can also also be found in the molecules of crude oil. Hydrocarbons are molecules of varying lengths and complexity made of mostly hydrogen and carbon atoms and a small number of oxygen atoms. The varying physical and chemical properties of the molecules are a result of the differences in their structures. This behavior means that crude oil finds its use in a broad area of applications. The major components of crude oil are hydrocarbons, including alkanes, cycloalkanes and aromatic hydrocarbons. Some nitrogen, oxygen and sulfur-containing organic compounds are also present in crude oil. Trace amounts of metals such as iron, nickel, copper and vanadium may also be present in the oil. The exact molecular composition of crude oil varies widely from source to source but the proportion of chemical elements vary over a fairly narrow range as shown in Table 9.1.

Four main hydrocarbon groups are found in crude oil. These are paraffins, naphthenes, aromatics and asphaltics. Their relative amounts in crude oil vary depending on the source. The average presence and range depending on the source are shown in Table 9.2.

Table 9.1 Elemental composition of crude oil by weight

Elements	Percentage range (% wt)
Hydrogen	10.0–14.0
Carbon	83.0–87.0
Nitrogen	0.1–2.0
Oxygen	0.1–1.5
Sulphur	0.5–0.6
Metals	<0.1

Table 9.2 Composition of hydrocarbon groups in crude oil

Hydrocarbon	Average composition (% wt)	Composition depending on source (% wt)
Paraffins	30.0	15.0–60.0
Napthenes	49.0	30.0–60.0
Aromatics	15.0	3.0–30.0
Asphaltics	6.0	Remainder

Paraffins or alkanes are saturated hydrocarbons described by the general formula, C_nH_{2n+2}. The group comprises both straight and branched chained alkanes. In crude oil, they generally have from 5 to 40 carbon atoms per molecule, although trace amounts of shorter or longer molecules may be present in a mixture. The alkanes from pentane (C_5H_{12}) to octane (C_8H_{18}) are the main components of petrol, the ones from nonane (C_9H_{20}) to hexadecane ($C_{16}H_{34}$) are refined into diesel fuel and kerosene which is the primary component of many types of jet fuel. The ones from hexadecane upwards are the main components of fuel oil and lubricating oil. At the heavier end of the range are paraffin wax and asphalt. The former is alkane with approximately 25 carbon atoms while the latter has 35 carbon atoms or more. They are usually cracked in modern refineries into more valuable products. The shortest molecules, those with four or fewer carbon atoms such as methane, ethane, propane and butane, are in a gaseous state at room temperature. These categories of molecules, also called petroleum gases, are flared off, sold as liquified petroleum gas under pressure, or used to power a refinery's own burner during production. The option to take depends on the demand for the product and the cost of recovery.

The next group of hydrocarbons found in crude oil, the naphthenes or cycloalkanes, are saturated hydrocarbons that have one or more carbon rings to which hydrogen atoms are attached. Cycloalkanes are described by the general formula, C_nH_{2n}. Besides the boiling points which are higher than that of the alkanes, cycloalkanes have similar properties to that of the alkanes. Aromatics are unsaturated hydrocarbons that have one or more benzene rings to which hydrogen atoms are attached according to the formula C_nH_n. Many of these aromatic molecules have a sweet aroma, however, they all burn with a sooty flame. Finally, the asphaltics are aromatic-type substances, which are classified on the basis of their solubility. The asphatenes are soluble in carbon disulfide but insoluble in light alkanes such as pentane or heptane. Asphatene molecules carry a core of attached, flat sheets of fused aromatic rings linked at their edges by chains of aliphatic and/or naphthalenic-aromatic rings.

9.2 Classification and Characterization of Crude Oil

Each crude oil variety has a unique mix of molecules, which define its physical and chemical properties, for instance, color and viscosity of the crude, and ultimately its market value. Market forces use three indicators to classify crude oil. These are the geographical location of its production, its specific gravity according to the American Petroleum Institute's (API) system and its sulfur content. In addition to these, the unique molecular characteristics of crude oil are determined by molecular assay analysis. The location where the crude oil is assayed and classified is used as a reference in pricing the oil. The geographic location is important because it affects the cost of transportation of the oil to its destination. Crude oil is termed light if it has low density or heavy if it has high density. According to the API system, each crude oil is given an API degree rating in relation to its specific gravity compared to that of water. On this scale, water is rated $10°$ API. Crudes lighter than water have a higher API. Crudes with API less than $10°$ API are so heavy that they will sink in water. All other things being equal, lighter crudes are more valuable as they are easier to refine and produce more petrol than heavier crudes. The higher the percentage of sulfur present in crude, the less attractive the crude oil is because of its corrosive qualities in refinery equipment and because of the environmental regulations that exist in many markets. Sulfur when present in fuel burns to produce emissions of hydrogen sulfide and the oxides of sulfur that pollute the environment. Generally, the sulfur content of crude can vary from as little as 0.1% (wt) to as high as 7.0% (wt). When the sulfur content is low the crude is termed sweet and has a high market value. On the other hand, when the sulfur content is high, it is termed sour and has a low market value.

Crude may also be characterized by its thermal properties. These characteristics include heat of combustion, thermal conductivity, specific heat and latent heat of vaporization, which are explained below.

i. Heat of combustion (Q_v)

$$Q_v = 12,400 - 2,100d^2 \qquad (9.1)$$

where Q_v is measured in cal/gram and d is the specific gravity at 16 °C.

ii. Thermal conductivity (K)

$$K = \frac{0.813}{d}[1 - 0.0203(t - 32)] \tag{9.2}$$

where K is measured in BTU $hr^{-1}ft^{-2}$, t is measured in °F and d is the specific gravity at 16 °C.

iii. Specific heat (c)

$$c = \frac{1}{\sqrt{d}}[0.388 + 0.00045t] \tag{9.3}$$

where c is measured in BTU/lbm-°F, t is the temperature in Fahrenheit and d is the specific gravity at 16 °C. In units of kcal/(kg·°C), the formula is:

$$c = \frac{1}{\sqrt{d}}[0.402 + 0.00081t] \tag{9.4}$$

where the temperature t is in Celsius and d is the specific gravity at 15 °C.

iv. Latent heat of vaporization (L)

$$L = \frac{1}{d}[194.4 - 0.162t] \tag{9.5}$$

where c is in kcal/kg, the temperature t is in Celsius and d is the specific gravity at 15 °C.

9.3 Crude Oil Refinery

By itself, crude oil has little commercial value. However, when separated into its various hydrocarbon components the resultant products can be put to a host of commercial uses. The process of turning crude oil into useful commercial products is termed 'refining' and is carried out in very complex plants called *oil refinery*. Prior to the refinery process, water and sediments (particulates) that accompany the oil from oil wells are removed by sedimentation. Sedimentation is carried out at or near the wellhead or at a refinery where the oil is transported in huge tanker ships or pipes into huge holding tanks. Water and contaminants settle and separate out of the crude, which can subsequently be refined into its components. The crude oil refinery is carried out in three main steps. In the first step, the oil is separated into the various hydrocarbon components in a process called *distillation*. The second step transforms the products from the first step into other products that are commercially more useful. This step is called *conversion*. The conversion may be chemical or physical. Almost all the products from the distillation process must undergo some form of conversion before use. In the final step, also called *purification*, hydrogen sulfide from the distillation and conversion steps are converted to sulfur. Not all refineries operate this last step even though in most countries regulation mandates this, in part to protect the environment from hydrogen sulfide pollution. However, the economics of most plants also make this last step necessary as the sulfur is usually sold to fertilizer manufacturers.

A crude oil value is related to the yield of valuable products each barrel will produce. Indeed, tens of thousands of products can be obtained from crude oil as a result of the refinery process. The number of economic products that can be extracted from crude oil depends on the quality of the oil and the configuration of the refinery (processing facilities of the refinery) plant. Generally, three groups (fractions) of products are obtained from the refinery process (see Table 9.3). These are the *light fraction* made up of refinery gas, LPG, gasoline, aviation fuel, motor fuel and naphtha; *middle distillates*, i.e. a medium fraction, are mostly kerosene,

Table 9.3 Major refinery products

Product	Composition	Application
Bitumen (asphalt)	Alkanes >C34	Road surfacing, roofing, hydraulic uses, printing ink, electrical cables, duplex paper manufacture
Paraffin wax	Alkane >C25	Candles, insect bait, explosives and pyrotechnics, medicine, foods
Lubricating oil	Alkanes >C16	Lubricating engines, machinery parts
Fuel oil	Alkanes >C16	Heating, furnace fuel
Aviation fuel	Alkanes of C9–C16	Aviation/jet engines
Diesel	Alkanes of C9–C16	Diesel generators, vehicular diesel engines
Kerosene	Alkanes of C9–C16	Domestic lighting, aviation engines
Gasoline	Alkanes of C5–C8	Ignition engnines for vehicles
Refinery gas	Alkane <C5	Refinery burners/generators
LPG	Alkane of C3–C4	Domestic heating, and ignition engines of vehicles
Petrochemicals	–	Plastics, detergents, solvents, fibers, monomers, elastomers, paints, lacquers, fertilizers, insecticides and pharmaceuticals
Petroleum coke	–	Fuel, steel manufacture, heat protection, electrodes
Sulfur	Elemental sulfur	Fertilizer, insecticides, fibers, steel manufacture, heat protection, electrodes, personal care products, cosmetics, parmaceuticals

lubricating oil, light gas oil, heating oil and diesel oil; and the *heavy fraction* or bottom residue is made up of fuel oil for power stations and boilers, asphalt and bitumen, and waxes. The quantities of the light, medium and heavy fractions vary considerably from one crude oil to the other and depend on the source and grade of the crude oil.

9.3.1 Distillation of Crude Oil

Distillation is the first step in the crude oil refinery process. It is also generally called the *primary refining process* because it is the first in the series of processes that finally deliver the refined crude product. Distillation separates the naturally occurring hydrocarbons in crude oil into a number of different cuts or fractions. A fraction is a mixture of hydrocarbons whose boiling points fall into a narrow range. Distillation is based on the principle that when a mixture of different components is heated, the vapor that emerges from the mixture will be richer in the component with the lower boiling points. Therefore, when this vapor is condensed the condensate will be richer in the lower boiling point component. The residual liquid in the mixture will therefore be poorer in the lower boiling point liquid and richer in the higher boiling point liquid. When the mixture is made up of more than two components, it is possible to control the passage of the vapor through a long column, generally called a *fractionating tower*, such that the different liquids will condense at different levels along the length of the tower.

During the distillation, the crude oil is first pre-heated (heated in a furnace) to a pre-determined temperature and subsequently fed into a fractionating tower. The tower contains several trays which are suitably spaced and fitted with vapor inlets and liquid outlets. An illustration of a fractionating column used in the distillation of crude oil is shown in Fig. 9.1. The pre-heated crude oil upon entering the fractionating tower separates into vapor and liquid phases. The vapor passes up the column, and the liquid flows down to the bottom of the column where it is drawn off as a bottom residue. As the vapor moves along the length of the column it becomes cooler higher up the column and partially condenses into a liquid that collects onto the trays. The liquid collections on the trays are drawn off as products, a fraction of a mixture of hydrocarbons, whose boiling points fall into a very narrow range. Excess liquid on a tray overflows and passes through the liquid outlets onto the next lower tray. The bottom of the column is kept very hot but temperatures gradually reduce toward the top so that each tray is a little cooler than the one below it. Hot vapors moving up the column mix with cooler liquids flowing down the column on the trays. The result of this activity is that a temperature gradient is established throughout the length of the column. When a fraction reaches a tray where the temperature corresponds to its own particular boiling range, it condenses and changes into a liquid. Consequently, it is possible to separate the different fractions from each other on the trays of the column.

The fractions that form at the top of the column are called the light fractions while the fractions that form on the lowest trays of the column are called the heavy

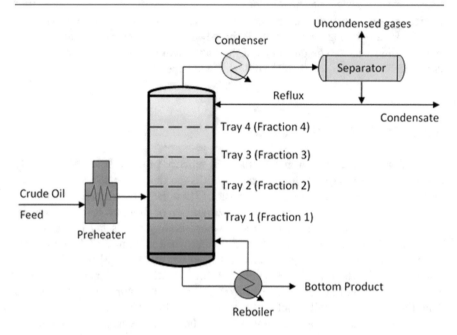

Fig. 9.1 An illustration of fractional distillation

fractions. The refinery gas refers to the very lightest fraction drawn from the top most trays of the column. It is often used as a fuel in refinery furnaces. Next in order of volatility comes gasoline (used for making petrol) followed by kerosene, light and heavy gas oils, and finally the bottom residue in that order. The fractions from the distillation unit are called *straight runs*. They range from gas, gasoline and naphtha, to kerosene and light diesel, and to lubricating oil and bitumen or asphalts at the bottom. In the case of some crude oils, the diesel fuel and heavy fuel fractions from the distillation process can be marketed directly. For most of the products from the distillation process, however, the quality has to be improved or transformed into other products before marketing. These changes are carried out in subsequent refinery processes.

9.3.2 Conversion of Straight Runs

All the products from the distillation process must be treated or modified to improve the quality and value before use or sale. The modification may be to purify the product to rid it of obnoxious compounds or elements as for instance, in the extraction of compounds of sulfur from a product. Such obnoxious impurities may be harmful to refinery equipment, the environment or may violate statutory regulations. Other impurities, besides sulfur, that are often removed from crude oil include nitrogen, oxygen and salt as well as small quantities of metals such as vanadium and nickel, which are found in certain types of crude oil. Some crude also contains arsenic and chlorine in small quantities, which may be removed. Sometimes, rather than extract an unwanted substance from crude, it may be necessary to sweeten the crude by adding special compounds to react with an inherent corrosive or a bad-smelling chemical to produce a harmless and odorless product.

Conversion of crude oil may be a chemical modification that converts an intermediate product, as in, for instance, the conversion of straight-chain paraffins into branched chains to increase the octane rating of gasoline. In certain instances too, products containing long-chain hydrocarbons are broken down into shorter chains to increase the value of the product, as in the case of the cracking of the bottom residues into gasoline and LPG. Whatever the nature of the modification, the object may be one of three, namely, to make an otherwise useless product useful, to improve the quality of a product and make it commercially more useful, or to purify a product to make it desirable and in order to meet an environmental regulation. The conversion processes include desulfurization, catalytic cracking, thermal cracking, hydrocracking, catalytic reforming, polymerization, alkylation, bitumen manufacture, crude sweetening, lube oil formulation and blending, among others. Table 9.4 lists some of the major crude oil modification processes.

Table 9.4 Major crude oil modification processes

Conversion	Description of process	Final product
Desulfurization	Amine absorption, hydrotreatment, caustic wash, Merox process	Sweetened fuel
Catalytic cracking	Catalysts are used to breakdown long-chained hydrocarbons	Light fuel products
Thermal cracking	Heat energy is applied to break down long-chained hydrocarbons	Light fuel products
Hydrocracking	Steam is employed to break down long-chained hydrocarbons	Light fuel products
Catalytic reformation	Platinum catalysts are used to convert naphtha boiling range molecules into into branched, cyclic and aromatic compounds (increase the octane numbers of gasoline)	High octane rating gasoline
Crude sweetening	Sulfur compounds in straight runs are converted into less objectionable forms that remain in the fuel	Sweetened fuel
Blending	Fractions are mixed or additives are added to specific products to improve quality	Improved product
Polymerization	Hydrocarbon monomers are polymerized	Polymers
Alkylation	Branched chain are introduced into hydrocarbons in crude oil	High octane-rating product
Bitumen manufacture	Vacuum distillation of bottom residue	Butimen
Lubrication oil formulation	Distillation, removal of aromatics, dewaxing and finishing of the base oil, and addition of additives	Lubrication (or lube) oil

9.3.2.1 Desulfurization

Literally, all the intermediate products from the distillation unit contain sulfur which must be removed from the products before use. Sulfur is present in the fuel as hydrogen sulfide (H_2S), mercaptans (organic sulfur compounds, i.e. R-SH) or sulfur dioxide (SO_2). When present in a fuel product, these sulfur-based compounds are undesirable for four reasons. One, they are corrosive and therefore harmful to refineries and other equipment. Secondly, fuel containing sulfur when burnt releases sulfur dioxide (SO_2) into the environment which is undesirable as it adversely affects communities, flora and fauna in diverse ways. Thirdly, sulfur poisons the catalysts that are used in the other treatment process following the fractionation. Finally, when present in the air in significant quantities hydrogen sulfide and sulfur-containing organic compounds in oil has a very undesirable smell. However, the sulfur in crude oil could be extracted and sold in commercial quantities and thereby improve the economy of a refinery plant. For these reasons, the compounds of sulfur present in crude are almost always extracted or reduced to insignificant levels before the products are finally used or marketed. Another way to deal with the sulfur in crude oil is to convert it into less objectionable forms in the product before use. Removal of sulfur through the desulfurization processes can add significance to the refinery costs, both financially and in terms of energy consumption. Indeed, the sulfur content of crude oil is one of the factors that determine the configuration of the refinery that can be used to process it.

There are two main protocols for removing sulfur-containing compounds from crude oil products. When present as hydrogen sulfide or sulfur dioxide it is removed by *absorption* into an appropriate medium. However, when present as mercaptan it is removed in a process called *mercaptan oxidation* (*Merox*) during which the sulfur compounds are first oxidized before extraction. A third approach to dealing with the sulfur in the fuel is the *Milnak process* that converts the sulfur into a less objectionable product that remains in the product.

i. Removal of the sulfur present as H₂S gas by absorption

The refinery fuel gas of the fractionation process often has associated H_2S which must be removed before burning the fuel. Otherwise significant quantities of SO_2 would be produced from the combustion of the fuel. When the SO_2 escapes into the environment it causes a lot of environmental nuisance. In particular, it combines with moisture in the atmosphere to cause undesirable acid rain. The H_2S in the refinery fuel gas is removed by absorption during which the fuel gas is contacted with a medium with a component that reacts with the H_2S and thus removes it from the fuel. Aqueous alkanol amine solutions are the frequently used media for the absorption process. Two commonly used alkanol amines in refinery industries are di-isopropanol amine and diethanol amine. A number of processes based on these alkanols are in existence in the industry. One such protocol is the Shell ADIP process which uses an aqueous medium containing di-isopropanol amine (DIPA) that reacts with the H_2S in the fuel upon contacting the streams containing the two different substances. In practice, the two streams are brought into contact in an absorption tower in a counter-current process. The ADIP solution flows down from the top of the tower while the fuel gas flows upwards from the bottom of the tower. Inside the tower are perforated trays which facilitate intimate contact between the fuel gas and the amine solution and thereby enhancing the reaction between the amine and the H_2S. The product of the reaction, a tetra ammonium cation is transferred into the aqueous ADIP solution and in this fashion removes the H_2S from the fuel gas. The reversible reaction between the H_2S and the amine occurs as follows:

$$H_2S \; + \; CH_3CHCH_2 \quad CH_2CHCH_3 \; \rightleftharpoons \; HS^- \; + \; CH_3CHCH_2 \quad H$$

| Hydrogen sulfide | Di-isopropanol amine (DIPA) | Quaternary ammonium salt |

$$(9.6)$$

A flowchart for the Shell ADIP process is shown in Fig. 9.2.

The Shell ADIP process is a regenerative absorption process. As such it enables the regeneration and recirculation of the amine solution. The process consists of two main steps, the H_2S absorption steps and the amine regeneration step. The concentration of DIPA in the amine solution to the absorption tower is about 4 mol/L. The H_2S concentration in the recycled solution (lean ADIP solution) is typically about 80 ppm (wt). The absorption occurs at high pressures and low temperatures. Under these conditions, the forward reaction is favored (i.e. the extraction of the H_2S is favored). The fuel gas leaves at the top of the adsorption tower depleted of H_2S and are ready to be used. The bottom product, the now quaternary ammonium salt-laden aqueous solution, is sent to the regeneration tower where at high temperatures and low pressures the reaction is reversed so that the amine solution is regenerated and recycled back as a lean ADIP solution to the absorption tower.

The H_2S generated at the regeneration tower is sent to a sulfur recovery plant to be processed into sulfur. The reaction between the H_2S and DIPA is an exothermic process, therefore the ADIP solution leaving the absorber has a relatively high temperature. It is therefore used to heat the lean ADIP solution recycled to the absorber.

The protocol using diethanol amine (DEA) is in many respects similar to the di-isopropanol process except in place of DIPA, and DEA is used in the aqueous medium. The DEA-based process is also a regenerative absorption process and removes CO_2 and SO_2 that might be present in the gas alongside the H_2S. The reactions occurring during the absorption process are as follows:

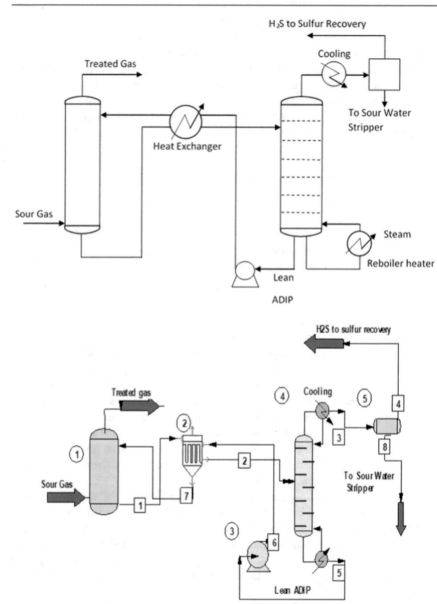

Fig. 9.2 A schematic of the Shell ADIP process

(a) The reaction with H_2S

$$H_2S \ + \ 2R_2NH \rightleftharpoons (R_2NH_2)_2S \tag{9.7}$$

$$(R_2NH_2)_2S \ + \ H_2S \rightleftharpoons 2R_2NH_2HS \tag{9.8}$$

(b) The reaction with CO_2

$$2R_2NH \ + \ H_2O \ + \ CO_2 \rightleftharpoons (R_2NH_2)_2CO_3 \tag{9.9}$$

$$(R_2NH_2)_2CO_3 + H_2O + CO_2 \rightleftharpoons 2R_2NH_2HCO_3 \tag{9.10}$$

or

$$2R_2NH \ + \ CO_2 \rightleftharpoons R_2NCOONH_2R_2 \tag{9.11}$$

An example of the DEA process is used in Tema Oil refinery in Ghana. At the Tema plant, the sour gas first flows through a lean gas knock-out drum to remove any entrained liquids before being routed to the absorption tower. The sour gas enters the absorption tower at a pressure of 11.5 kg/cm^2g (1127.8 kPa) where it is contacted with a 20% (wt.) DEA solution at 55 °C. The treated fuel gas emerging from the tower may contain entrained DEA solution which is removed in a treated gas knockout drum. The DEA-rich solution leaves at the bottom of the absorber and is first treated in a fresh drum to remove light hydrocarbons absorbed into the solution before a subsequent treatment to remove heavy hydrocarbons in a gravity separator. The amine solution containing the absorbed H_2S and CO_2 is finally sent to the regenerator to recover the amine solution.

ii. Removal of the sulfur present as a mercaptan (R-SH)

When sulfur is present as an organic compound (generally mercaptans) it is usually dealt with in one of three ways. It may be converted into less objectionable disulfides and allowed to remain in the product in a process called *Merox sweetening* (also called *Milnak process*). Alternatively, it may be extracted from the fuel distillates by contact with an alkaline medium in the presence of a catalyst. This latter process is called *Merox extraction*. Finally, the sulfur-containing fuel may be hydro-treated with hydrogen to convert the organo-sulfur compound to hydrogen sulfide (H_2S) which is then subsequently removed as the liquid fuel product cools. The Merox extraction process was described in Chapter 8 under LPG processing. The Merox sweetening uses oxygen from the air to convert mercaptans in fuel distillates into organic disulfide compounds, which are allowed to remain in the product as these are less objectionable. Post-treatment is usually not required of the product after a Merox sweetening. The oxidation is facilitated by the use and presence of an organometallic catalyst. The oxidation takes place in an alkaline environment and the catalysts often employed are iron group metal chelates. The reaction proceeds at an economically normal rate at normal temperatures for refinery run-down products. For light stocks, the operating pressure is kept slightly above the bubble point to assure liquid-phase operation. For heavier stocks, the operating pressure is normally set to keep air dissolved in the reaction medium. The oxidation proceeds according to the following reactions:

$$\underset{\text{Mercaptan}}{2RSH} + \underset{\text{Oxygen}}{1/2O_2} \xrightarrow{\text{NaOH}(aq)} \underset{\text{Disulfide}}{RSSR} + \underset{\text{Water}}{H_2O} \qquad (9.12)$$

Or

$$\underset{\text{Mercaptan}}{2R'SH + 2RSH} + \underset{\text{Oxygen}}{O_2} \xrightarrow{\text{NaOH}(aq)} \underset{\text{Disulfide}}{2R'SSR} + \underset{\text{Water}}{2H_2O} \qquad (9.13)$$

In the above reactions, R and R′ may be straight, branched or cyclic hydrocarbon chains. The chain may be saturated or unsaturated.

The use of an aqueous medium serves a dual purpose during the process. One, it facilitates the transfer of the mercaptan to the aqueous phase. Secondly, an aqueous environment is needed for the reaction to proceed in the desired direction. A flowchart of the Merox sweetening process is shown in Fig. 9.3.

During operation, the air and caustic are injected into the hydrocarbon feed stream, which then enter the reactor downflow. The sweetened product, essentially freed of entrained caustic, is withdrawn directly from the side near the bottom of the

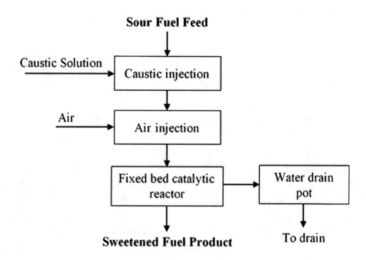

Fig. 9.3 A schematic of the Merox sweetening process

reactor. The aqueous caustic phase coalesces, collects at the bottom and is discharged to a sewer or a spent caustic disposal system.

For successful operation of the Merox sweetening process, substances (impurities) that can potentially adversely affect the activity of the catalyst should be removed before feeding to the catalytic bed. Elemental sulfur, hydrogen sulfide and carbon dioxide are some of the more common substances found in the feed which have also been identified to detract from the activity of the catalyst. They may also cause the degradation of the treating chemicals and solvents if not removed from the feedstock. Elemental sulfur, once formed, is very difficult to remove. In industry, therefore, every precaution is usually taken to prevent its formation. Hydrogen sulfide concentration in the Merox feedstock must not exceed 5 wt-ppm as the system cannot handle it. Therefore, when the H_2S concentration in the feedstock is greater than 5 wt-ppm, it is usually removed by pre-washing with a dilute caustic solution before entry into the reactor. Dilute caustic solutions are preferred so as to prevent the precipitation of sodium sulfide crystals. Hydrogen sulfide removal by this method is only economically viable if the hydrogen sulfide concentration does not exceed 100 wt-ppm. Carbon dioxide in the feedstock is often removed alongside H_2S during caustic washing.

iii. Desulfurization by hydrotreatment of distillates

Another approach to removing sulfur compounds present in gasoline, kerosene, diesel oils and naphtha is by reacting the sulfur compounds in the distillate with hydrogen. The process is called *hydrotreatment* and forms hydrogen sulfide as a product, which is then removed as a gas from the cooled oil liquid. The hydrotreatment process includes facilities for the capture and removal of the resulting hydrogen sulfide (H_2S) gas, which is then subsequently converted into by-product elemental sulfur or sulfuric acid (H_2SO_4). Data from 2005 indicate that a greater percentage of the 64,000,000 metric tons of sulfur produced worldwide was by-product sulfur which was used in hydrocarbon processing plants.

The process is carried out over a cobalt-molybdenum catalyst, which is selective for sulfur removal at a pressure of about 20 atmospheres and a temperature of about 350 °C. The overall reactions occurring can be simplified to:

$$2H_2 + R-\underset{\overset{|}{H}}{C}{=}S \rightleftharpoons R-CH_3 + H_2S \tag{9.14}$$

$$H_3C - CH_2 - CH_2 - CH_2 - SH + H_2 \rightleftharpoons CH_3CH_2CH_2CH_3 + H_2S \tag{9.15}$$

The above reactions are both reversible. The direction of equilibrium depends on the conditions of the reaction and is determined by Le Chatelier's principle. The forward reaction (desired direction) is favored by high pressures and by using an excess of hydrogen feed. On the other hand, at high temperatures, the hydrocarbons decompose to lighter hydrocarbons and finally to carbon. The catalyst, after operating for a while, does become deactivated by laydown of carbon. The catalyst can be regenerated by passing air and steam through the catalyst, a process that results in the burning away of the carbon from the surface of the catalyst. The use of pure air alone is not recommended as it leads to uncontrolled combustion, and the resulting high temperature adversely affects the catalyst.

9.3.3 Sulfur Recovery Unit

Sulfur recovery from refinery plants serves a number of purposes. First, the sulfur recovered can be sold to fertilizer manufacturers as sulfur is a major ingredient of many fertilizer products. The economics of the refinery plant could be significantly improved by the recovery and sale of sulfur. The sulfur if unrecovered ends up in the environment where it is harmful to flora and fauna and surrounding communities. Also, regulatory agencies require that refinery plants reduce the sulfur contents of their emissions to an acceptable level of about 5 ppm. For these reasons, sulfur recovery is a very important process in refinery plants.

As previously discussed, sulfur occurs in crude oil principally as sulfur dioxide, hydrogen sulfide and as mercaptans. Following the crude distillation, the reforming processes that convert the straight runs into usable products also produce hydrogen sulfide and small amounts of sulfur dioxide as by-products. These are the sources of hydrogen sulfide from which the sulfur is recovered. The process takes place in a sulfur recovery unit and is based on the Claus principle according to which hydrogen sulfide is converted into elemental sulfur in a two-step process by reaction with oxygen. The two reactions that convert the hydrogen sulfide into elemental sulfur proceed as follows:

$$H_2S + \frac{3}{2O_2} \rightarrow SO_2 + H_2O \tag{9.16}$$

$$2H_2S + SO_2 \rightarrow 3S + 2H_2O \tag{9.17}$$

The overall reaction proceeds according to Eq. (9.19).

$$H_2S + 1/2O_2 \rightarrow S + H_2O \tag{9.18}$$

Equation (9.17) is accompanied by the oxidation of any ammonia and hydrocarbons in the hydrogen sulfide stream according to Eqs. (9.20) and (9.21), and dissociation of a small quantity of the hydrogen sulfide according to Eq. (9.22).

$$C_nH_{2n+2} + (3_{n+1})/2O_2 \rightarrow nCO_2 + (n+1)H_2O \tag{9.19a}$$

$$C_nH_{2n+2} + (2_{n+1})/2O_2 \rightarrow nCO + (n+1)H_2O \tag{9.19b}$$

$$C_nH_{2n} + (3_n/2)O_2 \rightarrow nCO_2 + nH_2O \qquad (9.19c)$$

$$C_nH_{2n} + nO_2 \rightarrow nCO + nH_2O \qquad (9.19d)$$

$$NH_3 + 3/2O_2 \rightarrow 1/2N_2 + 3/2H_2O \qquad (9.20)$$

$$H_2S \rightarrow S + 1/2H_2 \qquad (9.21)$$

In practice, the sulfur recovery unit is made up of a thermal reactor and two catalytic reactors, which are inter-spaced with condensers. A schematic flowchart of the process is shown in Fig. 9.4.

In the first of the three stages of the Claus process, enough air is supplied to the thermal reactor (T1) to convert about a third of the hydrogen sulfide into sulfur dioxide as described in Eq. (9.17). The sulfur formed in the thermal reactor (see Eq. 9.22) is condensed and drained from the process gas stream into the sulfur pit. The two catalytic reactors follow the thermal reactor sequentially. Each of the catalytic reactors in turn is followed by a condenser. These latter units are provided for further conversion of the remaining hydrogen sulfide and sulfur dioxide to elemental sulfur. The sulfur in a liquid state is channeled into the sulfur pit from where it is sent to storage.

After the second catalytic reactor, the sulfur content of the process stream is less than 5%. This sulfur is oxidized in the catalytic incinerator at a temperature of approximately 400 °C. This high temperature is achieved by burning fuel gas in addition to the process gas. The hydrogen sulfide and the sulfur vapor in the process stream are at this stage oxidized almost completely into water and sulfur dioxide. The reactions going on in the incinerator are represented by Eqs. (9.23) and (9.24).

$$H_2S + O_2 \rightarrow H_2O + SO_2 \qquad (9.22)$$

$$S + O_2 \rightarrow SO_2 \qquad (9.23)$$

Finally, before releasing into the atmosphere, the gas mixture from the incinerator is first passed through a sulfur dioxide scrubber where the sulfur dioxide is removed into a solution.

9.3.4   Catalytic Reforming

Catalytic reforming represents one of the most important processes in the modern refinery. Through reforming it is possible to alter straight runs (intermediate distillates) into new compounds through a combination of high temperatures and pressures in the presence of appropriate catalysts. The most important of the reforming reactions is the production of high-quality gasoline. In this process, gasoline from the straight run is converted into high octane rating gasoline by conversion of straight-chain paraffins in the gasoline into branched-chain compounds.

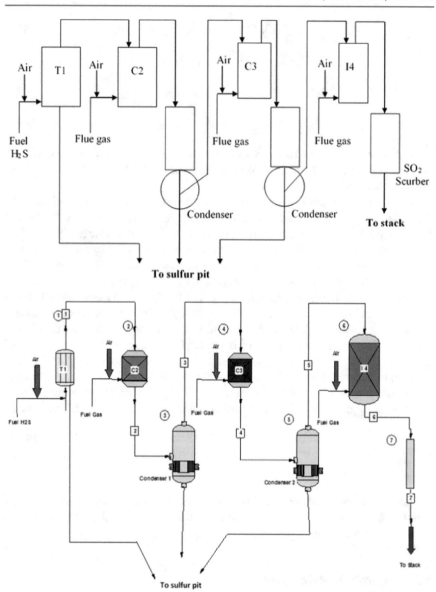

Fig. 9.4 The sulfur recovery unit (T1 = Thermal reactor; C2 = Catalytic reactor; C3 = Catalytic reactor; I4 = Incinerator)

When an unsuitable fuel is used in an internal combustion engine, 'knock' and 'pinking' occur in the engine. The extent or measure of the 'knock' and 'pinking in the engine is called the octane number of the fuel. Continual operation of an engine in which such knocking is occurring causes rapid destruction of bearings. To avoid this, fuels with high octane ratings are the preferred ones for use in internal combustion engines. The octane number is defined by an arbitrary scale which allocates zero octane number to *n*-heptane and 100 to 2,2,4-trimethyl pentane (*iso-octane*). The octane number of a fuel is defined as the percentage by volume of *iso-octane* in a mixture with *n*-heptane that gives the same degree of knocking as the fuel in question in a special test engine. Octane numbers of hydrocarbons vary with their oxidation stability. Straight-chain paraffins have the lowest octane numbers (e.g. *n*-butane has an octane number of 25) with branched-chain paraffins being higher (2-methyl pentane has an octane number of 70) followed by naphthenes (cyclohexane has an octane number of 83) and then aromatics (the octane rating of benzene is 100). What this means is that reforming conditions are created to convert straight-chain groups in gasoline into branched, cyclic or aromatic groups and thus increase the octane number of the gasoline.

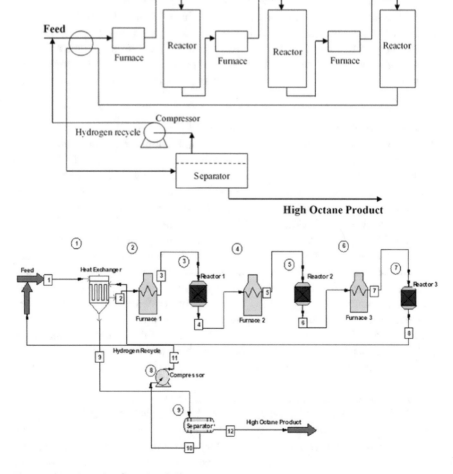

Fig. 9.5   A schematic of a reforming unit

Reforming of gasoline takes place at a temperature of about 500 °C and a pressure of 25 atmospheres in the presence of a platinum catalyst. The process is endothermic and is therefore favored by the high temperatures. A schematic of the reforming operation is shown in Fig. 9.5.

The important features of the reforming process are: (i) the use of three furnaces to provide the heat needed for the endothermic reaction; (ii) the use of the product heat to preheat the feed prior to entry into the first furnace; (iii) hydrogen recycle to the feed to provide the excess hydrogen required to prevent coking resulting from the high temperatures of the reaction; (iv) the deactivation of the catalyst with time as a result of coke lay-down on the catalyst; and (v) reactivation of the catalyst by periodic removal of coke by passing a mixture of air and nitrogen through the catalyst, thus burning off the coke. A high temperature also ensures that the reactions occur at a reasonably fast rate. However, a high temperature favors the cracking reactions which are undesirable as it leads to a lay-down of carbon on the catalyst. To suppress coking, a high hydrogen pressure is used by recycling some of the light gases from the separator even though this adversely affects the formation of the desirable aromatic products. During the catalyst regeneration process, nitrogen is used to control the rate of burn-off, and if pure air is used the efficiency of the catalyst would be destroyed by the resulting high temperature. Notice that in contrast with the desulfurization process, steam is not used in this instance as water adversely affects the platinum catalyst.

9.3.4.1   The Fluid Catalytic Cracker (FCC)

The heavier fractions of the straight runs—the high boiling, high molecular weight hydrocarbons—namely fuel oils, bitumen, lubrication oil and waxes can be turned into lighter more valuable products by application of appropriate processes. Fluid catalytic cracking (FCC) is one such process where the heavier fractions are cracked into lighter products. Thermal cracking, hydrocracking and visbreaking are other processes used to break down heavier distillate fractions into lighter fractions. However, FCC is preferred compared to other cracking processes in part because it produces gasoline with high octane rating. Also, it produces by-product gases that are more olefinic and hence are more valuable than those produced by other cracking processes. In fact, the FCC is the most important conversion process used in the petroleum industry. The FCC uses a finely divided silica/alumina catalyst, high temperatures and moderate pressures to break down the heavier distillate fractions into lighter hydrocarbons. A unique feature of the FCC is that the active catalyst is suspended within the hydrocarbon phase and flows with it. (The FCC unit is a fluidized bed reactor.) A schematic of the FCC process is shown in Fig. 9.6.

The pre-heated high-boiling petroleum feedstock at a temperature of about 315–430 °C is combined with recycled slurry from the distillation unit and injected into the catalyst riser where it is vaporized and cracked into smaller molecules by contacting and mixing with the very hot catalyst from the regenerator. The mixture flows upward to enter the reactor at a temperature of about 535 °C and a pressure of about 1.72 bar. In the reactor, the spent catalyst is separated from cracked vapor

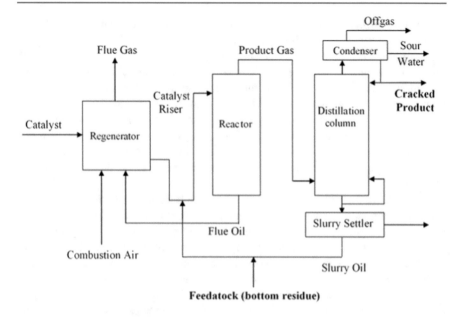

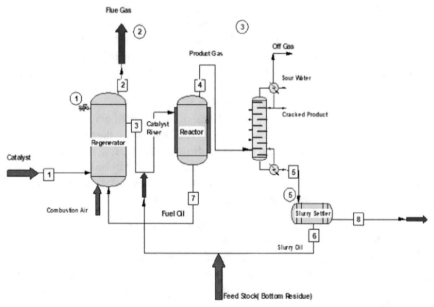

Fig. 9.6 A schematic of the FCC unit

products in a set of two-stage cyclones. The spent catalyst flows downwards through a steam stripping section to remove any hydrocarbon vapors before it is returned to the catalyst regenerator. The reaction products, mostly in the vapor phase, flow from the top of the reactor and are fed to the bottom of a distillation column where they are distilled into products of cracked naphtha, fuel oil and an off-gas LPG. The naphtha is further processed to remove sulfur and blended with other components to produce high octane gasoline. Many FCC units have two sidecut strippers and produce light fuel and heavy fuel. Similarly, the main fractionators produce light-cracked naphtha and heavy-cracked naphtha.

FCC units come in two basic types: the 'stacked' type and the 'side-by-side' type. In the stacked type, the reactor and the catalyst regenerator are contained in a single vessel with the reactor above the regenerator. The side-by-side type has the reactor and the catalyst regenerator occupying different vessels and located beside each other.

The FCC reactor only serves as a vessel in which the cracked product vapors are separated from the 'spent' catalyst. The cracking/reaction itself does not really occur in the reactor. It takes place in the catalyst riser where the catalyst first contacts and vaporizes the feed into vapor. At the prevailing conditions of temperature and pressure in the catalyst riser, the high molecular weight hydrocarbons in the feedstock get broken down into lighter products. Upon contact with the high temperature regenerated catalyst the hydrocarbon feed is vaporized and the vapor, in turn, carries the catalyst in its medium as it flows up the catalyst riser. Along the way, the hydrocarbons are broken down into smaller molecules. By the time the mixture reaches the reactor vessel, the reaction would have been completed.

The cracking reaction produces coke, a carbonaceous material, that deposits on the catalyst and causes the activity of the catalyst to deteriorate very quickly. Consequently, the catalyst must be reactivated through regeneration. The regeneration takes place in the catalyst regenerator which uses air to burn away the deposited coke at a temperature of about 715 °C. The combustion of the coke is an exothermic reaction and generates a great deal of heat; part of which is used to feed the cracking reaction which is endothermic. The heat of the combustion also provides energy for the vaporization of the feedstock. The catalyst is a finely divided powder of silica/alumina. The average particle size is 60–100 μm with a size distribution of 10–150 μm. The density of the catalyst is in the range of 800–900 kg/m^3. The catalyst has four main components:

i. A zeolite, which is the active component and comprises about 20% of the catalyst. The zeolite is a molecular sieve composed of silica and alumina. It does not allow molecules larger than 8–10 nm to enter its lattice.
ii. A matrix made up of amorphous alumina also provides catalytic activity, particularly in larger pores that allow entry for larger molecules than does the zeolite.
iii. The binder is usually a silica sol, which provides strength and integrity to the catalyst
iv. A filler, which serves a similar role to the binder.

Thermal cracking and visbreaking also use higher temperatures and pressures to break down the heavier hydrocarbons molecules into smaller molecules, except that visbreaking is capable of breaking much heavier molecules than thermal cracking. Hydrocracking, on the other hand, uses hydrogen to upgrade heavier fractions into lighter fractions.

9.3.5 Bitumen Manufacture

The term bitumen is used to refer to a non-crystalline solid or viscous material derived from petroleum either by natural or refinery processes. Bitumen is black or brown in color and has adhesive properties. It can be found in nature but is usually made as an end product during the distillation of selected crude oil. Natural asphalts are found in Trinidad, Venezuela, Cuba and some other countries. But natural asphalts have now been replaced by the refinery product as the latter is increasingly the preferred ones for many applications for which bitumen is required. In the USA, bitumen and asphalt are used synonymously. Outside of the USA, the word asphalt is used to refer to a mixture of bitumen and solid mineral matter. Bitumen has many uses, the chief among which is in road construction. Bitumen is valued because of its various properties. It is waterproof, ductile, adhesive, chemically inert and resistant to atmospheric exposure and the effects of acids and alkalis. The important quality indicators of bitumen are its penetration number which determines the hardness of bitumen at 25 °C, the softening point which determines the temperature at which the material softens under test conditions, flashpoint which determines the temperature at which sufficient vapor from the substance develops to give a flame flash and specific gravity which is used for quantity reckoning and identification purposes.

Bitumen is made from the bottom residue of crude oil fractionation. When this heavier residue is further distilled under vacuum and at lower temperatures in a second fractionating column, lighter fractions are separated and drawn off from the residue leaving a vacuum residue, which is removed from the bottom of the column. This residue is bitumen which for heavy crudes is a commercial product. However, for lighter crudes, the vacuum residue requires further processing before use. Such further processing includes air blowing, which consists of reacting the bitumen feedstock in a reactor at a temperature of about 220–300 °C. Oxidation and condensation reactions take place resulting in the formation of high molecular weight compounds, which give harder and less temperature susceptible bitumen. Another treatment is thermal conversion where temperatures up to 450 °C and high pressures are applied to bitumen feedstocks and followed by vacuum distillation to remove volatiles leaving a residue that can be used as a component of bitumen. Solvent precipitation is yet another treatment whereby solvents like propane or propane/butane mixture are used to dissolve the oil. In the process, the asphaltic fraction precipitates and is drawn off and used in the manufacture of bitumen.

9.3.6 Manufacture of Lubrication Oil

Lubrication oils are just one of the fractions that can be derived from crude oil during distillation. It is one of the middle distillates of the straight run and is also one of the products of the vacuum distillation of the bottom residue from the straight runs. Lube oil fractions from these two sources are often treated to rid them of impurities and subsequently blended with additives before use or sale. As a first step in the processing of lube oil, the fractions from the distillation process are passed through several ultrafine filters, which remove residual impurities from the oil. This is followed by solvent extraction to remove aromatics from the oil as these aromatics when present adversely affect the viscosity of the oil. Finally, to achieve the desired physical properties for the lubricant, additives are added to the oil. Common additives include lead and metal sulfides, which enhance the lube oil's ability to prevent galling and scoring when metal surfaces come in contact under extremely high pressures. High molecular weight polymers are also sometimes added to the lube oil to improve its viscosity and counteract the tendency of the oils to thin at high temperatures. Nitrosoamines are employed as antioxidants and common inhibitors by neutralizing acids and forming films on metal surfaces. Lube oil is characterized by its viscosity, specific gravity, color, flashpoint and fire point. The viscosity determines the oil's resistance to flow at a given temperature and pressure. There is too much resistance to the relative movement of metal parts if the oil is too viscous. On the other hand, if it is not viscous enough, it will be squeezed out from between the mating surfaces and hence not be able to lubricate them sufficiently. The lube oil's color indicates the uniformity of a particular grade or brand. The flashpoint indicates the temperature at which the lube oil will give sufficient flammable vapor when heated such that it will flash when brought into contact with a flame. The temperature at which the oil vapor continues to burn when ignited is called the fire point. The fire point temperature is higher than the flashpoint temperature and both are safety properties of the oil. The specific gravity of lube oil is a measure of the heaviness of the oil compared to water. It is only significant in determining the volumes and weight of the oil. The specific gravity of the oil depends on the refining method and the types of additives present in the oil.

Most applications of lube oils require that they be non-resinous, pale-colored, odorless and oxidation-resistant. Common lube oils are classified by viscosity and performance factors such as wear prevention, oil sludge deposit formation and oil thickening.

9.4 Refinery Waste Treatment

Crude oil refinery plants generate a lot of waste which necessarily must be treated before disposing of into the environment. The need to treat waste generated by refinery plants is dictated by a number of considerations. The principal among which is the requirement of environmental regulatory agencies is that the waste can only be disposed of into the environment if it meets certain criteria. When the criteria are not met sanctions are imposed on the plants, and in some cases, the plant may be closed down until measures are put in place to deal with the waste to warrant the plant resumption of operations. But the need to treat waste is sometimes born of a plant's self-interest. For instance, when waste is not treated it may affect the health of the workers of the plant which in turn would affect the production of the plant. Communities affected by the operations of a plant often agitate that measures be put in place to deal with the pollution generated by a plant which in turn brings pressure to bear on plant managers to institute waste treatment measures. Some international funding institutions and banks like the International Monetary Fund and the World Bank require, as part of funding arrangements to refinery plants, to build a waste treatment plant to treat wastes at plants funded by these agencies. Non-waste treatment of certain refinery wastes before disposal contributes to the general degradation of the environment, and general increased awareness of the need to protect the environment has caused refinery managers to react to treat these wastes to protect the environment. Finally, the activities of environmental pressure groups have made it difficult for plants to dump their waste into the environment without notice. Often, the pressure groups would harass polluting companies by use of tactics such as image denting picketing and public denouncements, and sometimes even by legal actions. The result is refineries would rather treat the waste they generate than attract the kind of adverse public exposure brought on their companies by environmental pressure groups.

The processes designed and operated to remove naturally occurring contaminants from the crude oil in refineries often produce wastes, including water, sulfur and compounds of sulfur, nitrogen compounds and heavy metals. The nature of the wastes generated at a refinery, therefore, depends primarily on the characteristics and source of the crude oil to be refined. However, the configuration of the refinery plant also plays a key role in the eventual waste that a plant churns out. Solids, liquid effluents and gaseous emissions are all generated at refinery plants and must be treated to meet the requisite standards before discharge into the environment. Environmental regulatory agencies in the countries where the plants are sited set the standards which the plant management is required to comply with. But there are internal standards and sometimes international standards that companies sometimes are required to abide by. The need for international standards arises especially where the plant was funded through loans contracted from international banks or funding agencies. For instance, the World Bank and the International Monetary Fund (IMF) often require plants with loans or grants provided by them to abide by certain guidelines. Table 9.5 lists the type, specific waste, source of wastes, and the methods of treatment of the wastes.

9.4.1 Wastewater Treatment Unit

Huge volumes of wastewater are produced on a daily basis in a refinery. The sources of wastewater in a refinery are many and varied. Many of the treatment

Table 9.5 Sources of wastes and methods of treatment in refinery plants

Type	Waste	Source	Method of treatment
Solids	Spent catalysts	Reforming, cracking and Merox units	Extraction of catalyst metals or dumping to land fills
	Sludge	API separation, chemical precipitation and biological units	Filteration followed by drying and sale or disposal
	Scrap metals	Container tanks, maintenance scraps	Sale to scrap dealers
	DAF floats and suspended solids	DAF unit and effluent ponds	Incinerated or dumped
Liquid effluents	Desalter effluent	Crude oil	Chemical treatment
	Oil water	RFCC, storm drain, crude storage, and product storage	API separator, DAF equalization basin
	Sour water	Residual fluid catalytic unit	Stripping/absorption
	Spent caustic	Neutralization tank	Disposal into drains
Air emissions	CO_2 and CO	Flaring, incinerator and boiler	Discharged into the environment
	H_2S and NH_3	Sour water treatment unit	Sulfur recovery unit or biological digestion
	NO_x and SO_x	Incinerator, boiler, flaring units	Absorbed in stacks
	Hydrocarbons	Fugitive emissions	Flared

processes require water to carry out and as such invariably produce wastewater after the valuable oil product has been extracted. The crude oil desalter, sour water stripping and amine treatment units are some examples of process units where water is extensively used. Also, crude oil and petroleum products storage units are another source of wastewater. During storage, the water separates out of the oil and must be separated and discharged. During rainfall, water collects oil from the contaminated land area of the plant, especially the loading bay which is channeled for treatment. Not all the wastewater in a refinery plant is contaminated with oil. For instance, sewage water and utility discharges may not at all be contaminated with oil yet requires to be treated before disposal. In view of the variety of sources of refinery wastewater, different process steps may be required for their treatment. Figure 9.7 shows a schematic flowchart of wastewater treatment protocols used in the oil industry. The major sources of wastewater in a refinery are listed in Table 9.6 together with the modes of treatment.

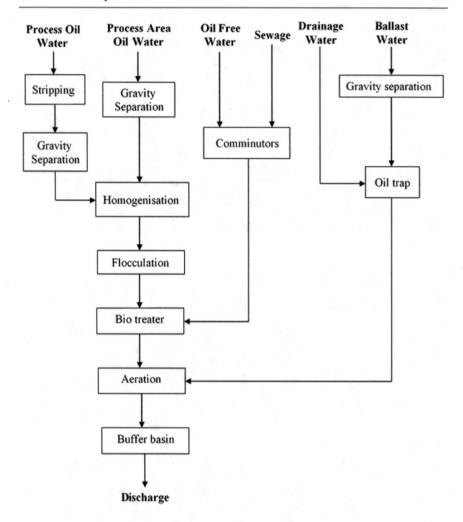

Fig. 9.7 A generalized flowchart of the wastewater treatment processes in the refinery

9.5 Other Refinery Processes

Other refinery processes that upgrade the straight runs from the fractionator before use are listed in Table 9.7.

Table 9.6 Refinery wastewater sources and modes of treatment

Type	Source	Mode of treatment
Major oily water	Process units and process areas	Sour water stripping/gravity separation, homogenization, flocculation, biotreatment, aeration, pH adjustment and finally discharge
Minor oily water	Drains	Oil separation, aeration, pH adjustment followed by discharge
Oil-free water	Utility units	Comminution, biotreatment, aeration, pH adjustment and finally discharge
Sewage	Domestic	Comminution, biotreatment, aeration, pH adjustment and finally discharge
Ballast water	Ships	Gravity separation followed by aeration, pH adjustment and then dicharge into drains

Table 9.7 Other refinery process

Refinery process	Definition
Vacuum distilation	Unit further distills residual bottoms after atmospheric distillation and produces additional light and middle distillates
Naphtha hydrotreater	Hydrogen is employed in an atmospheric distillation process to desufurize naphtha. The naphtha must be hydrotreated prior to processing in the catalytic reformer unit
Coking units	(Delayed coking, fluid coker and flexicoker) Process very heavy residual oils into gasoline and diesel fuel, leaving petroleum coke as a residual product
Alkylation unit	Produces high-octane components for gasoline blending
Dimerization unit	The process by which two olefins molecules are joined together by bonds into higher-octane gasoline blending components. For instance, isooctene can be formed from the dimerization of butenes. The isooctene can thereafter be hydrogenated to form isooctane
Isomerization unit	This process converts linear molecules to higher-octane branched molecules. The higher-octane molecules are blended into gasoline or used as feed to alkylation units
Claus unit	Converts hydrogen sulfide from hydrodesulfurization into elemental sulfur
Solvent refining	**This process uses solvents to remove asphaltenic and other unwanted materials from lubricating oil stock or diesel stock. The solvents often used in this process include creasol and furfural** Units to remove the heavy waxy constituents petrolatum from vacuum distillation products
Solvent dewaxing	Units remove the heavy waxy constituents petrolatum from vacuum distillation products

9.6   The Downstream Oil Industry in Ghana

The oil industry is generally grouped into three areas, namely the upstream industry which deals with the exploration and production of crude oil, the midstream industry which deals with the transportation and storage of crude oil, natural gas and natural gas liquids, and the downstream sector, also called onshore operations, which focuses on the processing of the crude oil and natural gas into consumable products and the distribution, and marketing of petroleum and gas products. In Ghana, the Ministry of Energy and Petroleum has a supervisory role in the sector. Its core duties in the sector include policy formulation, planning, monitoring and evaluating energy sector policies. Under the Ministry, the Ghana National Petroleum Company (GNPC) regulates the upstream sector of the oil industry. The National Petroleum Authority (NPA) is mandated to regulate, oversee and monitor activities in the petroleum downstream industry. The storage and transportation of oil and petroleum products are the preserve of the Bulk Oil Storage and Transportation Company Ltd (BOST). BOST has the mandate to develop a network of storage tanks, pipelines and other bulk transportation infrastructure throughout the country and to keep strategic reserve stocks for Ghana. It has the additional mandate as the Natural Gas Transmission Utility (NGTU) entity. In this latter role, BOST is expected to develop the natural gas infrastructure throughout the country. The key downstream sector players include Tema Oil Refinery (TOR), Ghana Gas Company Ltd and Platon Gas and Oil Ghana Limited. These companies refine crude oil and natural gas into various petroleum and gas products. Established in 1963, TOR is the leading refinery in Ghana. It has a production capacity of 45,000 barrels per stream day of petroleum products. The major petroleum products of the company are liquefied petroleum gas (LPG), gasoline (petrol), kerosene, aviation turbine kerosene (jet AI), gas oil (diesel), premix, naphtha, and fuel oil. TOR operates two major plants—a Crude Distillation Unit (CDU) and a Residual Fluid Catalytic Cracking (RFCC) Unit that has the capacity to produce and store 10,560 tons of LPG per day. Further, the company provides storage facilities in excess of about 1,000,000 tons for crude oil and finished petroleum products, which it also rents to third parties that import and market oil products into the country.

Platon Gas and Oil Ghana Limited was established in 2014. The company is able to process raw materials with a density range from 0.7 to 0.9 g/cm^3. Depending on the density it can process anything from 230,000 m^3 to 340,000 m^3 of raw materials per year. Its key products are diesel, RFO and naphtha. Both TOR and Platon Gas and Oil Ghana Ltd are situated in Tema. The other important downstream player in the sector is the Atuabo Gas Plant owned by the Ghana Gas Company Limited. The Atuabo plant situated at Atuabo in the western region of Ghana processes raw gas from the Jubilee oil fields of Ghana. The plant which started operations in 2015 has a capacity to process 120 million cubic feet of gas daily. The processed gas from the plant is mainly fed to the Volta River Authority plants to generate electricity for the nation. In addition to natural gas, the plant also has the capacity to produce 180,000

Table 9.8 Some bitumen and asphalt-producing companies in Ghana

Company	Capacity	Products	Location
Mustek Bitumen Limited	–	Bitumen	Accra
Locaf Industries Ltd	–	Bitumen emulsion	Abelenkpe, Accra
Platinum Seal Ltd	420 TPH bitumen plant;150 TPH hot mix asphalt palnt; 120 TPH mobile hot mix asphalt plant	Bitumen and Asphalt	Gomoa Akotsi, Central Region
Locaf Industries Ltd	10 tons/hour	Bitumen emulsion	Kotwi, Ashanti Region
Gelcon Ltd	200 tons/hour	Hot mix asphalt	Ojobi, Central Region

tons of liquefied petroleum gas (LPG), 46,000 tons of condensate and about 15,000 tons of isopentane annually.

Another player in the downstream sector of the industry in Ghana is Tema Lube Oil also situated in Tema, which blends and packages quality lubricants for the Ghanaian market and export. The company operates three plants, including an 80,000 kg per 8 h shift blending plant, a filling plant that packages blended products, and an 80 drum per hour steel drum plant that produces steel drums for the filling plant. The key products of the blending plant are engine oils, hydraulic oils, gear oils, marine oils and industrial oils. These are mostly packaged into steel drums, plastic bottles, plastic pails, integrated bulk containers (IBC) and bulk trucks at the filling plant. A number of companies operate bitumen and asphalt plants also in the downstream industry. Table 9.8 lists some of the companies operating in the sector.

Further Readings

Coker, A. K. (2018). *Petroleum refining, design and applications handbook* (Vol. 1). Wiley.

Fahim, M., Al-Sahhaf, T., & Elkilani, A. (2009). *Fundamemtals of petroleum refining* (1st ed.). Elsvier Science.

Fanchi JR, Christiansen RL (2017) Introduction to petroleum engineering. John Wiley and Sons, Ltd., Chichester, UK.

Favennec J. P. (Edn) (2001). Petroleum refining: refinery operation and management, Vol. 5. Editions Technip, Paris.

Gupta, O. P. (2019). *Elements of petroleum refinery engineering*. Khanna Publishers.

Kaiser, M. J., Gary, J. H., & Glenn, E. (2007). *Handwerk, petroleum refining: Technology and economics* (5th ed.). CRC Press.

Leffler WF (2008) Petroleum refining in nontechnical language, 4th edn. PennWell, Tulsa.

Meyers, R. (Ed.). (2016). *Handbook of petroleum pefining processes* (4th ed.). McGraw-Hill Education.

Nag A (2015) Distillation & hydrocarbon processing practices. PennWell, Tulsa.

Nasr, G. G., & Connor, N. N. E. (2014). *Natural gas engineering and safety challenges: Downsrteam process, analysis, utilization and safety.* Springer International Publishing.

Cocoa Processing and Chocolate Manufacture

10

Abstract

Chocolate is a vital ingredient in several foods and drinks. Chocolate and other cocoa products are produced from cocoa beans by different companies all over the world. Different types of chocolate are produced using different blends of ingredients, namely plain dark chocolate, milk chocolate and white chocolate. Many processing steps are required, and the procedures are complex. Some of the processes are carried out by the farmers namely cocoa harvesting, fermentation and drying of the cocoa beans. The refined cocoa beans are transported to the chocolate manufacturing industries for cleaning, roasting, grinding and other processing steps. The processing steps such as fermentation, drying and roasting influence the characteristics flavor and taste of the final chocolate. Quality control is essential in the chocolate industry. Consistent control and management of the cocoa butter composition, contamination and bacteria (*Salmonella*) across all areas of the process are required. However, each processing step generates a large amount of waste that causes environmental problems. This chapter covers the detailed processing methods required to produced chocolate and other cocoa products. The health and nutritional benefits of chocolate are included. Quality control in the cocoa processing and chocolate manufacturing industry and the environmental and pollution concerns have been covered.

10.1 Introduction

Chocolate is made from cocoa beans, which are grown on cacao trees (*Theobroma cacao* L.). It is believed that chocolate consumption originated in the Pre-Columbian societies of Central America some thousands of years ago, in about the seventh century A.D. During this era, cocoa beans were used as currency, and that the upper class of the native populations drank *cacahuatl,* a frothy beverage consisting of roasted cocoa beans blended with red pepper, vanilla and water. However, currently, the production and consumption of chocolate is a complex world trade network with an estimated global market worth of $98.3 billion in 2016.

Most of the major cocoa-producing countries come from warm and wet climates similar to where the bean originated. Surprisingly, the largest producing continent is Africa, but not the Americas where it originated from, with four of the top five nations found in Africa. As shown in Table 10.1, Ghana is the second-largest cocoa-producing country after Cote d'Ivoire.

In Ghana, the cocoa crop was introduced to the country by Tetteh Quarshie upon returning to Ghana in 1876 from the Island of Fernando Po (now Bioko in Equatorial Guinea) and planted on his farm at Akuapim Mampong. Although Tetteh Quarshie is credited with bringing cocoa into the then Gold Coast, it is in the record that Dutch missionaries planted cocoa in the coastal belt of Ghana as far back as 1815 (61 years earlier). Also, in 1857, the Basel missionaries planted cocoa in Aburi, the eastern region of Ghana. Following the cultivation of the plant in his farm, cocoa farming and its production spread from the eastern region to Ashanti, Brong Ahafo to central and the western regions of Ghana.

Table 10.1 Top cocoa-producing countries in the world in 2016 (reproduced with permission from WorldAtlas by Miklos Mattyasovszky. @2018 worldatlas.com)

Rank	Country	Production (tonnes)
1	Cote d'Ivoire	1,448,992
2	Ghana	835,466
3	Indonesia	777,500
4	Nigeria	367,000
5	Cameroon	275,000
6	Brazil	256,186
7	Ecuador	128,446
8	Mexico	82,000
9	Peru	71,175
10	Dominican Republic	68,021
11	Colombia	46,739
12	Papua New Guinea	41,200
13	Venezuela	31,236
14	Uganda	20,000
15	Togo	15,000
16	Sierra Leone	14,850
17	Guatemala	13,127
18	India	13,000
19	Haiti	10,000
20	Madagascar	9,000

Chocolate is a key ingredient in many foods and beverages such as milkshakes or chocolate drinks, candy bars, cookies and cereals. Complex procedures are required to produce chocolate. The process involves harvesting the cocoa, refining cocoa to cocoa beans and transporting the cocoa beans to the manufacturing factory for cleaning, roasting and grinding. Different companies over the world produce chocolate and other cocoa products and have developed different flavors over the years.

10.2 Cocoa Processing

10.2.1 Harvesting

Chocolate production starts with harvesting the cocoa in a forest. Cocoa beans grow in pods that sprout off the trunk and branches of cocoa trees (as shown in Fig. 10.1). The ripe cocoa pods (which are usually orange or yellow) are hack gently off the trees using hooked blades or cutlasses. The pods are collected and taken to a processing house or a gathering place on the farm. The beans are then removed from pods by either manually splitting the pods using knives, cutlasses or machetes, or by using a cocoa pod splitting machine. A pod can contain up to 50 cocoa beans. The fresh cocoa beans do not have the characteristic chocolate flavor and are not brown. The flavors and the brown color are produced during fermentation and drying of the pods.

Ripe and unripe pods on the plant

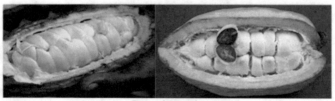

Open ripe pods

Fig. 10.1 Ripe and unripe cocoa pods

10.2.2   Fermentation

The beans are fermented, which enables chemical compounds to develop inside the beans, which are the precursors of the flavor in the final chocolate. Poor fermentation affects the quality of the beans, which cannot be rectified at a later stage. Also, poor control can cause molds, which results in very unpleasant flavored products, even if the beans are fermented correctly. After fermentation, pulps are removed from the beans before being dried. As shown in Fig. 10.2, the cocoa beans are removed from the pods and heaped in a pile and covered with plantain or banana leaves in small farms or wooden boxes in plantations for typically 5–7 days after which the pulp is removed. Platforms, baskets and trays are among the other used fermentation methods. The heap method results in better quality cocoa beans than the box method, due to a more uniform fermentation. Also, the box method affects the pH value, tannins and sugars contents, and causes the presence of purple beans. The platform method has a low fermentation rate and it can be applied adequately to beans, which require only a short fermentation period (2 or 3 days). Many factors such as the method, duration, pod storage and cocoa bean genotype influence the fermentation and lead to differences in the quality of the cocoa beans.

During the fermentation, natural yeasts and bacteria multiply in the pulp, causing the breakdown of the sugars and mucilage. Much of the pulp then drains away as a liquid. In the fermentation stage, complex biochemical reactions take place producing cocoa flavor precursors such as nitrogenous compounds and reducing sugars. Also, the development of color, as well as the reduction in bitterness and astringency, starts during fermentation. The concentrations and the ratio of flavor precursors at the end of the fermentation are vital to the optimal development of flavor volatiles during bean roasting.

Fully fermented cocoa beans have brown color due to enzymatic browning catalyzed by polyphenol oxidase, but unfermented cocoa beans are dark gray and are more astringent. Partially fermented beans are purple-colored and have a bitter and harsh flavor, which leads to loss of flavor in the final chocolate. Overfermentation produces beans with undesirable characteristics. It produces cocoa beans with high pH, causes blackening or darkening of the beans, and gives the beans a characteristic hammy off-flavor.

Fig. 10.2 Fermentation of cocoa beans

10.2.3 Drying

The next stage after the complete fermentation of the cocoa beans is drying. This is required before the beans can be put into sacks for transporting or shipping to chocolate industries. To dry the fermented beans, the beans are spread on trays or mats to dry in the sun. Artificial drying is an alternative drying method; however, sun drying is preferred because it gives a more pronounced chocolate flavor. Artificial drying can lead to off-flavors such as smoky, hammy, rubber or gasoline notes. It takes approximately a week to properly dry the cocoa beans, which reduces the mass of the beans to about half the weight of the fresh beans. During drying the beans are manually sorted to remove defective beans and foreign materials that may cause contamination during chocolate manufacture. Drying continues the oxidative processes that started during fermentation. During drying, the transformation of polyphenols to quinones is catalyzed by polyphenol oxidases. These subsequently undergo further condensation with free amino and sulfhydryl groups forming brown polymers (Fig. 10.3).

Drying reduces the moisture content of the beans to about 7.0–7.5% to prevent overfermentation, mold contamination and bean damage during storage. Well, sun-dried, good-quality beans have the characteristic flavor and brown color. Also, good sun drying reduces bitterness, astringency and acidity, and the absence of off-flavors such as smoky notes and excessive acidity. Incomplete drying and rain soaking produce high concentrations of strong-smelling carbonyl compounds and hammy off-flavors. The drying can be done directly in the open sun (i.e. open sun drying) or a greenhouse sun drying (first picture on the left-hand side). The latter prevents rain contact and/or soaking, while the open sun drying requires monitoring to prevent rain contact.

After drying, the beans are packed into sacks for storage and transportation to chocolate processing factories. It is important to note that besides correct fermentation and drying of the beans, correct transport conditions are also required when moving the beans from the place of drying and/or storage to the chocolate manufacture, or from the country of growing to that of chocolate manufacture. A typical composition of cocoa beans is shown in Tables 10.2 and 10.3. As traded, dried

Fig. 10.3 Drying of cocoa beans

Table 10.2 Composition of
cocoa beans as traded

Parameter	Composition (%)	
	Unfermented beans (dry, fat-free)	Fermented beans (dry, fat-free)
Moisture		5.0–8.0
Shell (dry basis)	–	10.0–16.0
Nib (dry basis)	–	78–82.0
Total polyphenols	150.0–200.0	–
Procyanidins	61.0	23.0
Epicatechin	–	3.0–16.0

Table 10.3 Typical
composition of cocoa bean
nib and shell

Parameter	Composition (%)	
	Nib	Shell
Moisture	2.0–5.0	4.0–11.0
Fiber (crude)	2.1–3.2	13.0–19.0
Fat (cocoa butter and shell fat)	48.0–57.0	2.0–6.0
Protein	11.0–16.0	13.0–20.0
Starch	6.0–9.0	6.5–9.0
Ash	2.6–4.2	6.5–20.7
Theobromine	0.8–1.4	0.2–1.3
Caffeine	0.1–0.7	0.04–0.3

cocoa beans have a typical moisture content of 7.5%, which varies depending on storage conditions and the degree of drying. Also, the composition depends on the type of beans and the method of analysis.

10.3   Chocolate Manufacturing Process

The chocolate production process starts with the primary processing steps, which are harvesting of the cocoa and removal of the beans from the pods, fermentation and drying of the beans. The generalized flowchart for the manufacturing of chocolate and other cocoa products is shown in Fig. 10.4.

10.3.1   Cleaning of Cocoa Beans

On arrival at the processing factory, chocolate manufacturers evaluate cocoa beans for bean size, mold, infestation, filth and degree of fermentation to ensure that only the finest beans make it into their facilities. Hence, it is necessary to clean the beans to remove non-cocoa components such as metals and stones and other extraneous

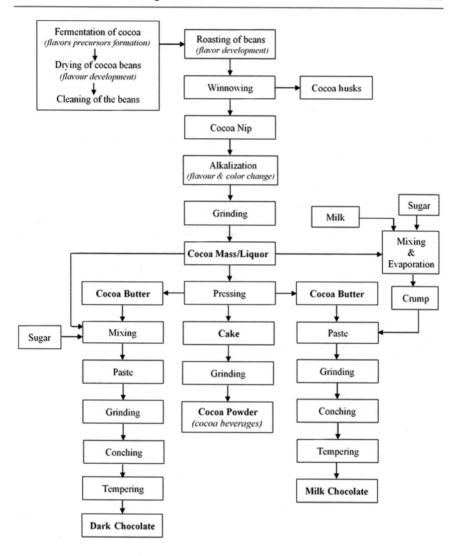

Fig. 10.4 A generalized flow diagram for the manufacture of chocolate from cocoa

material that might contaminate the product. Cleaning is the first critical step in ensuring that the finished chocolate will meet the quality criteria of confectioners. However, no processing steps can improve upon inferior beans caused by, for example, poor fermentation. Also, impurities may cause wear and damage to the processing machinery.

Cleaning of the beans is mostly carried out in several stages as explained below.

(i)   **Removal of coarse and fine impurities**: Both coarse and fine particles are removed by sieving using a set of vibrating screens. Cocoa beans pass through coarse screens while clustered beans, strings, large debris and pieces of wood are collected by the coarse screens. Smaller particles pass through finer screens.

(ii) **Removal of ferrous matter (metals)**: Metals are removed by magnets or metal detectors during further processing.

(iii) **Destoning and removal of other high-density particles**: Stones and other heavy particles are removed using destoners, which operate on the principle of a fluidized bed. The stones and other heavy particles are vibrated toward the top of a slopping screen with air passing through the screen, and the lighter particles remain at the lower end of the screen.

(iv) **Dust collection**: Several cleaning steps using counter air flow techniques are utilized to separate low density and dust particles from the beans during the cleaning operation. Removal of dust and sand from the beans is very important because their abrasive nature causes rapid wear of the balls in ball mills or the surface of refiner rolls.

10.3.2 Roasting of Cocoa Beans

The first step of chocolate manufacturing after cleaning the dried cocoa beans is to roast them. The process involves, first roasting the beans on screens and then in revolving cylinders through which heated air at 110 to 140 °C is blown for a period of 20 min to 2 h. The roasting reduces the moisture of the beans from about 7.0% to around 1.0–2.0%. Over roasting (over 160 °C) causes the development of burnt taste, as well as off-flavors. Roasting is frequently performed by three methods, namely *whole bean roasting*, *nib roasting* and *liquor roasting*. In liquor roasting, the first step is heat pre-treatment, which is followed by removal of the shells, and then finally, transferring the nibs into liquor before roasting. In nib roasting, nibs are roasted in a batch system that uses humidity to control the removal of volatile substances, which enhances flavor development and aids in the microbiological quality of the product. In whole bean roasting, the beans are usually passed over a fluidized bed at the roasting temperature and residence time. Bed depth, bean flow rate and internal bean temperature are the key variables that need to be optimized.

The roasting process causes a browning reaction, in which about 600 different compounds (alcohols, carboxylic acids, aldehydes, ketones, esters and pyrazines) present in the cocoa beans react, developing the color and the rich aroma/flavor of the beans, which palate experts expect from chocolate. Roasting also eliminates undesired volatiles (acetic acid). Various factors influence the specific cocoa aroma. These include the genotype of the cocoa, farming practices, environmental conditions, the complex chemical and biochemical reactions during the post-harvest processing of the raw beans such as fermentation and drying, chemical composition of the raw seeds and the chocolate manufacturing stages.

The aroma of fine cocoa is made up of three types:

1. *The base aroma*, already present in the fresh seed (e.g. the chocolate characters and some wine-like flavor in Trinitario cocoa from West Indies, South America, and Central America; the Arriba flavor with aromatic, floral, spicy, and green notes in Nacional cocoa from Ecuador).
2. *The fermented aroma* caused by the products produced during microbial fermentation.
3. *The thermal aroma*, clearly chocolate, is produced during the roasting process.

10.3.3 Winnowing of Cocoa Beans

The roasting process makes the shells or hulls of the cocoa brittle, which breaks away from the meat of the bean (or cotyledons) into small pieces of cocoa bean meat called 'cocoa nibs'. The process of removing the outside cocoa shells from the cocoa beans is called 'winnowing'. During the winnowing process, the roasted beans are separated into shells and cocoa nibs, which then pass down through a series of sieves through vigorous shaking, which sorts the nibs according to size. This separation process can be completed by blowing air across the beans as they go through a giant winnowing machine called a cracker and fanner, which loosens the hulls from the beans without crushing them. The shells which are separated from the nibs are used as commercial boiler fuel and as a mulch or fertilizer (Fig. 10.5).

10.3.4 Alkalization

Alkalization is the treatment of cocoa nibs, cocoa mass, liquor or powder with an alkaline solution, usually potassium carbonate, that raises its pH from 5.5 to 7 or 8. Also, it can be achieved before roasting. Alkalization is also carried out to improve the flavor of the cocoa by decreasing its bitterness to make it mild and darken the color of the cocoa. In addition, it reduces the tendency of the nib particles to form

Fig. 10.5 Crushed cocoa nibs

clumps in the liquor. Furthermore, alkalization reduces astringency by complex polymerization of polyphenols and increases the dispersibility of cocoa powder in beverages.

This process can be batch (when pressurized conditions are required) or continuous (under atmospheric conditions). Each process has advantages and disadvantages and will result in specific cocoa powders with individual flavor and color characteristics.

10.3.5   Grinding of Cocoa Nibs

After the cocoa beans are roasted to the desired flavor profile and alkalized, the nibs are then ground into chocolate liquor by a variety or combination of grinding or milling equipment. These include ball mills, shear mills and stone mills. The choice of grinding equipment depends on its ability to achieve the suitable particle size reduction, fat release, flow rate capacity, energy consumption, design and cost. The grinding process generates heat, and the dry granular consistency of the cocoa nib is then turned into a liquid as the high amount of fat contained in the nib melts. Another name for cocoa liquor is cocoa mass or unsweetened chocolate. The cocoa mass is used in two ways: (i) to make cocoa powder or cocoa beverage or chocolate drinks such as Milo (produced by Nestle), Cadbury drinking chocolate (produced by Cadbury), etc.; (ii) pressed to produce cocoa butter, which is mixed with cocoa mass to produce chocolate.

10.3.6   Pressing of Cocoa Liquor

The liquid cocoa liquor obtained after grinding is stored in large storage tanks and kept at a temperature of about 70 °C to keep the liquor in its liquid form. To obtain cocoa butter, the liquid cocoa liquor is pumped (at between 15 and 20 bar) to liquor conditioning tanks mounted on each pressing to achieve optimum conditions required when it is pressed into cocoa butter and cocoa cake. The liquor is pressed using high-speed stirring gear to ensure quick heat transfer and homogenization of the product as well as reducing the viscosity. This enhances the flow and pressing properties of the product as it results in a thin-fluid consistency. Depending on the type, a press can hold between 165 kg (12 pots) and 220 kg (14 pots) of liquor. Usually, the first butter produced immediately after the hydraulic pump is started contains very fine cocoa particles that are not filtered out by the filter screen, which makes it impure. However, coarse cocoa particles accumulate on the screen to form a filter cake or additional filter, which helps to aid in obtaining clear pressed butter.

The fat content decreases as the pressure increases. At between 100 and 150 bar, the fluid liquor is pressed into cakes, while at about 350 bar the residual fat content of the cake is reduced to between 22 and 24%. To obtain fat content of about 10–12%, the pressure must be above 500 bars and maintained at this pressure for about

10 min. At this pressure, about three to four pressing cycles per hour can be achieved.

The cocoa butter produced is filtered and deodorized. The main use of cocoa butter is for making chocolate, which is performed by adding it to the cocoa mass paste (see Sect. 10.3). Due to its melting point (at 34 °C), cocoa butter can also be used in cosmetic or medicinal products, but it is increasingly being replaced by cheaper synthetic materials.

Apart from pressing, an expeller can be used to produce expelled cocoa butter from cocoa beans, nibs, mass (reduced fat), press cake or any combination of these. In addition, cocoa butter can be obtained from cocoa beans by solvent extraction. Pressed, expelled or solvent extracted cocoa butter can be refined by neutralization with an alkaline solution and decolored with, for example, bentonite, activated carbon, etc.

10.3.7 Grinding and Refining

The next step is refining, during which the cocoa liquor or cake obtained after pressing is further ground between sets of revolving metal drums. The purpose is to make the cake finer or to reduce the size of the particles in the liquor to about 0.001 inch (0.00254 cm).

10.3.8 Making Cocoa Powder

The mass which remains in the press after the butter has been extracted is called the 'presscake', which still contains 10–20% of fats. To produce cocoa powder, the cake is broken, crushed and then sieved. Pin or hammer mills are usually used to pulverize the cocoa cake. After pulverization, the cocoa powder is tempered to crystallize the cocoa butter into its stable forms. Incorrectly tempered cocoa powder may result in changes in its visual appearance, called 'external' color, and lumps may be formed.

To produce cocoa powders with fat contents of less than 10%, further extraction is needed, and CO_2 and/or other solvents can be used. To produce cocoa powders with fat contents above 30%, other techniques such as cryogenic grinding of the cocoa mass are used. Cocoa powder with specific characteristics can be produced by blending different cakes in the desired ratio. The advantage of blending properties such as the color of the powder can be standardized. Additives such as sugar or other sweeteners, lecithin and milk powder may be blended with the cocoa powder. Lecithin is added to increase the wettability of cocoa powder for specific applications, for example, (agglomerated) in dry drink mixes.

10.3.8.1 Types of Cocoa Powder
A variety of cocoa powders can be produced, and the final characteristics may be different in:

(i)   **Fat content**: Normally low-fat cocoa powders are less expensive, while high-fat powders give improved flavor characteristics in certain applications.
(ii)   **Flavor**: Customized flavor can be developed depending on alkalization and roasting conditions.
(iii) **Physical characteristics**: The addition of lecithin to cocoa powder increases the wettability properties.
(iv)   **Color**: High-fat powders have darker 'external' color. 'Intrinsic' color can vary from light brown–non-alkalized to dark brown and dark reddish colors.

10.3.9 Mixing: Producing Chocolate

Production of chocolate involves mixing the cocoa butter back with the cocoa liquor or cocoa powder in varying quantities and then refining. The formulation of milk chocolate depends on the individual manufacturer's formula and manufacturing methods. Generally, fresh whole milk, sweetened condensed milk or roller-dry low-heat powdered whole milk is added to milk chocolate. A refiner is a series of rolls that reduce the particle size of the ingredients to the required fineness using a differential roll speed causing sheer. Usually, pre-refiners are used to condition the mass prior to the five roll refiners. The mass is introduced to the lower roller and moves up the rolls increasing the surface area and causing a dry paste or flake to come off the refiner as very fine particles (20 microns). This helps to create an agreeable texture in the mouth.

The different types of chocolate are produced using the following basic blends with ingredients roughly in the order of highest quantity first:

- *Plain Dark Chocolate*: cocoa powder, cocoa liquor, cocoa butter, sugar, lecithin and vanilla.
- *Milk Chocolate*: sugar, milk or milk powder, cocoa powder, cocoa liquor, cocoa butter, lecithin and vanilla.
- *White Chocolate*: sugar, milk or milk powder, cocoa liquor, cocoa butter, lecithin and vanilla.

The function of the cocoa butter in the chocolate is to provide texture and consistency. The amount of cocoa butter depends on the type of chocolate, as different quantities of cocoa butter are needed for different types of chocolate.

10.3.9.1 Conching

The mixture, after mixing and refining, now undergoes a process known as 'conching', in which it is continuously turned and ground in a huge open vat. Conching is a mixing and heat treatment step, which helps to develop the smooth texture and the final flavor of chocolate. The conching process improves the flavor profile, decreasing off-flavors. It reduces or eliminates the concentration of free acids and other volatile by-products of the cocoa beans to give a more mellow flavor. It improves the rheology (reduction in viscosity). It removes moisture and thus reduces lumping and graining.

Conche is a large mixer that applies high mechanical energy to the chocolate mass. The friction created through the mixing generates a shearing action, which results in a temperature rise. The conching process is usually performed by stirring the mixture at a temperature above 40 °C, while dark chocolate is typically conched at 70 °C or up to 82 °C. The process can last from between three hours to three days, although more time is not necessarily the best. During the conching cycle, cocoa butter, emulsifiers (lecithin) or flavorings may be added depending on the chocolate texture required. The chocolate must contain between 31 and 35% of fats to ensure that all the particles are coated in an unbroken fatty film. Conching is the most important step in making chocolate. The crucial aspect of conching is the time and rate at which the ingredients are added, and the temperature of mixing. These factors determine the quality of the final product.

Conching takes place in three stages or phases, although not all occur for every recipe and all types of conche. The three phases are:

Dry phase: chocolate mass is still crumbly and like a powder, moisture is removed.

Pasty phase: chocolate is a thick paste, much of the moisture has been release, high work input is required by the conche.

Liquid phase: high-speed stirring to mix in the final fat and emulsifier additions, the chocolate becomes liquid.

The dry and pasty phases are the most important, as most of the changes take place during these stages.

10.3.9.2 Tempering

After conching, the chocolate is tempered. Tempering chocolate is a process of using mechanical operation and temperature control to induce crystallization in the correct quantity of the cocoa butter into its stable beta form (β shape) of crystals so that when molded it forms a stable solid with a smooth, shiny surface. The beta crystals are the finest and most stable. Tempering is an important step in chocolate manufacture. If the chocolate is not properly tempered, dull-looking pieces with poor texture will be produced. Tempering is performed by reheating and cooling the chocolate around the 34 °C melting point.

After tempering is complete, molding is the final procedure for chocolate processing. The paste is poured into molds, cooled to create the chocolate, cut and finally, wrapped or packaged. Chocolate with '70% cocoa' contains 30% of sugar and 70% of the mass mixture plus cocoa butter (not including the additives such as lecithin or vanilla which make up less than 1%). Each manufacturer's mass to butter ratio is highly confidential.

10.4 Quality Control in Chocolate Manufacture

Cocoa processing and chocolate production require consistent control and management of the fat (or cocoa butter) composition across all areas of the process. For
instance, the quality and price of cocoa liquor depend on the fat content. Hence,
processing the liquor involves maximizing the production of the expensive fat while
producing cocoa powder to specification. Typical cocoa powder contains 10–12%
cocoa butter. Accurate analysis of these products and tight control of the process
will produce in-spec cocoa powder, more expensive cocoa butter, and quality
chocolate. In chocolate processing, the quality of the ingredients can be evaluated
before mixing to reduce recycling and rejected batches. The chocolate can be
analyzed for sugar and fat contents to precisely control the quality of the final
product and maximize the use of expensive cocoa butter.

The different aspects of quality required in cocoa processing and chocolate
manufacture are presented in the following sections.

10.4.1 Cocoa Bean Physical Characteristics

Chocolate manufacturers aim to produce chocolate of high quality. The physical
characteristics of cocoa beans affect the quality of chocolate and other cocoa
products. These include yield of the cocoa nib, bean size and weight (should at least
weigh 1.0 g), shell percentage, and foreign matter. Others include damaged beans,
moisture content and fat content.

10.4.1.1 Shell Percentage

The shells of cocoa beans are required to be strong to remain unbroken during
normal handling, but loose for easy removal during processing. The most contaminated of the bean is the shell that, for example, can be polluted with pesticides
residues, polycyclic aromatic hydrocarbons (PAHs) and ochratoxin A (OTA).
Hence, the first step in cocoa/chocolate processing is careful de-shelling. The
Codex Standard for Cocoa Mass and Cocoa Cake specifies that the shell and
embryo (germ) of cocoa cake must be more than 5.0% w/w (between 11.0 and
12.0%) calculated on the fat-free dry matter, or not more than 4.5% calculated on an
alkali-free basis (for cocoa shell only). For cocoa mass (cocoa liquor), the shell and
germ should be less than 5.0% w/w of the cocoa mass (cocoa liquor) calculated on
the fat-free dry matter or not more than 1.75% calculated on an alkali-free basis (for
cocoa shell only).

10.4.1.2 Foreign Matter

Foreign matter in cocoa beans affects the yield of the nib and hence reduces the
value of the cocoa to the chocolate manufacturer. It can also be a source of contamination to the products and affects the flavor of the products. Therefore, the
foreign matter must be removed during the cleaning of the beans. There are two

types of foreign matter: (i) one that has no commercial value to the chocolate manufacturer, e.g. stones, sticks, etc. (which are not cocoa-related) that can damage machinery as well as placenta, pod husk and flat or shriveled beans containing very little nib (which are cocoa-related) that can be detrimental to the flavor as well as reducing the yield of edible material. (ii) The foreign matter that has only a reduced value, called 'Cocoa residue', which includes broken beans and fragments or beans and shells.

10.4.1.3 Damaged Beans

Damaged beans are removed from bean samples during cleaning before processing as this reduces the usable nib and affects the wholesomeness of the beans. Double beans and clumped beans (clusters) are also rejected during cleaning and can cause a serious loss to manufacturers.

10.4.1.4 Moisture Content

Cocoa beans are required to have a moisture content of about 7.0%. Moisture content above 8% causes loss of nib, and risk of mold and bacterial growth, which cause potentially serious consequences to food safety, flavor and processing quality. Moisture content less than 6.5% results in a brittle shell and causes the beans to disintegrate into high levels of broken beans. If such cocoa beans are transported or stored in bulk, because there are less protected from damage if not bagged, they are likely to have higher levels of lipolysis that results in free fatty acids (FFAs).

10.4.2 Cocoa Flavor Quality

The flavor is a key quality criterion for manufacturers of cocoa products. The inherent potential chocolate flavor of a particular source of cocoa beans is determined principally by the variety of the trees. This means that flavor characteristics are inheritable. Hence, if two types of beans with contrasting flavor characteristics are crossed, the flavor quality of the resulting product will be the average of the two types.

Regardless of the genetic background of the cocoa beans, the flavor development of the beans and chocolate similarly depends on correct fermentation and drying procedures explained earlier in Sect. 10.2, as well as further processing steps such as roasting, alkalization or conching discussed in Sect. 10.3. Under-fermentation, overfermentation and taints cause flavor defects, which can affect all types of cocoa.

The 'cut-test', which is used in grading cocoa beans for the market, can be used to give an indication of gross flavor defects of the beans such as excessive bitterness and astringency from a high proportion of slaty beans, or moldy/musty notes from moldy or infested beans. However, apart from these examples, the cut test is not a reliable indicator of flavor quality.

The best way of evaluating the flavor of cocoa beans is to turn into cocoa liquor or made up fully into chocolate and tasted. This is usually performed by a taste panel of experienced tasters. Liquors can be tasted directly without the addition of cocoa butter, sugar and milk products, which dilute the taste impression and impart flavor notes unrelated to the cocoa beans being tested. To taste chocolates, time is required for the flavor to stabilize after preparation.

Samples of the cocoa beans can be assessed for strength of the cocoa or chocolate flavor, residual acidity, bitterness and astringency, normally present in liquors and chocolates, as well as the presence of any off-flavors and any positive ancillary flavors such as fruity or floral notes.

10.4.2.1   Moldy Off-Flavors

These flavors are caused by the presence of molds mostly inside the beans, which is caused mainly by germinated beans, prolonged fermentation, inadequate or too slow drying and adsorption of moisture during storage under highly humid conditions. Just about 3.0% of mold inside beans can cause a moldy flavor to liquor, and subsequently to chocolate. Moldy off-flavors cannot be removed during the chocolate-making process, hence always impact the moldy flavor of the products. Apart from off-flavors, mold growth results also in increased levels of free fatty acids (FFA) in cocoa butter and specific molds (fungi) such as *Aspergillus* molds could even lead to the formation of mycotoxins, including Ochratoxin A (OTA). Mycotoxins are carcinogenic; hence, it is important that measures are taken to minimize their formation during post-harvest processing, storage and transportation in cocoa beans. Effective de-shelling of the beans is essential to minimize mycotoxins since most of these compounds are found on the outside of the bean. Cutting the beans into halves is used to reveal the presence of moldy beans.

10.4.2.2   Earthy Off-Flavors

Earthy off-flavors in beans are caused by the accumulation of dirt and debris during storage of the beans in a silo and increase in the last beans. Beans with an earthy off-flavor at high levels should be rejected as this can be very unpleasant.

10.4.2.3   Acid Taste

This is caused by excessive amounts of acetic and lactic acids, which are formed during deep box fermentation. Proper drying reduces the acidity level in the fermented beans; however, the acidity will remain in the beans if drying is carried out hastily. During chocolate processing, the acetic acid present, which is volatile will be reduced to an acceptably low level, but the non-volatile lactic acid if in excess will remain, causing an off-flavor in chocolate. Also, excess acidity, usually with pH of 5.0 or less, in the dry beans results in poor development of chocolate flavor. The pH of liquors produced from properly fermented and dried beans is about 5.5. However, control of pH does not guarantee a good chocolate flavor. The pH can be raised by alkalization (see Sect. 10.3.3), but this alone cannot achieve an acceptable flavor note.

10.4.2.4   Bitterness and Astringency

Excessive bitterness and astringency are caused by poor fermentation of the beans and/or certain varieties. These are objectionable flavors, which cannot be removed by the normal processing method. Samples of cocoa beans containing more than 3.0% slaty or unfermented beans give extremely bitter and astringent cocoa liquors and chocolate.

10.4.2.5   Contamination

The high-fat content of cocoa beans makes it extremely effective for absorbing all manner off-flavors during the storage and transport of the beans. Contamination of cocoa beans with foreign matter does not only affects the quality and flavor of the products but may also cause damage to the plant and machinery (as mentioned in Sect. 10.4.1.2).

10.4.3   Cocoa Butter Quality

Cocoa butter is generally the most valuable part of the bean. Chocolate manufacturers require cocoa beans with a fat content of about 55–58% fat in the dry nib. Qualities used to characterize cocoa butter include free fatty acid (FFA) and hardness. Cocoa beans from Ghana mostly contain a higher percentage of fat (48.0%) than Côte d'Ivoire (about 46.4%), Brazil (44.6%) or Indonesia (39.9%).

The important factors for obtaining high-quality cocoa butter are the origin of the beans and the processing conditions used. High-quality cocoa beans are required to produce the best quality cocoa butter. The type of process used may affect the quality of the cocoa butter. For example, severe alkalization, or a higher deodorization temperature can totally change the butter solidification characteristics. The quality parameters of cocoa butter are presented in Table 10.4.

10.4.3.1   Free Fatty Acid

The hardness of cocoa butter, and thus its processing quality, particularly, the crystallization properties of the butter is affected by the FFA content. High FFA butter produces poor quality chocolate, affects bloom, tempering, and may affect the flavor. The FFA content of whole and healthy cocoa beans is between 1.0 and 1.3%. These are beans that have been fermented properly and dried properly without delay, stored properly and exported on time. Beans with higher FFA content above 1.75% may be caused by diseased pods, very slow drying of beans after fermentation (especially of a cluster of beans), long storage under humid

Table 10.4 Quality parameters of cocoa butter

Parameter	Value
Free fatty acids	1.75%
Iodine value	33.0–42.0
Refractive index	1.456–1.459
Moisture content	Max 0.1%
Unsaponified matter	Max 0.035%
Blue value	Max 0.05%

conditions or beans with a moisture content above 8.0%, broken beans in a higher percentage, or beans infested with inserts during storage. The FFA content of 1.75% is the legal limit for cocoa butter within the EU according to Directive 2000/36/EC and the Codex Standard for cocoa butter (86–1981, Rev.1–2001).

10.4.3.2   Hardness

The composition of fat in cocoa butter affects its texture and consequently affects the behavior of chocolate in the manufacturing process as well as the texture and appearance of the final product. Cocoa butter contains a mixture of triglycerides; mostly stearic acid, palmitic acid and/or oleic acid. However, the proportions of these fatty acids differ; hence, the fat of cocoa beans from different countries may have different properties. Cocoa beans containing high levels of FFA tend to be softer. Relatively hard and consistent cocoa butter is preferred by manufacturers. Cocoa beans from most West African countries give cocoa butter the desired physical properties. Cocoa beans from Cameroon and Brazil give softer cocoa butter, while butter from Southeast Asia is harder.

10.4.4   Cocoa Powder Quality

The quality of cocoa powder must conform to the specification in Table 10.5. To produce cocoa powder within the recommended specifications, it is important to use quality cocoa beans and apply optimal processing conditions. Before the cocoa powder is approved, it is recommended that the relevant quality parameters are determined. *Salmonella* may be present in the final cocoa powder because of the nature of the harvest and subsequent fermentation in the country of origin. The total bacterial load of cocoa beans can be between 1.0 and 10.0×10^6 cfu/g. This can be reduced by roasting and/or nib treatment; both of which can also kill any *Salmonella* present.

Table 10.5 Quality parameters of cocoa powder

Parameter	Value
Fat content	10.0–24.0%
Moisture	Max 4.5%
Fineness	Min 99.8% passing sieve
pH	5.0–8.2
Molds	Max 50 cfu/g
Yeast	Max 10 cfu/g
Total plate count	Max 5000 cfu/g
Enterobacteriaceae	Absent per gram
E. coli	Absent per gram
Salmonella	Absent per 750 g
Contaminants	Free from heavy metals, pesticides

Cocoa powder is very sensitive to high temperature and temperature fluctuations, which may result in the melting and re-crystallization of the cocoa butter present. This may lead to lump formation, which is a common problem that causes difficulties in processing cocoa powder. Cocoa powder is hygroscopic; hence, it absorbs moisture from the environment, which should be avoided, and appropriate packaging used.

10.4.5 Chocolate Quality

Careful selection of the ingredients and control of the processing method is necessary to produce chocolate of desirable flavor characteristics and smoothness in the palate. The fundamental chocolate characteristic is the continuous phase lipid composition, which influences melting properties and mouthfeel. Chocolate is solid at room temperature (20–25 °C) and melts during consumption at 37 °C (oral temperature). The main triglycerides in chocolate are saturated stearic fatty acid (34%), monounsaturated oleic acid (34%) and palmitic fatty acid (27%). Chocolate manufacturers use several characteristics such as appearance, particle sizes, polymorphism, rheology, aroma to produce quality chocolate to the customers.

10.4.5.1 Flavor Quality
The flavor properties of chocolate are mainly influenced by the cocoa beans' aroma and by the manufacturing process. The typical flavor of dark chocolate is a praline, chocolate note with malty, nutty and caramel notes, whereas milk chocolate has a sweet, milky and honey-like flavor with coconut notes. The main odor-active compounds in cocoa and chocolate are pyrazines, aldehydes, esters, alcohols, acids and hydrocarbons. Dark chocolate contains compounds such as pyrazines (tetramethylpyrazine), Strecker aldehydes (3-methylbutanal, 2-methylpropanal), pyrroles, alcohols (phenylethanol) and carboxylic acids (3-methylbutanoic acid). Other compounds in dark chocolate are nitrogen heterocycles: 2,3-dimethylpyrazine, trimethylpyrazine, tetramethylpyrazine, 3 (or 2), 5-dimethyl-2 (or 3)-ethylpyrazine), 3,5 (or 6)-diethyl-2-methylpyrazine), furfurylpyrrole and acetylpyrrole, all with praline, nutty and coffee notes. Lactones (decalactone), esters (phenylethylacetate), long-chain aldehydes (heptanal, octanal), ketones (2-nonanone) and sulfur compounds (dimethyl disulfide, trimethyl trisulfide) are the dominant compounds in milk chocolate.

10.4.5.2 Visual Characteristics
Visual characteristics include color and gloss. Color is often related to flavor characteristics, an important consideration when aiming for consistent flavor quality. Good-quality chocolate has a continuous light to dark brown color (depending on the product type) and a glossy appearance. White dots, patches or marks attributed to the migration of fat or sugar to the chocolate surfaces, known as 'blooming', are undesirable and affect the quality and acceptability of chocolates.

10.4.5.3   Rheological Properties

The rheological properties of chocolate are very important in the manufacturing process. If the viscosity of chocolate is too high, will form 'feet' or bubbles that will not come out of a molded tablet. Also, chocolate with high viscosity has a pasty mouthfeel persistent in the mouth. For a very low viscosity, the weight of the chocolate on an enrober-coated sweet will be too low. Incorrect viscosity of the chocolate will result in a poor-quality product, which may have to be sold cheaply as a miss-shape or possibly reprocessed.

The flavor of chocolate in the mouth is affected by viscosity. This is because as the chocolate melts it comes into contact with a lot of different flavor receptors in the mouth, each capable of detecting a single type of flavor; for instance, sweet on the tip end of the tongue and sour at the throat end. Therefore, the perceived taste of chocolate depends upon the order and rate of contact, which is related to the viscosity and the rate of melt.

The flow property of chocolate is very complicated because the viscosity is not a single value; that is, it is non-Newtonian. This means that if measured, the viscosity varies depending on the flow rate of the chocolate. Chocolate consists of solid particles in a suspension within a continuous fat system. At low flow rates, the solid particles within the chocolate collide with each other and resist the force, initially preventing movement and then making the apparent viscosity relatively high. When the flow rate of the chocolate is faster the particles can all move together along with the flow rather than resist it, so the chocolate has a lower apparent viscosity and behaves like a thinner liquid. It is therefore important to measure the viscosity at several flow rates so that the results can be presented as several single-point measurements or by using mathematical models to produce the two flow parameters, the yield values and plastic viscosity. Yield value relates to both pattern holding and formation of feet and tails; plastic viscosity is important in controlling the filling processes and pumping. The accepted method of measuring viscosity in the confectionery industry is the procedure by the Internal Office of Cocoa. In this method, a rotational viscometer fitted with a concentric cylinder is used to measure the chocolate flow curve between 2 and 50 s^{-1} at 40 °C. The viscosity of the tempered chocolate will be affected by the lower processing temperature (about 30 °C) and the presence of fat crystals produced during tempering, increasing the dispersed phase volume fraction.

10.4.5.4   Texture Quality

Texture is a quality parameter that refers to the feel of food in the mouth and the impression one has of its physical characteristics as a result of biting and chewing. The chocolate texture is the most complex of all its physical characteristics, and in addition to flavor, it is the quality that is mostly used during the selection of chocolate. The texture of chocolate can be described in several words depending on whether the emphasis is on structure, consistency or mouthfeel. These are hardness at first bite, smoothness, adhesiveness, thickness, chunky and melting in the mouth. Among these qualities, the three textural sensory properties of great importance are smoothness, meltiness and hardness. In terms of hardness, plain chocolates show greater bite firmness than milk chocolates, and concerning fineness, plain chocolates have superior quality. Also, plain chocolates melt more slowly than milk chocolates.

Desirable chocolate is a firm solid with a good snap at ambient temperature and a glossy appearance that melts easily in the mouth with a smooth mouthfeel. Defects in chocolate texture include excessive hard or soft, poor snap, sticky surface, not melting readily in the mouth and gritty mouthfeel. Particle size distribution, fat and lecithin content influenced the hardness and mouthfeel in dark chocolate. Therefore, to control chocolate texture, optimization of grinding, refining and conching processes as well as the composition are needed.

10.4.5.5   Taste Quality

The chocolate taste is one of the key qualities that determine the acceptance or rejection of chocolate. The different taste characters in chocolate are sweetness, saltiness, acidic, bitterness and umami. These taste characters are used as a critical factor for quality and also dictate their preferences and marketability by consumers. Ingredient composition influenced chocolate taste greater than the production process.

10.5   Nutritional and Health Benefits of Cocoa and Chocolate

10.5.1   Health Benefits of Cocoa and Chocolate

Cocoa and chocolate contain high polyphenol levels and have been acclaimed to have nutritional and medicinal/health benefits. Cocoa/chocolate shows promising antioxidant, cardioprotective, neuroprotective and chemopreventive potential. Cocoa polyphenols have strong antioxidant properties. Cocoa and dark chocolate may exert suppressive effects on low-density lipoprotein (LDL) oxidation and the associated development of atherosclerosis with cardioprotective implications because of their interference in many pathophysiological mechanisms.

Cocoa and dark chocolate lower blood pressure through the induction of nitric oxide (NO)-bioavailability, a molecule which play a crucial role in the maintenance of vascular homeostasis. The consumption of cocoa-based beverages with high flavanol concentrations attenuates the blood pressure response to exercise. Polyphenol in cocoa and dark chocolate reduces the occurrence of stroke, atherosclerosis, coronary artery disease, heart failure and cardiovascular disease-related mortality. Cocoa and dark chocolate have cardioprotective properties, which are attributed to an improvement in antioxidant status and endothelial function, modulation of the blood pressure, metabolic and anti-inflammatory effects, and inhibition of the platelet activation and aggregation.

Cocoa and dark chocolate consumption help to improve endothelial function, by improving the flow-mediated dilatation (FMD) of the brachial artery, a clinical marker of endothelial function. Furthermore, the polyphenol-rich in dark chocolate intake improves glucose metabolism and increases insulin sensitivity, decreased platelet activation and function, as well as modulation of immune function and inflammation. In addition, chocolate releases phenylethylamine and serotonin into the human system, when consumed, producing some aphrodisiac and mood-lifting effects. Consumption of cocoa or dark chocolate might also have beneficial effects on serum lipids.

10.5.2   Nutritional Benefits of Cocoa and Chocolate

Consumption of cocoa and chocolate contributes positively to human nutrition through the provision of the major constituents—carbohydrates, fat and protein for energy and other metabolic functions. Also, chocolates contain minerals, namely magnesium, potassium, iron and copper. The fat in cocoa and chocolate contains several fatty acids and triglycerides, mainly saturated stearic, palmitic and monounsaturated oleic, which do not raise blood cholesterol levels.

Chocolate contains the energy-providing nutrients, fat, carbohydrate and protein. Dark chocolate contains 5.0 g protein, 28.0 g fat and 63.5 g carbohydrate per 100 g serving. Milk chocolate, on the other hand, contains 7.7 g protein, 30.7 g fat and 56.9 g carbohydrate, while white chocolate contains 8.0 g, 30.9 g and 58. g of protein, fat and carbohydrate, respectively. The fat, which is mainly from cocoa butter contains 34% stearic acid (a saturated fatty acid), about 34% oleic acid (a monounsaturated fat) and 27% palmitic acid (a saturated fatty acid), and most of the remaining are polyunsaturated fatty acids.

Essential vitamins and minerals present in cocoa and chocolate include iron, copper, zinc, magnesium, phosphorus, manganese and potassium, which have important roles in the metabolism of the human body. These are presented in Table 10.6. The mineral content of chocolate varies and depends on the origin of

Table 10.6 Vitamin and mineral content in chocolate, per 100 g	Vitamin/Mineral (mg)	Chocolate Type		
		Dark (Plain)	Milk	White
	Iron	2.3	1.4	0.2
	Zinc	1.3	1.1	0.9
	Copper	0.71	0.24	Trace
	Manganese	0.63	0.22	0.02
	Calcium	33.0	220.0	270.0
	Magnesium	89.0	50.0	26.0
	Potassium	300.0	390.0	350.0
	Phosphorous	140.0	220.0	230.0
	Riboflavin	0.06	0.49	0.49
	Thiamin	0.04	0.07	0.08
	Niacin	0.4	0.4	0.2
	Vitamin E	1.44	0.45	1.14
	Vitamin B12	0.0	1.0	1.0

the cocoa, the amount of cocoa powder used and the milk in the final product. Chocolate with a high proportion of cocoa solids is richer in iron, manganese and copper, while milk chocolates contain riboflavin, vitamin B12 and more calcium. The cocoa butter provides vitamin E.

10.6   Environmental and Pollution Concerns from the Cocoa Processing Industry

The production of chocolate and other products starts from harvesting, removing the cocoa pods, fermentation, and drying of the beans before the beans are sent to the cocoa processing industry for roasting, de-shelling, grinding and pulverization. Each of these steps generates a large amount of waste that causes environmental problems. The cocoa processing industry contributes to about 2.1 million tons of greenhouse gases (GHGs) annually. A large amount of wastewater is generated during the production of chocolate and other cocoa products. About 10,000 L of water is required to produce a kilogram of chocolate and emits between 2.9 and 4.2 kg of CO_2.

The wastewater from the chocolate production plant mostly come from the production line clean-up and includes chocolate, cream and filling mass crap; and contains about 30–35% fat (cocoa butter, cocoa mass), and around 40–45% sugar (in the scrap mass) and flour.

The chocolate industry wastewater does not contain hazardous elements but it is complex with a strong organic load, and that has a high level of biochemical oxygen demand (BOD), chemical oxygen demand (COD), total solids as well as high color content. Hence, it has a high pollution potential. These contaminants may be in a suspended, emulsified or dissolved state. Although the wastewater contains detergents and sanitizers used for washing, it does not contain substantial obstinate or toxic chemicals. Hence, it can be treated by biological techniques. Aerobic and anaerobic treatment techniques are the commonest biological processes that are used for the treatment of chocolate manufacturing wastewater. These processes have been explained in detail in Chap. 11. The suspended solids are removed before the wastewater is treated aerobically or anaerobically. Oil separators can be installed at the end of every production line to remove the fats from the wastewater before the biological treatment.

Apart from wastewater, other wastes generated in the cocoa processing industry include cocoa bean shells produced during the processing step of winnowing, aluminum packaging, and plastic waste (packaging). The cocoa bean shell comprises a lignin-cellulosic complex, which contains about 18–60% dietary fiber, 17–23% carbohydrates, 15–18% total proteins, 2–7% lipids, ∼ 6% pectin and 6–11% ash. The cocoa bean shell can be used for environmental decontamination, for example as a low-cost adsorbent for heavy metals, gases or industrial dyes removal due to its high degree of porosity, adsorption capacity and mechanical strength. It can also be used as an alternative and environmentally friendly source of fuel with a calorific value (~ 17–20 MJ/kg) higher than that of wood.

10.7 Cocoa Processing Companies in Ghana

There are a number of companies in Ghana that process cocoa to chocolate and other intermediate products. Notable cocoa processing companies in Ghana are listed in Table 10.7. Cocoa Processing Company Ltd (CPC) is based in Tema in the Greater Accra region and was established in 1965. The company has three factories: two cocoa factories and a confectionery factory. The cocoa factory processes raw cocoa beans into semi-finished products—cocoa liquor, butter, natural/alkalized

Table 10.7 Cocoa processing and cholate manufacturing companies in Ghana

Company	Location	Capacity (metric tons/year)	Products
Cocoa Processing Company (CPC)	Tema	65,000	– Chocolate bars and couverture – Chocolate spread – Drinking chocolate (powder) – Chocolate dragee – Cocoa butter – Cocoa liquor – Cocoa cake – Cocoa powder
Cargill Ghana Limited	Tema	65,000	– Chocolate – Cocoa butter – Cocoa powder – Coatings and fillings
Niche Cocoa Industry Ltd	Tema	40,000	– Cocoa liquor – Cocoa butter – Cocoa cake
Cocoa Touton Processing Co. Ltd (CTPC)	Tema	32, 000	– Cocoa butter – Cocoa cake – Cocoa powder
Nestle Ghana Ltd	Tema	No available	– Drinking chocolate (powder) – Chocolate bar
Afrotropic Cocoa Processing Ltd (ACPL)	Accra	6,000	– Cocoa butter – Cocoa cake – Cocoa powder
Olam Ghana	Kumasi	30,000	– Cocoa liquor
PLOT GHANA	Takoradi	32,000	– Cocoa liquor – Cocoa butter – Cocoa cake – Cocoa powder

Chocolate bar

Milk chocolate   Dark chocolate

Drinking (powdered) chocolate

Fig. 10.6 Some chocolate products by Cocoa Processing Company (CPC), Ghana

cake or powder whilst the confectionery factory produces the chocolate bars (Golden Tree brand—see Fig. 10.6), couverture, chocolate-coated peanut (Pebbles), drinking chocolate powder (VITACO and ALLTIME), chocolate spread (Choco Delight) and natural cocoa powder (Royale brand). The cocoa factories have an annual throughput of 65,000 metric tons of premium Ghana cocoa beans.

Afrotropic Cocoa Processing Ltd (ACPL) was established in 2005 in Accra, with an annual throughput of 6,000 metric tons. Their products are pure prime pressed (PPP) cocoa butter, natural PPP cocoa cake/kibble, alkalized PPP cocoa cake/kibble, natural cocoa powder, alkalized cocoa powder and dark alkalized cocoa powder, and flavored cocoa powder. Their products are sold to traders and chocolate industries, and food manufacturers worldwide.

Cargill has been sourcing cocoa from Ghana for over 40 years and in 2008 opened its state-of-the-art cocoa processing facility in Tema with an annual capacity of 65,000 metric tons. Cargill manufactures the Gerkens® brand of cocoa powders, cocoa liquors, chocolate (dark, milk and white options), cocoa butter, coatings and fillings.

Olam Ghana started production in 2009 and produces natural cocoa liquor (the UNICAO Ghana brand) with three liquor recipes with a specific flavor profile. The products are exported to five different continents. Olam Ghana Ltd took over operations of Archer Daniels Midland (ADM) Ghana Ltd in October 2015, a cocoa paste manufacturing company in Kumasi. The company has an annual throughput of 30,000 metric tons.

Niche Cocoa Industry Ltd is in Tema and was established in May 2011. Niche produces a range of semi-finished cocoa products from 100% Ghana cocoa beans with an annual throughput of 40,000 metric tons. The brand name for their products is 'Ghana Prime Brand'. Niche produces conventional and specialized cocoa liquor, natural cocoa butter and natural cocoa cake.

PLOT Ghana Limited is a Cocoa Processing Company in Takoradi in the western region of Ghana that manufactures natural and alkalized cocoa liquor, natural cocoa butter, natural and alkalized cocoa cake and natural cocoa powder from Ghana Cocoa. The brand name is 'Gold Brand' and the plant capacity is 32,000 metric tons per annum. It was established in February 2011.

Cocoa Touton Processing Company Ltd (CTPC) was established in 2015 in Tema. The capacity of CTPC is 32,000 tons of beans and produces cocoa liquor, cocoa powder and cocoa butter.

Further Readings

Afoakwa, E. O. (2010). *Chocolate science and technology*. Blackwell Publishing.

Aprotosoaie, A. C., Luca, S. V., & Miron, A. (2016). Flavor chemistry of cocoa and cocoa products–An overview. *Compr Rev Food Sci Food Saf, 15*, 73–91. https://doi.org/10.1111/1541-4337.12180.

Beckett, S. T. (2008). *The science of chocolate* (2nd ed.). RSC Publishing.

Beckett, S. T. (Ed.). (2009). *Industrial chocolate manufacture and use* (4th ed.). Blackwell Publishing Ltd.

CAC, Codex Standard for cocoa butter (86–1981, Rev 2001). Rome: Food and Agriculture Organization (Codex Alimentarius, 2001).

CAC, Codex Standard for chocolate and chocolate products (87–1981, Rev 1–2003). Rome: Food and Agriculture Organization (Codex Alimentarius, 2003).

End M. J., Dand, R. (Eds). (2015). Cocoa beans: Chocolate and cocoa industry quality requirements. ECA-CAOBISCO-FCC Cocoa.

Ghana Cocoa Board. (2017) Regional Cocoa Purchases 2017.

Guehi, S. T., Dingkuhn, M., Cros, E., Fourny, G., Ratomahenina, R., Moulin, G., & Clement Vidal, A. (2008). Impact of cocoa processing technologies in free fatty acids formation in stored raw cocoa beans. *African Journal of Agricultural Research, 3*(3), 174–179.

International Office of Cocoa. (2000). Viscosity of cocoa and chocolate products. Analytical Method 46. CAOBISCO, rue Defacqz 1, B- 1000, Bruxelles.

Schwan, R., & Fleet, G. (Eds.). (2014). *Cocoa and coffee fermentations*. CRC Press.

Sundara R, Rasburn J, Vieira J (2013) Quality sentries: Some trends in chocolate manufacturing. Nestlé Product Technology Centre.

The World Atlas of Chocolate. (2019) The production of chocolate. https://www.sfu.ca/geog351fall03/groups-webpages/gp8/prod/prod.html. Accessed 20 Sep 2019.

Mattyasovszky, M. (2018). Top 10 Cocoa Producing Countries. WorldAtlas. https://www.worldatlas.com/articles/top-10-cocoa-producing-countries.html. Accessed 20 Oct 2019.

Vásquez, Z. S., de Carvalho Neto, D. P., Pereira, G. V. M., et al. (2019). Biotechnological approaches for cocoa waste management: A review. *Waste Management, 90*, 72–83.

Wood, G. A. R., & Lass, R. A. (1985). *Cocoa, Tropical Agricultural Series* (4th ed.). Longman Group.

Milk and Dairy Products Manufacture

11

Abstract

This chapter covers the various milk and dairy products produced from cow's milk. The quality control measures in the dairy industry and the testing techniques have been presented. Typical quality analysis includes taste and smell made at the farm, bacteria count, protein content, fat content, and freezing point. Processes for the manufacture of the different milk and dairy products have been included in detail such as pasteurized milk, Extended Shelf-life (ESL) milk, long shelf-life milk, evaporated milk, condensed milk, fermented milk products (e.g., yoghurt, cultured/sour cream, cheese), ice cream, milk powders, whey products (e.g., demineralized whey, whey power, lactose, whey protein concentrates, whey protein isolates,) using different membrane techniques, chromatographic techniques, ion exchange and electrodialysis. Environmental effects associated with the dairy products manufacture and the treatment methods have been elaborated. These include types of waste generated in dairy processing and minimization strategies, the types of wastewater treatment technologies in the dairy industry covering the primary, secondary for treatment of organic matter (e.g. aerobic and anaerobic processes), and tertiary for removal of phosphorous (e.g. the Struvite process) as well as advanced treatments for reuse such as membrane filtration and reverse osmosis.

11.1 Introduction

Milk, as defined by the U.S. Code of Federal Regulations (CFR), 21 CFR 131.110, is the lacteal secretion, practically free from colostrum, obtained by the complete milking of one or more healthy cows. Although the CFR states that milk is obtained from cows, milk can also be produced from other dairy animals such as goats, sheep, and water buffalo. Milk contains solids, fat, and water. The 'non-water' components of milk are the milk solids, and consist of lactose, protein, and minerals, which is also referred to as the *solids not fat content*. It is called *total solids content* when the fat is included. After milking, the fluid milk is taken through further processing steps such as standardization, homogenization, pasteurization to produce fluid milk (whole milk). Several milk products can be produced from the whole milk. These include other fluid milk products (e.g. semi-skimmed, skimmed, evaporated, and condensed milk), ice creams, yoghurt, cheese, powdered milk, milk proteins, casein, whey powder, among others. The various milk and dairy products from cow milk will be covered in this chapter.

11.2 Milk Composition and Nutrients

Cow's milk composition can vary widely depending on different breeds, animal's feeds and the stages of lactation. As shown in Table 11.1, cow milk consists of about 87% water and 13% dry matter. The dry matter is suspended or dissolved in the water and are distributed differently in the water phase, depending on the type of solids and size of particles (show in Table 11.2).

11.2.1 Milk Fat

The primary fat in milk is a complex combination of lipids called *triglyceride* or *triacylglycerol* (esters of three fatty acids with one molecule of glycerol). There are many different fatty acids that can be attached to the glycerol molecule. Hence, there are many different types of triglycerides or fats. Fat compounds can also be

Table 11.1 General milk composition

Component	Composition (%)
Water	87.3 (85.5–88.7)
Milk fat	3.9 (2.4–5.5)
Proteins	3.3 (2.3–4.4)
Casein	2.6 (1.7–3.5)
Serum (whey) proteins	0.6
Minor proteins	
Carbohydrates (Lactose)	4.6 (3.8–5.3)
Minerals (ash)	0.7 (0.5–0.8)
Cationic: Na, K, Ca, Mg, …	
Anionic: Cl^-, PO_4^{3-}, $C_6H_5O_7^{3-}$, CO_3^{2-}	
Organic acids	0.18 (0.13–0.22)
Citric, lactic, formic, acetic, oxalic	
Enzymes – peroxidase, catalase, phosphatase, lipase	

Table 11.2 Relative sizes of particles in milk

Types of particles	Size (mm)
Fat globules	10^{-2}–10^{-3}
Casein-calcium phosphates	10^{-4}–10^{-5}
Whey proteins	10^{-5}–10^{-6}
Lactose, salts and other substances in true solutions	10^{-6}–10^{-7}

monoglycerides that have one fatty acid or diglycerides that have two fatty acids on the glycerol molecule. Mono- and diglycerides are used as emulsifiers, which help to keep the fat and water from separating in products such as ice cream.

There are more than 400 fatty acids in cow's milk. However, approximately 15 to 20 fatty acids make up 90% of the milk fat. Milk contains both saturated and unsaturated fatty acids. The major fatty acids in milk fat (about 65%) are straight chain fatty acids that are saturated and have 4 to 18 carbons. The remaining 35% are unsaturated fatty acids, which are further classified as monounsaturated (about 30%) and just 5% polyunsaturated (depending on the number of double bonds in the carbon chain of the fatty acid molecule). The polyunsaturated fats in milk contain the omega-6 essential fatty acid (linoleic acid) and the omega-3 fatty acid (linolenic acid). These names refer to the position of the double bond in the carbon chain of the fatty acid molecule. These are called essential fatty acids because they are essential to health but cannot be made within the body and so must be obtained from the diet.

11.2.2 Proteins

Proteins are chains of amino acid molecules connected by peptide bonds. All the nine essential amino acids required by humans are found in milk proteins. Total milk protein content and amino acid composition varies with cow breed and individual animal genetics. Caseins are the primary group of proteins in cow's milk, making up around 86% of the total protein content. The remaining portion is made up of whey proteins. The casein family contains phosphorus and will coagulate or precipitate at pH 4.6. The whey proteins do not contain phosphorus, and these proteins remain in solution in milk at pH 4.6. There are four types of casein (alpha-s_1, alpha-s_2, beta and kappa casein) combine to make up a structure known as a *casein micelle*. The micellar structure of casein, which forms the principle of coagulation, or curd formation, at reduced pH is the basis for cheese curd formation.

Milk serum proteins, commonly called '*whey protein*' is a group of proteins that remains when the casein is removed from skim milk by precipitation, usually by the addition of mineral acid. The whey proteins are: α-lactalbumin, β-lactoglobulin, serum albumin, immunoglobulin. The whey protein derivatives are widely used in the food industry. Whey proteins in general, and particularly, α-lactalbumin, have very high nutritional values.

11.2.3 Carbohydrates

The major carbohydrate in milk is a disaccharide of glucose or galactose (or sugar) called *lactose*. Milk also contains trace amounts of monosaccharides and oligosaccharides. Digestion of lactose is achieved by breaking it down by the enzyme lactase in the intestine to glucose and galactose, which are the component monosaccharides of lactose. Many people have a digestive problem, called *lactose intolerance* when they consume cow's milk and dairy products because they are unable to digest lactose. Symptoms of lactose intolerance may include flatulence (wind) diarrhoea.

11.2.4 Vitamins and Minerals

Milk contains good sources of vitamins, which are present in varying amounts (see Table 11.3). Milk contains the water-soluble vitamins thiamine (vitamin B1), riboflavin (vitamin B2), niacin (vitamin B3), pantothenic acid (vitamin B5), vitamin B6 (pyridoxine), vitamin B12 (cobalamin), vitamin C, folate, pantothenate and biotin. The amounts of niacin, pantothenic acid, vitamin B6, vitamin C, and folate are small and hence, milk is not considered as a major source of these vitamins in the diet.

Milk also contains the fat-soluble vitamins A, D, E, and K. The content of fat-soluble vitamins in dairy products depends on the fat content of the product. Reduced fat (2% fat), low-fat (1% fat), and skim milk must be fortified with vitamin A to be nutritionally equivalent to whole milk. Fortification of all milk with vitamin D is optional. Other vitamins in milk are E and K, but these are in small amounts, so milk is not a major source of these vitamins in food. Minerals found in cow's milk include sodium, potassium, calcium, magnesium, phosphorus and chloride, zinc, iron (although at extremely low levels—total concentration is less than 1%), selenium, iodine and trace amounts of copper and manganese.

Table 11.3 Vitamins in milk

Vitamin	Amount per liter of milk (mg)
A	0.2–2.0
B$_1$ (thiamin)	0.4
B$_2$ (riboflavin)	1.7
C	5.0–20.0
D	0.002

11.3   Physical Properties of Milk

11.3.1   Appearance

The color of milk depends on the carotene content of the fat and varies from white to yellow. Proteins, suspended particles of fat, and certain minerals cause the opacity of milk. Skim milk has a slightly bluish tinge color, and it is more transparent.

11.3.2   Density

The density of cows' milk normally ranges between 1.028 and 1.038 g/cm^3, depending on its composition. The density of milk at 15.5 °C can be calculated by using the following formula:

$$\rho_{15.5°C}(g/cm^3) = \frac{100}{\frac{F}{0.93} + \frac{SNF}{1.608} + \text{Water}} \tag{11.1}$$

where F is the fat content (%), SNF is the solid non-fat content (%), and water $= 100 - F - SNF$.

11.3.3   Acidity

Normal milk is slightly acidic with a pH between 6.6 and 6.8. The typical pH is 6.7 at about 25 °C.

11.4   Collection and Storage of Milk

Healthy cows are milked using a milking device that is safe and comfortable for the cows. The milk is chilled to 4 °C within two hours after milking and stored in bulk in tanks on the farm before being transported to a dairy processing factory using insulated road tankers. The tanks have a capacity of 300–30,000 L (or larger up to 500,000 L) and are fitted with an agitator and the cooling facility. The agitation prevents separation of the whole milk by gravity. Gentle agitation is required as extreme mixing causes aeration of the milk and fat globule disintegration. This results in the lipase enzymes in the milk attacking the fat. To avoid mixing warm milk from the cow with already chilled contents of the tank, mostly, separate plate coolers are installed for chilling the milk before it is transferred to the tanks. The milk room should have a facility for cleaning and disinfecting the pipe system, equipment and bulk cooling tank.

11.5 Milk Quality

Milk from sick animals and milk which contains antibiotics or sediment are not accepted by the dairy industry. Traces of antibiotics in milk can make it unsuitable for manufacturing dairy products, which are obtained by acidification with bacteria cultures, such as cheese and yoghurt. General assessment of milk quality is usually made at the farm, which are taste and smell, and cleaning checks inside the surfaces of farm tanks. The composition and hygienic quality are usually determined on arrival at the dairy. Typical tests carried out on milk supplies are explained below.

11.5.1 Taste and Smell

Samples of the milk are taken at the farm for testing. Milk that deviates in taste and smell from normal milk is rated low. Milk with significant deviations in taste and smell is rejected by dairy manufacturers.

11.5.2 Hygiene or Resazurin Tests

The hygienic quality of milk is assessed by determining the bacteria content of the milk. The Resazurin tests are mostly used to analyze the bacteria content of milk. Resazurin is a blue dye which becomes colorless when it is chemically reduced by the removal of oxygen. When it is added to the milk sample and there are bacteria present, the metabolic activity of the bacteria changes the color of the dye at the same rate as the number of bacteria in the milk sample.

Two tests are used; one is a quick-screening test and the other is a routine test. The quick test forms the basis for rejecting a bad consignment. If the milk sample changes color immediately, the batch is considered unhealthy for human consumption. The routine test involves storing the sample in a refrigerator overnight before a Resazurin solution is added. The sample is then incubated in a water bath and held at 37.5 °C for 2 h.

11.5.3 Bacteria Count

The Leesment method can be used to analyze the bacteria content of milk. In this method, the bacteria are cultivated at 30 °C for 72 h in a 0.001 ml milk sample with a nutritive substrate. The bacteria count is determined with a special screen.

11.5.4   Somatic Cell Count

The cell content is determined with specially designed particle counters (e.g. a Coulter counter). A large number (more than 500,000 per ml of milk) of somatic cells in the milk indicates that the cows are suffering from udder diseases.

11.5.5   Protein Content

In milk, nitrogen is an important component used to characterize proteins, and each protein has a unique nitrogen content. Protein nitrogen derived from amino acids represents about 95% of nitrogen. Non-protein nitrogen, such as urea, exists in minor quantities at approximately 5%. Therefore, nitrogen determination has always been used as a standard method for the estimation of the protein content of milk. The protein content of the milk is analyzed by means of instruments operating with infrared rays. Three methods can be used to determine the proteins in milk: (i) determination of total nitrogen (ii) direct protein determination; (iii) indirect protein determination. Each of these methods has advantages and disadvantages.

i **Determination of protein by total nitrogen**

The analysis of protein by determining total nitrogen content may involve chemically digesting or, alternatively, combusting the sample, and converting the resulting material to a form that allows the total nitrogen to be measured. The two international standards are the Kjeldahl method and the Dumas method, which respectively use the chemical digestion and combustion methods. Because the total nitrogen analysis does not directly measure protein, the total nitrogen value is converted to total crude protein using a nitrogen conversion factor of 6.38 for milk proteins and 6.25 for milk based infant foods. True protein nitrogen is obtained by subtracting from the total nitrogen content of the filtrate of a sample treated with trichloroacetic acid. The result (true protein nitrogen) is multiplied with the conversion factor (6.38) and gives the true protein content.

These methods have high accuracy and reliability, which is their major advantage. A disadvantage is that they require dedicated laboratory equipment and skilled staff which makes them expensive and time-consuming to perform.

a. **Total Kjeldahl Nitrogen analysis**:

The Association of Analytical Communities (AOAC) 991.20 method for Total Kjeldahl Nitrogen (TKN) analysis in milk involves three stages:

A. **Digestion**

This is the initial digestion stage, in which a mixture of potassium sulfate, copper sulfate, and sulfuric acid are added to a digestion flask containing a pre-weighed sample of milk, pre-heated at 38 ± 1 °C. The digestion solution is heated and kept at a rolling boil for about 1.5 to 2 h. After cooling, purified water is added to the digest. The digest is then transferred to a distillation flask, where sodium hydroxide is added to neutralize the solution and ultimately convert ammonium sulfate to ammonia gas.

B. **Distillation**

In the distillation stage, the solution is again heated until all the ammonia gas is liberated and captured in a boric acid solution.

C. **Titration**

The analysis stage begins with titration of a blank sample, containing only digested reagents, followed by titration of the sample-containing distillate with a standard sulfuric or hydrochloric acid solution.

b. **The Dumas method**:

In this method, organic and inorganic nitrogen is converted into nitrogen gas, which is analysed by gas chromatography, volumetrically or by thermal conductivity.

The method involved combusting the sample in an oxygen atmosphere at a high temperature. By means of subsequent oxidation and reduction tubes, organic and inorganic nitrogen is quantitatively converted to nitrogen gas (N_2), which is analysed by gas chromatography, volumetrically or by thermal conductivity. Other volatile combustion products are either trapped or separated. Using thermal conductivity, a Thermal Conductivity Detector measures the nitrogen gas. Results are given as % or mg nitrogen, which may be converted into protein by using the conversion factor mentioned above.

ii. **Direct protein determination**

Specific protein components in milk can be quantified or determined by using a variety of methods explained below.

a. **Dye-binding assays**: These utilise dyes that specifically bind to proteins. The protein content is determined by measuring the intensity of the dye colour, which depends on the protein concentration.

b. **Immunology**: This method relies on using the interaction between an antigen and its corresponding antibody. This methodology is very suitable for the quantification of minor proteins.

c. **Chromatographic methods**: These involve separating the intact proteins based on their physical properties such as size, electrical charge or hydrophilic/hydrophobic properties and then measuring the relative amounts of each. There are several different types of chromatography methods available such as ion-exchange chromatography, size-exclusion chromatography, and reverse phase chromatography. The proteins when separated can be detected by various techniques, e.g., by using mass spectroscopy, fluorescence, or ultraviolet detectors. Calibration with appropriate commercially available protein standards allows quantitation of the protein in the milk.

d. **Electrophoresis**: This is the separation of proteins by charge using an electric field. In this method, the proteins are heat-denatured in the presence of sodium dodecylsulfate (SDS), which enables the alkaline and acidic proteins to migrate in the same direction in the electric field. The SDS binds to the hydrophobic parts of the proteins exposed via denaturation, resulting in an almost equal relation between negative charge and molecular mass for every protein. In a polyacrylamide gel, the rate of movement of the proteins is affected by the pore size of the gel and the electrical field's strength. During application of an electric current, the distance migrated by each protein in a mixture of proteins applied to gel is determined by its molecular mass. Therefore, the smaller proteins move through the gel faster than larger proteins.

iii. **Indirect protein determination**

Protein in milk can be determined using spectroscopic methods but rely on calibration to a chemical or reference method to ensure consistency. Near and mid infrared spectroscopy are used for both liquid and solid dairy products in production laboratories and independent testing laboratories.

11.5.6   Fat Content

Various methods can be used to analyze the fat content in milk. The most used method for whole milk is the Gerber test. In this test, the milk fat is separated from proteins by adding sulfuric acid (H_2SO_4) and amyl alcohol. The H_2SO_4 increases the specific gravity of milk serum, which makes greater difference between milk serum (specific gravity 1.43) and fat globules (specific gravity 0.9). It also destroys stickiness of milk by dissolving all the solid non-fat (SNF). The amyl alcohol helps the separation of fat from the milk acid mixture and prevents the charring of fat and sugar by the H_2SO_4. Mixing of the milk and sulfuric acid produce an exothermic reaction, (producing heat) which completes digestion of the SNF (i.e. disintegrating the emulsion structure in the milk). The free fat globules rise to the surface by subsequent centrifugation of this mixture, which enables the fat particles to release freely to the surface. The free fat is collected in the graduated portion of the neck of a special calibrated butyrometer (or Gerber tube) for measurement.

11.5.7 Freezing Point

The Freezing Point of milk is determined by the water-soluble components in milk such as salts and lactose. The freezing point of the milk is measured by the dairies to determine if it has been diluted with water. Milk of normal composition has a freezing point of −0.54 to −0.59 °C. The freezing point will rise if water is added to the milk. The official method recommended by the Association of Official Analytical Chemists (AOAC) for determining freezing point depression of milk is the **Thermistor Cryoscope method**. The cryoscope uses thermistor probes to measure the changes in electrical resistance with variation in temperatures. The most important cryoscope instruments are the "Fiske Cryoscope" and "Advanced Milk Cryoscope".

11.6 Pasteurized Milk Production

Production of fluid milk starts from the milking stage to obtain the whole milk. The simplest process for producing fluid milk is to pasteurize the whole milk. This process consists of a pasteurizer, a buffer tank and a filling machine. However, the process becomes more complex if several types of fluid milk are to be produced, such as whole milk, skim milk, semi-skim milk, standardized milk of varying fat content, and cream of different fat content. Typical flow diagram for producing (pasteurized) fluid milk is shown in Fig. 11.1.

Fig. 11.1 Flow diagram for production of pasteurized fluid milk with partial homogenization

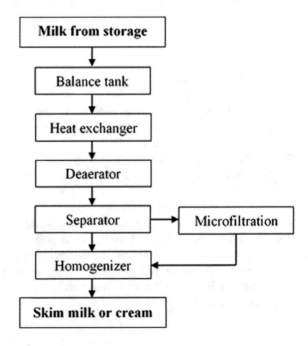

The fresh milk from the storage tank enters the process unit through the balance tank and is pumped to a plate heat exchanger, where it is pre-heated before it continues to the separator, which produces skim milk and cream.

11.6.1 Standardization

The fat content of milk varies depending on species (cow, sheep, goat, water buffalo), animal breed, feed, stage of lactation, among others. To provide the consumer with a consistent product, milk is usually standardized by processing the milk through centrifugal separators to create a skim portion and a cream portion of the milk. The skim portion has less than 0.01% fat, whilst the cream portion is typically 40% fat, even though the separator settings can be altered to achieve the desired fat content of the cream portion. In order to produce the desired fat content for the product, the cream portion is added back to the skim portion. Common products following standardization are whole milk (3% fat), semi-skim milk (between 2 and 1% fat), and skim milk (<0.1% fat).

11.6.2 Homogenization

Homogenization is a process by which the fat droplets in milk are emulsified to prevent the cream from separating. The fat in milk is in globules of non-uniform size between 0.20 and 2.0 μm. The non-uniform size of the globules causes them to float, or cream, to the top of its container. Homogenization reduces the milk fat globules size to less than 1.0 μm, which allows the fat to distribute evenly in milk. It is a high-pressure process that forces the milk at a high velocity through a small orifice to break up the globules. This results in the creation of many fat globules of a smaller size. Homogenization may be partial or total. A smaller homogenizer can be used for partial homogenization, which makes it more economical.

Several factors affect the fat globule size: namely the geometry of the gap and the applied pressure (i.e. higher pressure gives smaller fat globules), milk fat content, homogenization temperature, the type of homogenization device.

Natural enzymes in milk can deteriorate fat (lipases) causing rancidity. This produces off-flavors and reduces the shelf life in milk. Therefore, homogenized milk needs to be pasteurized to inactivate the enzymes. However, unlike homogenized milk, pasteurized milk does not necessarily require homogenization.

11.6.2.1 Homogenization Efficiency

Homogenization must be efficient to prevent creaming. The degree of creaming of homogenized milk can be measured by using the homogenization index, called NIZO (Netherlands Institute for Dairy Research) value, which is commonly accepted by the industry. It is measured using a pipette. The pipet is filled with milk to an upper line marking on the tube, about 25 mL. The tube is sealed and then spun in a heated centrifuge (40 °C) for 30 min. The milk is drained from the pipet after centrifuging until the top level reaches a line at the bottom level of the tube. This leaves about 5 mL of milk in the pipet (about the top 20% of the sample) which is discarded. The bottom portion of the sample is analyzed for percent butterfat, and this percentage is divided by the butterfat percentage for the original milk sample and multiplied by 100 as shown in Eq. (11.2). The larger the number is, the better the homogenization.

$$\text{Homoginization efficiency } (\%) = \frac{\text{fat content in bottom 20 ml of centrifuged sample}}{\text{fat content in non} - \text{centrifuged sample}} \times 100$$

$$(11.2)$$

The required NIZO value of homogenized milk is normally between 85 and 90%, which varies depending on the expected shelf life of milk, for example 70% for pasteurized milk and 80% for extended shelf life (ESL) milk.

11.6.3 Pasteurization

The process of heating a liquid to a temperature below the boiling point to destroy microbes is called pasteurization. In dairy processing, the purpose of pasteurization is to: (i) Increase milk safety for the consumer by destroying disease causing microorganisms that may be present in milk; (ii) Maintain the quality of milk products by destroying spoilage microorganisms and enzymes that contribute to reduce quality and shelf life of milk. This means, pasteurization gives milk a longer shelf life.

Pasteurization conditions require heating the milk to 62.8 °C for 30 min for a batch pasteurization process, or 71.7 °C for 15 s for a continuous process in order to inactivate *Coxiella burnetii*, which causes Q fever in humans and it is the most heat resistant pathogen recognized in milk. However, milk can be pasteurized at higher temperatures and longer processing times than these required minimums.

The most common process used for pasteurization of fluid milk is the continuous process. Batch pasteurization is used in smaller milk processing plants. In the continuous process, the milk is pumped from the raw milk storage tank to a holding tank that feeds into the continuous pasteurization system. This consists of a series of thin plates, which heat up the milk to the appropriate temperature. It is ensured that the pasteurization temperature of the milk is maintained for the suitable time before it flows through the cooling area of the pasteurizer. The cooled milk is pumped to the other processing lines, for example to the bottling station.

The pasteurization process may differ from one country to another, according to their national legislation. In all countries, a common requirement is that the heat treatment must guarantee significant reduction of spoiling microorganisms and destruction of all pathogenic bacteria, without causing damage to the product.

11.6.3.1   Quality of Pasteurized Milk

The composition of milk makes it highly susceptible to chemical (iron, copper, etc.) and bacterial contamination. Therefore, it is very important to ensure that processing lines have high standards, good cleaning facilities and water of high quality.

Exposure to light also affects the quality of milk, such as nutrients and taste, particularly when it is homogenized. Hence, once packed, the product must be protected from sunlight and artificial light.

11.6.3.2   Shelf Life of Standard Pasteurized Milk

The shelf-life of pasteurized milk dependents on the quality of the raw milk. An ordinary pasteurized milk produced from very high-quality raw milk and under good hygienic conditions should have a shelf-life of 8–10 days at 5–7 °C in an unopened package. However, the shelf-life can be shortened considerably if the raw milk is contaminated with heat-resistant bacilli such as *Bacillus cereus* and *Bacillus subtilis* which survive pasteurization in the spore state and/or microorganisms such as species of *Pseudomonas* that form heat-resistance enzymes (lipases and proteases). After pasteurization, re-contamination of milk with Gram-negative bacteria also reduces the shelf-life; hence, it must be avoided.

11.6.4   Extended Shelf-Life (ESL) Milk

Extended shelf-life (ESL) treatment helps to extend the shelf-life of milk beyond its conventional shelf-life by reducing the major sources of re-contamination or reinfection and maintaining the quality of the product. The three main processing technologies for ESL treatment are:

- Pasteurization combined with bactofugation or double bactofugation;
- Pasteurization combined with microfiltration;
- High heat treatment (HHT).

The choice of technology depends on the type of the desired product. Pasteurization combined with single or double bactofugation can prolong the shelf-life of milk by additional 2–5 days. Bactofugation use centrifugal process to separate the microorganisms. Double bactofugation can reduce bacteria spores up to >99%, but this is not sufficient to extend the shelf-life of pasteurized milk beyond 14 days. In order to achieve a shelf-life of 2–4 weeks, microfiltration using a 1.4 μm filter or HHT is required. This can achieve reduction of up to 99.99% of bacteria pores, and the reduction can be increased to up to 99.9999% by application of 0.8 μm membranes.

The HHT treatment requires heating the cream at up to 130 °C for some seconds and remix with microfiltered skim milk through the standardization unit. The standardized milk following remixing is homogenized and finally pasteurized at 72 °C for 15–20 s and cooled to +4 °C.

11.6.5 Packaging of Pasteurized Milk

Milk is a sensitive product; hence, exposure to daylight or artificial light affect the taste and destroys some essential vitamins. Packing protects and preserve the food value and vitamins in milk. It also protects milk from mechanical shock, light and oxygen. Packages use for fluid milk include tin cans, glass bottles, plastic bottles, plastic pouches and milk carton (or paperboard packages). Glass bottles are heavy and fragile, and must be cleaned before re-use, which causes problem for diary industries.

11.7 Production of Long Shelf-Life Milk

As explained in Sect. 11.6.3, pasteurization is the most common type of heat treatment, which is the least heat treatment needed to destroy most pathogenic microorganisms and most spoilage organisms. However, a few bacteria remain after pasteurization and packaging, which can grow during storage. Pasteurized milk is always kept refrigerated but can last for only about two weeks.

The best way of extending the shelf-life of milk for ambient storage is to heat it at temperatures high enough to destroy almost all microorganisms, which should be followed by storing in sealed containers to prevent contamination by bacteria. Two methods are used to produce long-life milk: (i) In-Container Sterilization; and (ii) Ultra High Temperature (UHT) Treatment. Both processes produce a 'commercially sterile' milk which does not contain microorganisms, and so can be stored at room temperature.

11.7.1 In-Container Sterilization

In-container sterilisation (in cans or bottles) involves two processes: (i) batch processing in autoclaves; and (ii) continuous processing systems such as in vertical hydrostatic towers and horizontal sterilizers. In-container sterilization in a batch operation involves heating the final containers of milk in an autoclave at 110–125 °C for 3–40 min.

The batch system can be operated by three methods: (i) in stacks of crates in a static pressure vessel, autoclave; (ii) in a cage that can be rotated in a static autoclave; and (iii) in a rotary autoclave. In autoclave sterilization, the milk is usually pre-heated to about 80 °C and then transferred to clean, heated bottles. The bottles are capped, placed in a steam chamber and sterilized, normally at the above stated conditions. The batch is then cooled, and a new batch filled in the autoclave. The same principle is used for cans. The rotary methods have an advantage over the static method such as: (i) quicker uptake of heat from the heating medium, and (ii) greater uniformity of treatment with respect to bacterial kill and color of the finished product.

Continuous sterilization systems operate at higher capacities and are generally favored. The continuous operation uses a pressure lock system through which the filled containers pass from low pressure/low temperature conditions into a relatively high pressure/high temperature zone. At the end of the sterilization process, the temperature/pressure is decreased gradually, and the containers are finally cooled with cold or chilled water. There are two types of continuous sterilization machines, which are the hydrostatic vertical bottle sterilizer and the horizontal rotary valve-sealed sterilizer. The main difference between them is the type of pressure lock the system uses.

11.7.2   UHT Treatment

UHT processing involves heating the milk in a continuous flow system at about 140 °C for a very short time—about five seconds. During the process, the milk is pumped through a closed system, which is then pre-heated, high heat treated, homogenized, cooled and packed aseptically without any re-infection. The major steps in UHT process are shown in Fig. 11.2. Hot milk from post-sterilization is used as a heating source in a tubular or plate heat exchanger to preheat the milk from ~ 5 to ~ 90 °C. This heat regeneration step is very important for energy efficiency of the UHT plant. Depending on the plant type, more than 90% of the heat can be regenerated.

The final heating step to the required sterilization temperature in UHT systems are divided into two: direct and indirect. In the direct systems, there is direct contact between the product and the heating medium, which is followed by flash cooling in a vacuum vessel, homogenization and finally further cooled indirectly to packaging temperature. There are two types of direct systems: steam injection systems (steam injected into product) and steam infusion systems (product introduced into a steam-filled vessel).

In the indirect systems, there is a partition (plate or tubular wall) that separates the product from the heating media, so during heat transfer, there is no direct

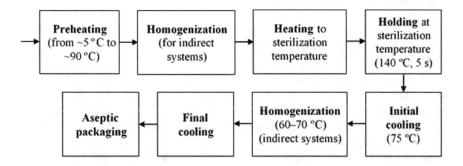

Fig. 11.2 Generalized flow diagram for UHT processing of milk

contact between the media and the product. The indirect systems can use the following heating media: plate heat exchangers, tubular heat exchangers, and scraped surface heat exchangers. The heat exchangers can be combined in the indirect systems based on process and product requirements. Also, a combination of direct heating and indirect cooling is possible without subsequent flash flooding. In treatments, the condensate created by the steam while heating remains in the product. The rate at which the milk is heated is the major difference between the direct and indirect systems. In the direct systems, it takes less than one second to heat the milk from preheat temperature to sterilization temperature, but the indirect systems can take quite a few seconds to minutes. The major consequence of this difference is that, for the same bactericidal effect, the direct systems produce much less chemical change in the milk constituents than the indirect systems.

Once the sterilisation temperature is reached, the milk enters a holding tube where it is maintained at 140 °C for five seconds as the milk passes through the holding tube. In a direct system, initial cooling of the product is achieved very quickly as it passes through a vacuum chamber, which removes water condensed into the product during the steam heating. This returns the product temperature to about 75 °C.

For a product containing fat, such as whole milk, a homogenization step at 60–70 °C is included. This is carried out either before or after the sterilization step. If the homogenization is carried out after sterilization, it must be aseptic so that no bacteria can be introduced after sterilization. Even though this puts a high demand on plant operators, milk process by a direct heating process must be homogenized after the sterilization step to break up aggregates of protein, which form during the heating and cause astringent taste in the milk.

In the final cooling step in direct systems, and in both cooling steps in indirect systems, the heat from the hot milk is transferred to the cold milk in the preheating/heat regeneration stages.

Compared to traditional sterilization in hydrostatic towers, UHT treatment of milk saves time, labor, space and energy. Also, UHT has less effect on the flavor and color of the milk. However, UHT-treated milk is tasteless compared with autoclave sterilized milk, which has a "cooked" or caramel flavor. For this reason, many consumers prefer pasteurised milk.

11.7.3 Packaging of UHT Milk

UHT milk has a long non-refrigerated shelf-life; hence, the package must also give almost complete protection against light and atmospheric oxygen in order to protect the sensory quality of the product and the nutritional value. Therefore, carton used for long-life milk packaging must be of high-quality carton board inserted between layers of polyethylene plastic. Shown in Fig. 11.3, the packaging process involves sterilization of the packaging material or container, filling with the UHT (sterile) milk in an aseptic environment and using containers that are tight enough to prevent recontamination, i.e. that are hermetically sealed. The most common aseptic packages used for UHT milk are paperboard and multi-layered plastic. The sterilization of the packages is usually done using hot hydrogen peroxide, which is followed by hot air to remove residual peroxide. The sterilization is done before the packages are filled.

The term "aseptic" refers to the absence of any unwanted organisms from the product, package, or other specific areas. "Hermetic" implies suitable mechanical properties to exclude the entry of bacteria into the package or, more strictly, to prevent the passage of microorganisms and gas or vapor into or from the container.

11.8 Concentration of Milk

In the dairy industry, evaporation is used to concentrate whole milk and other milk products. Indirect heating is used to evaporate water in the milk. A steel separates the product and the heating medium (steam) from one another. The partition (steel) serves as the medium through which the heat released during the condensing of the steam is transferred to the product. The time required for evaporation is determined by the properties of the product such as heat stability and viscosity. To avoid heat damage, and thus minimize the impact of heat on the products, evaporation takes place in a vacuum at pressures of 16–32 kPa.

11.8.1 Evaporated Milk

Evaporated milk, also known as 'unsweetened condensed milk', is produced by removing about 60% of the water from fresh milk. Evaporated milk is canned. It is a sterilized product, light in color and has a creamy appearance. The slightly caramelised colour of evaporated milk is the results of the heating during the evaporation process. Almost all the nutrients found in fresh milk, such as buttermilk, lactose, minerals and vitamins are retained in the evaporated milk. It can have a shelf-life of months or even years without refrigeration, depending upon the fat and sugar content. This makes evaporated milk a safe and reliable substitute for perishable fresh milk.

The process for making evaporated milk is shown in Fig. 11.4, in which 60% of the water is evaporated from the milk. Evaporation takes place under vacuum at a

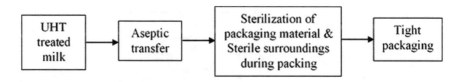

Fig. 11.3 Packing of UHT milk under aseptic conditions

temperature between 65 and 70 °C. Temperatures below 65 °C is an ideal growth conditions for spores and heat-resistant bacteria, which could result in the entire process being spoiled. Accurate control of the bacteria in the process is therefore an essential requirement in the manufacture of evaporated milk.

Another important condition in the manufacture of evaporated milk is the ability of the milk to tolerate intensive heat treatment without coagulating (i.e. thermal stability). This depends greatly on the acidity of the milk, which should be low, and on the salt balance in the milk. The thermal stability of the milk can be improved by addition of additives or standardization of fat content and solids-non-fat or pre-heating.

Prior to evaporation, the raw milk is standardized to the required content of fat and dry solids, which vary with the applicable standard, but are usually 7.5% fat and 17.5% solids-non-fat. This produces double concentrated milk with 25% total solids. Triple concentrated milk is another common standard with 33% total solids (often 4–10% fat content).

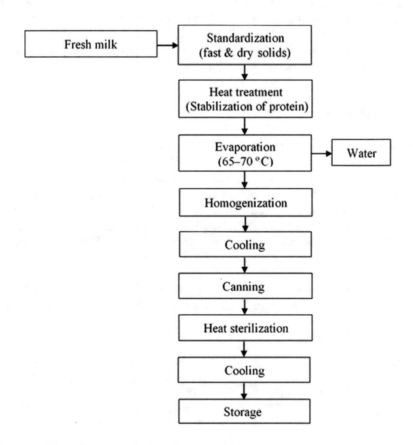

Fig. 11.4 Process line for production of evaporated milk

The standardized milk then undergoes intensive heat treatment to destroy microorganisms and to improve its thermal stability. The heat treatment is often integrated in the evaporation plant, which takes place in a tubular or plate heat exchanger at a temperature of 100–120 °C for 1–3 min, followed by chilling to about 70 °C before the milk is sent to the evaporator. The heat treatment denatures a great part of whey proteins, while calcium salts are precipitated. This stabilizes the protein complex of the milk so that it can withstand subsequent sterilization without coagulation during the process or successive storage. Heat treatment is very important for the product quality as this determines the color and viscosity of the produced milk.

During evaporation, the milk passes through steam-heated tubes under vacuum. Boiling takes place at 65 and 70 °C. The dry matter content of the milk increases as the water is evaporated. Continuous monitoring of the density is required. Multistage falling-film type evaporator is usually used.

The concentrated milk following evaporation is pumped to a homogenizer, which operates at a pressure between 50 and 250 bar. Homogenization disperses the fat and prevents the fat globules from coalescing during subsequent sterilization. For in-can sterilization, the pressure is generally in the range 125–250 bar (two-stage). It must be noted that too intensive homogenization might impair the stability of the protein, which will cause the milk to coagulate during sterilization. Hence, an optimal pressure is required. The pressure should be high enough to produce the needed fat dispersion, but low enough to eliminate the possibility of coagulation. After homogenization, the milk is cooled to about 14 °C before packaging, or to 5–8 °C, if it will be stored for sample sterilization. Usually, final check of the fat content and the solids-non-fat is made at this stage. Also, this is the stage for any addition of vitamins. The concentrated milk can now be stored, but not for so long to prevent bacteria growth and the tendency of age gelation of the final product.

The cooled condensed milk following homogenization is filled in cans and sealed using automatic canning machines before sterilization. The temperature of canning is carefully chosen so that the possibility of froth formation is at the lowest. The filled and sealed cans are transferred to the autoclave, which for either batch or continuous sterilization (see Sect. 11.7) are sterilized at a temperature of 100–120 °C for 15–20 min, and then cooled to room temperature. During this operation, any protein precipitated during the heat treatment is uniformly distributed throughout the milk.

11.8.2 Condensed Milk

Condensed milk is produced from either whole milk, or skim milk, or recombined milk added with anhydrous milk fat, skim milk powder and water. The increased dry matter is achieved by either evaporation of fresh milk or by addition of milk powder. Sweetened condensed milk is a concentrated milk, with added sugar. It has a yellowish color and high viscosity. The high sugar content (between 62.5% and

64.5%) preserves it for a long shelf-life, by increasing the osmotic pressure to a very high level, which inhibits bacterial growth and destroys most of the microorganisms. Therefore, sweetened condensed milk does not require heat treatment after packaging and requires less processing time. The manufacturing process is shown in Fig. 11.5.

During the process, the fat and solids-non-fat contents of the milk are standardized to predetermined levels in the same way as for evaporated or unsweetened condensed milk. The milk is heated to stabilize the protein complex and to destroy enzymes and microorganisms that might cause problems. The heat treatment is an important step as it determines the viscosity of the product during storage. A gel can

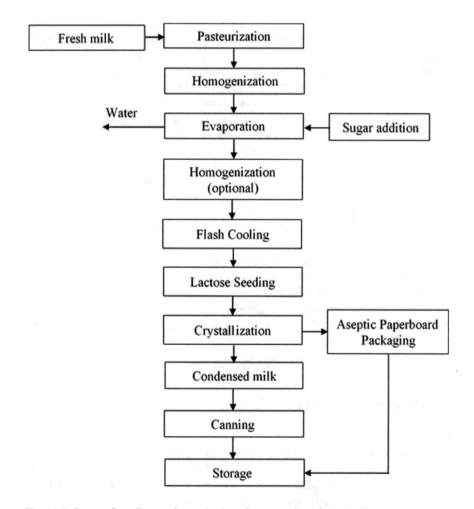

Fig. 11.5 Process flow diagram for production of sweetened condensed milk

form if a severe heat treatment is applied. To produce milk with a relatively high viscosity, it is usually heat-treated at 82 °C for 10 min.

If a low-viscosity milk is required, a higher temperature for a shorter time is required, for example 116 °C for 30 s.

The heat-treated milk is then pumped to the evaporator, where it is concentrated. The evaporation is carried out the same was as for evaporated or unsweetened condensed milk. Sugar syrup is added to the concentrate during evaporation, but the sugar can also be added in solids form, in the correct proportion calculated on dry substance, before evaporation. It is important to add the correct proportion of sugar (at least 62.5% in the aqueous phase), because the shelf-life of the milk depends on sufficiently high osmotic pressure that can inhibit the growth of bacteria. The dry matter content in the milk is measured indirectly by determining the density of the concentrate, and evaporation continues until the required dry matter content has been achieved. In some processes, the concentrate is homogenized at 50–75 bar immediately after evaporation to control the viscosity of the produced condensed milk.

After concentration, the product is cooled in such manner that the lactose forms very small crystals in the supersaturated solution. This is achieved by cooling the mixture rapidly under robust agitation, without air being entrapped, and usually by flash cooling. The crystals form must be less than 10 μm, so that they cannot be detected by the tongue. The crystals should remain dispersed in the milk under normal storage temperatures, 15–25 °C. Finely ground lactose crystals are added at a rate of about 0.05% of the total mix, either as a slurry or powder, once the milk has cooled to crystallization temperature (about 30 °C), the supersaturation temperature of the sugar solution. At this temperature, the seed lactose does not dissolve. However, a too slow temperature must be avoided in order to prevent spontaneous nucleation before mixing the seed crystals.

After cooling and crystallization, the sweetened condensed milk is packed, usually in cans, although it is possible to pack in aseptic paperboard packages. In the case of cans, they must be cleaned and sterilized before filling, because no sterilization takes place after canning.

11.9 Fermented Milk Products

Fermented milk products also known as cultured dairy products, or cultured milk products, are dairy products that have been fermented with lactic acid bacteria such as *Lactobacillus*, *Lactococcus*, and *Leuconostoc*. During the fermentation, the milk is inoculated with a starter culture which converts part of the lactose to lactic acid. Depending on the type of lactic acid bacteria used, carbon dioxide, acetic acid, diacetyl, acetaldehyde and several other substances are formed in the conversion process, and these give the products their characteristic fresh taste and aroma. The fermentation process increases the shelf-life of the product, while enhancing the taste and improving the digestibility of milk because of lactose breakdown.

Cultured milk products have low pH, which inhibits the growth of putrefactive bacteria and other detrimental organisms, thereby prolonging the shelf-life of the product. In contrast, acidified milk provides favorable conditions for growth of yeasts and molds, which cause off-flavors in the product, blown packages, etc.

Production of fermented milk products require the best possible growth conditions for the starter culture. These conditions are achieved by heating the milk to destroy any competing microorganisms. Also, the milk must be held at the optimum temperature relevant for the starter culture. The fermentation time is very vital; shorter or longer time will impair the flavor and affect the consistency and appearance of the product.

A variety of fermented milk products can be found around the world. Some of the most common fermented milk products include cheese, yoghurt and cultured (or) sour cream, which are described below. The production techniques of the fermented products have many similarities; for example, pre-treatment of the milk is almost the same. The general processing steps for the manufacture of yoghurt, sour cream and cheese are illustrated in Fig. 11.6.

Fermented dairy products are normally made from a standardized whole milk or a mix from whole, partially defatted milk, or condensed skim milk, cream and/or non-fat dry milk. All the raw materials should be selected for the products to have high bacteriological and sensory quality.

11.9.1   Manufacture of Yoghurt

Yoghurt is the most popular of all fermented milk products worldwide. It is a semi-solid fermented product made from a heat-treated fat-standardized milk fortified with non-fat milk solids by the activity of a characterizing symbiotic blend of *Streptococcus thermophilus* and *Lactobacillus delbrueckii* subspecies *bulgaricus* cultures. To make yoghurt, milk is added with non-fat dry milk or condensed skim milk to increase the solids-not-fat (SNF). In the industry, up to 12% SNF is used in the yoghurt mix to produce a thick, custard-like consistency. However, the most typical fat content is between 0.5–3.5%. As classified by FAO/WHO, yoghurt (full-fat) contains minimum of 3% milk fat, whilst milk fat for partially skimmed yoghurt contains 0.5–3%, and fat level not exceeding 0.5% for non-fat or skimmed yoghurt.

The consistency, flavor and aroma vary from one manufacture or country to another. Yoghurt can be produced as a drink, a softer gel, a highly viscous liquid or in a frozen form as dessert. Sweeteners may be used, such as invert sugar, sucrose, honey, refiner's syrup, brown sugar, high fructose corn syrup, fructose, fructose syrup, molasses (other than blackstrap), maltose, maltose syrup, dried maltose syrup, malt extract, dried malt extract, malt syrup, dried malt syrup, maple sugar, except table syrup. Normally, commercial yogurts contain an average of 4.06% lactose, 1.85% galactose, and 0.05% glucose.

In addition, stabilizers, flavoring ingredients, and color additives may be added. Stabilizers when added in yoghurt produce smoothness in body and texture and

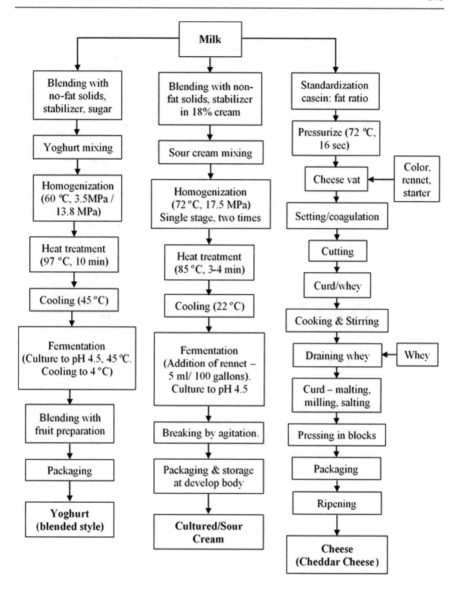

Fig. 11.6 General processing steps for manufacture of yogurt, sour cream, and Cheddar cheese

impart gel structure. Stabilizers increase shelf-life and provide some degree of uniformity of the product. A good yogurt stabilizer should not impart any flavor, should be easily dispersed in the normal working temperatures of the yoghurt, and should be effective at low pH values. Typical stabilizers used in yoghurt are modified starch, gelatin, whey protein concentrates and pectin. Generally, a

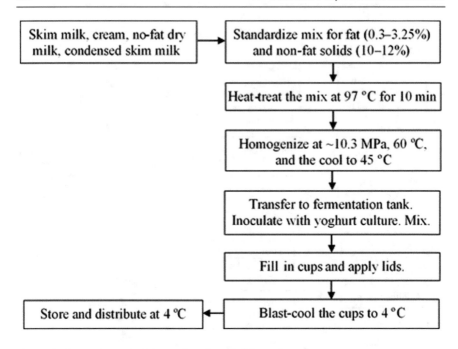

Fig. 11.7 Processing steps for manufacturing of plain yoghurt

combination of starch and gelatin is used in yoghurt mix. Also, whey protein concentrates at 0.5–1% level is usually used for their water-binding attributes.

11.9.1.1 Processing Steps

(a) **Blending**:

Milk of highest bacteriological quality is required to produce a high-quality yoghurt. The milk must have a low content of bacteria and substances which may delay the development of the yoghurt culture. The milk must not contain bacteriophages, antibiotics, residues of clean-in-place (CIP) solution or sterilizing agents. The milk must therefore be very carefully analyzed at the dairy.

The milk is blended with additives (non-fat solids, stabilizer, sugar, etc.) prior to heat treatment.

Standardization of the milk for fat and milk solids-non-fat content reduces the fat content and increase other solid components (lactose, protein, mineral, and vitamin content) to approximately 30–35%. Specific gravity changes from 1.03 to 1.04 g/mL at 20 °C. Nutrient density of yogurt is increased in relation to that of milk.

(b) **Homogenization**:

Homogenization prevents creaming during the incubation period and assure uniform distribution of the milk fat. It also improves the stability and consistency of fermented milks, even those with low fat contents. It is recommended that the milk should be homogenized at 65–70 °C and 20–25 MPa to obtain optimum physical properties in the yoghurt. Homogenization followed by heating at a high temperature, usually 90–95 °C for about five minutes improves the viscosity and stability of the final yoghurt.

(c) **Heat treatment**:

The milk is heated before it is inoculated with the started. Plate heat exchanger or high-temperature, short-time (HTST) pasteurization system is used to heat the mix at 85 °C for 30 min or 95–99 °C for 5–10 min. This condition denatures about 70–80% of the whey proteins (99% of the β-lactoglobulin) in particular. The β-lactoglobulin, which is the principal whey protein, interacts with the K-casein, thus helps to give the yoghurt a stable body.

The heat treatment:

- Produces a relatively sterile medium for the exclusive growth of the bacteria culture.
- Causes thermal breakdown of the milk proteins, releasing peptones and sulfhydryl groups, which provide nutrition and anaerobic conditions for the yogurt culture.
- Removes air from the medium to produce a more conducive medium for micro-aerophilic lactic cultures to grow.
- Denatures and coagulates whey proteins (β-lactoglobulin), which improves water-binding properties, thus creating a smooth consistency, high viscosity and stability of yogurt.
- Ensures that the coagulum of the finished yoghurt will be firm.
- Reduces the risk of whey separation in the produced yoghurt.

The heat treatment during yogurt processing inactivates milk enzymes inherently present and destroys all the pathogenic flora and most vegetative cells of all microorganisms contained in it. The destruction of competitive organisms produces favorable conditions for the growth of desirable yogurt bacteria.

(d) **Fermentation**:

The starter is generally inoculated at 5% level and yoghurt is maintained at a temperature of approximately 43 °C for 4–8 h. Longer fermentations can yield tangier flavor and fuller digestion of lactose. Appropriate fermentation with yogurt culture leads to the formation of the typical flavor compounds of yoghurt; lactic acid, acetaldehyde, acetoin, diacetyl, and other carbonyl compounds. The activity of the yoghurt culture causes the milk coagulum which results from the drop in pH. Streptococcus thermophilus dominates the early stage of yoghurt fermentation. As redox potential of the milk medium is reduced, the pH decreases from 6.5 to 5.5. This stimulates the growth of *Lactobacillus* (LB), which produces acetaldehyde and lactic acid, which give the green apple flavor characteristics of yoghurt. Continued acid production lowers the pH of yoghurt to 4.6, which results in clotting. Fermentation is completed at pH 4.5.

The fermentation tank normally has a cone bottom to facilitate draining of viscous yoghurt after incubation. The tank is often provided with swept surface type agitator for optimal mixing of the viscous cultured yoghurt. Triple-tube or plate heat exchangers are used in order to achieve efficient cooling after culturing.

(e) **Blending yoghurt with fruit preparation**:

Blending yoghurt with various fruits, flavoring and aroma additives is very popular in the market and are usually designed for marketing requirements. The contents of fruit preparations in the final yoghurt are generally in the range 10–15%, of which 50% are sugar. Fruit preparations mostly contain natural flavors to improve aroma and flavor. Additives usually added to yoghurt are fruits and berries in the form of a syrup or as a purée. The fruit is normally mixed with the yoghurt either before or during the packing. It can also be placed at the bottom of the pack before being filled with the yoghurt. Sometimes yoghurt is also flavored with vanilla, honey, coffee essences, etc. Colors are normally added to improve the appearance. Alternatively, the fruit can be separately packed but in the same package. The fruit mix should have zero mold and yeast population to prevent spoilage and to extend the shelf life of the yoghurt. The pH of the fruit mix should be compatible with the pH of the yoghurt (about 4.4). In addition, stabilizers may also be added to modify the consistency.

Typically, fruit yoghurt contains: 12–8% fruit mix, 0.5–3% fat, 3–4.5% lactose, 11–13% Milk solids non-fat (MSNF), and 0.3–0.5% stabilizer (if used).

11.9.1.2 Some Types of Yoghurt

(a) **Plain yoghurt**:

Plain yogurt is the basic type and forms an essential component of fruit-flavored yogurt. Plain yogurt normally contains no added sugar or flavors. This offers the option of adding flavoring of the consumer's choice. Plain yoghurt can also be used for cooking or for salad preparation with fresh fruits or grated vegetables. Full-fat yoghurt has a minimum fat content of 3.25%.

However, the fat content can be standardized to consumer demands. Typical packaging materials used in the industry are polystyrene and polypropylene cups and lids. The processing steps in Fig. 11.7 are involved in the manufacturing of plain yogurt.

(b) **Stirred style or blended yoghurt**:

Stirred style/blended yogurt is produced by incubating the warm cultured milk in a fermentation tank, cooled and then stirred for a creamy texture. Often fruits and other flavorings are added. Stirred style yogurt is usually thinner than plain yogurt and can be consumed as-is, added to cold beverages or desserts. Typical processing steps for production of stirred yoghurt is shown in Fig. 11.8.

The incubation tanks can be fitted with pH meters to check the level of acidity. When using highly concentrated culture (about 0.02% Inoculum), the incubation period of about 4–5 h can be used at 42–43 °C. The generation period for typical yoghurt bacteria is approximately 20–30 min. To achieve optimum quality conditions, cooling to 15–22 °C (from 42–43 °C) should be attained within 30 min after the desired pH has been reached, to stop further growth of bacteria.

(c) **Light yoghurt**:

Light yogurt is stirred/blended type, which is made without any added sugar. The sugar in stirred/blended type is replaced with high-intensity synthetic sweeteners such as sucralose of aspartame, which may be added during the fruit preparation. The fruit preparation contains 30–60% fruit. Light yoghurt consists of lower sugar content and calories.

(d) **Drinking yoghurt**:

Drinking yoghurt is a low viscosity made from low solid mixed with and contains a low-fat content. It is made in a similar process as used for stirred/blended yogurt, but the coagulant is broken down to a liquid before being packed. It can have the same composition as for stirred/blended yoghurt but can be reduced in dry matter by dilution with water. That is a mixture of 30% yoghurt and 70% water. Also, fruit preparations in drinking yoghurt mostly consist of purees and juices. Other additives added to drinking yoghurt are sugar and aroma. The steps for production of drinking yoghurt are shown in Fig. 11.9.

To produce a stable drinking yoghurt without sedimentation, a stabilizer should be added before cooling. Stabilizers usually used are pectin, modified starch or carboxymethyl cellulose (CMC). Addition of the stabilizer prevents sedimentation and whey separation. It also improves the viscosity and mouth feel of the yoghurt following heating. After addition of the stabilizer, especially, pectin, the yoghurt is homogenized before cooling to obtain the best stabilizing effect.

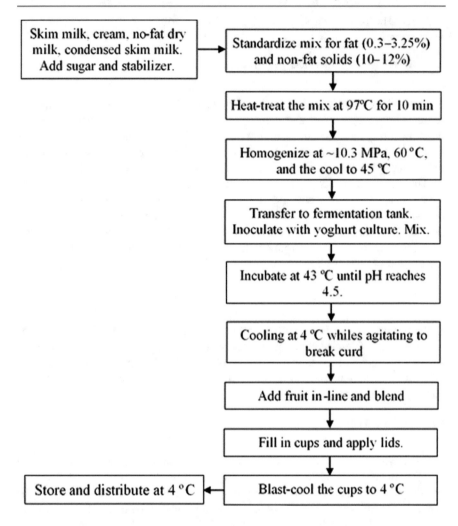

Fig. 11.8 Processing steps for manufacturing of stirred style or blended yoghurt

(e) **Frozen yoghurt**:

Frozen yogurt is a low-acid product, which has similar flavor and texture as that of a low-fat ice cream, with only 0.3% titratable acidity. There are two types of frozen yoghurt: soft-served and hard-frozen (Table 11.4). Usually, frozen yoghurt is produced by blending skim milk fermented with yoghurt culture with low-fat ice cream mix. An alternative procedure is to ferment low-fat ice cream mix to a pH of ~ 5.9–6.0 before further processing. In this case, a conventional line for production of stirred type/blended yoghurt can be used. About 4–6% starter culture is added to the mix as it is being pumped to the incubation tank. Yoghurt mix contains

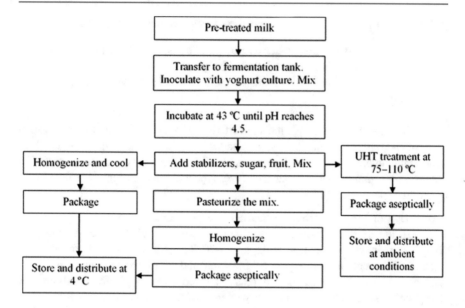

Fig. 11.9 Processing steps for production of drinking yoghurt

more carbohydrates than normal yoghurt; hence, the incubation time is much longer than for other yoghurt production. The incubation time for frozen yoghurt is about 7–8 h. This is the ideal time required to attain the typical acidity (pH 4.5) of yoghurt, for saccharose content of 10–12%.

Sometimes the blend is pasteurized to destroy pathogens and yoghurt bacteria in the low-acid product. To produce yoghurt containing live and active culture, a frozen culture concentrate is blended with the pasteurized yoghurt.

11.9.1.3 Production of Long-Life Yoghurt

Shelf-life of yoghurt depends on bacteriological defects and on other factors such as changes in viscosity, whey separation, color, acidity and aroma. To avoid the difficulty in maintaining the integrity of the cooling chain of 'normal' yoghurt, a means to sterilize yoghurt so that it can be stored at room temperature is essential. The shelf-life of all types of yoghurt can be prolonged by heating.

Table 11.4 Typical composition of frozen yoghurts

Ingredients	Soft-served (%)	Hard-frozen (%)
Fat	4.0	6.0
Milk solids non-fat	10.0–11.0	12.0
Sugar	11.0–14.0	12.0–15.0
Stabilizer, emulsifier	0.85	0.85
Water	71.0	66.0

There are two ways in which shelf-life of yoghurt can be extended:

(i) Production and packing under aseptic conditions; by taking measures to prevent the yoghurt from being infected by yeast and molds.
(ii) The finished yoghurt is heat treated either instantly before packing or in the package.

The heat treatment prolongs the shelf-life by: (a) inactivating contaminants such as yeast and molds; and (b) inactivating starter bacteria and their enzymes. The heat treatment temperature depends on factors as milk quality, milk pre-treatment, pH of yoghurt, fruit quality, particle size, stabilizer type and microbiological requirements of the final product.

Production of long-life stirred type/blended yoghurt:
The shelf-life of stirred type/blended yogurt can be extended by heat-treating the coagulum from the incubation tanks at 60–70 °C for a few seconds during production. This will minimize post-acidification and gives the yoghurt a bacteriological shelf-life of about 1 to 2 months if packed under hygienic conditions and stored cold. To be able to store the yoghurt under ambient conditions, a higher temperature in the range 75–110 °C for some seconds should be applied. The temperature to be applied depends on factors such as milk quality, milk treatment, and pH of yoghurt.

The heat treatment can be applied in different ways: (i) heat-treat and cool yoghurt and fruit mixed together; (ii) heat-treat yoghurt and fruit together, and cool separately before mixing; (iii) heat-treat and cool yoghurt, heat-treat fruit, and mix warm fruit with cold yoghurt.

In all cases, the yoghurt should be packed in an aseptic manner to prevent reinfection.

Common problems related to heating of fermented milk are decrease in viscosity and whey separation. However, adding stabilizers help to solve these problems by re-building the rheological properties of the yoghurt.

11.9.2   Manufacture of Cultured (or Sour) Cream

Cultured cream (or sour cream) is made by maturing pasteurized cream of 18% fat content with lactic acid and aroma-producing bacteria. Cultured cream contains not less than 18% milk fat. Cultured cream products containing bulky ingredients may have lower fat content but cannot be below 14.4%. Cultured cream must have a titratable acidity of at least 0.5%, expressed as lactic acid. Cultured cream is relatively viscous and has a uniform structure. It has a mild and slightly acidic taste. Cultured cream has a limited shelf-life; therefore, strict hygienic practices should be followed during production to ensure product quality. Extended storage causes the lactic acid and bacteria enzymes to break down β-lactoglobulin and become active resulting in the cultured cream becoming bitter. Packages must be airtight to prevent development of yeast and molds on the surface of the product. Also, poor packaging causes the cultured cream to lose its flavor because carbon dioxide and other aromatic substances diffuse through the packaging. Similar to yoghurt, the long-life of cultured cream can be achieved by heat treatment of the product before packing. Also, stabilizers may be added to improve texture, avoid whey separation, maintain body and enhance shelf-life. Salt, rennet, flavorings, nutritive sweeteners and fruit products may be added. Cultured cream is often used in cooking.

11.9.2.1 Processing Steps

The process steps for production of cultured cream involve standardization of the fat content, homogenization, heat treatment of the cream, inoculation and packing (Fig. 11.6).

(a) **Raw materials**:

Raw cream separated from milk for production of cultured cream should be fresh with low bacteria count and of high organoleptic and microbiological standards. The raw cream should not be mixed vigorously as this leads to activation of lipase and disruption of milk fat globule membrane, which result in rancid flavor. Condensed skim milk or non-fat dry milk is usually added to increase the solid-non-fat in cultured cream. The non-fat milk solids are used to standardize cultured cream mix to 9.0–9.5%.

Cream is mixed with non-fat dry milk (or condensed skim milk) at 4 °C to standardize its composition to 18.5% milk fat and 27.5% total solids. After this, a stabilizer is added during which care is taken to ensure hydration of the gums preventing incorporation of air and excessive foam formation. For cultured cream containing 18% milk fat, a blend of stabilizers at a level of 1.5–1.8% is used comprising of gelatin, modified starch, guar gum, locust bean gum, Grade A whey products, sodium phosphate and calcium sulfate. For low-fat cultured cream, the level of the stabilizer combination is increased to 1.75–2.0%. The blend of stabilizers used for non-fat cultured cream are modified starch, propylene glycol monoester and additional water-binding agents such as cellulose gum, microcrystalline cellulose, and gum acacia.

(b) **Homogenization**:

The cream mix is normally homogenized by two single-stage passes at a pressure of 17.2 MPa. This helps to produce firm-bodied sour cream. For cream with 10–12% fat, the homogenization pressure is usually 15–20 MPa at 60–70 °C. Typically, an increase in homogenization temperature improves the consistency of the cream, although this is up to a certain point. For cream with 20–30% fat, a lower homogenization pressure (10–12 MPa) is required, because there is not enough protein (casein) to form membranes on the enlarged total fat surface.

(c) Heat treatment:

After homogenization, the cream is heat treated at about 74–79 °C for 30 min or at 82–90 °C for 3–5 min. Other combinations of temperature and time can be used if the homogenization process is carefully matched to the heat treatment. The heat treatment denatures whey proteins, resulting in improved water absorption and leads to higher viscosity as well as a smooth thick body after culturing. After heat treatment, the cream mix is cooled to about 21–24 °C and then transferred to the fermentation tank.

(d) Fermentation:

The cooled heat-treated (or pasteurized) cream is inoculated with direct set or bulk cream culture, just like in yoghurt production. About 0.01% concentrated culture or 1–2% of bulk starter culture is added. The fermentation time is 18–20 h, and it is continued until the pH drops to 4.5 before breaking by agitation. After fermentation is completed, the cultured cream is cooled quickly, to prevent any further pH reduction. For low fat (10–12%) cream, because the viscosity is moderate the cooling can take place in a heat exchanger. In the case of higher fat cream, the fermented product is directly fed to the packaging and cooled in the packages as cooling in a plate heat exchanger will be difficult due to the high pressure drop. Alternatively, for high fat cream products, inoculation can also take place in individual-size packages. In this option, packages are filled with the inoculated mix and incubated to develop the required acidity to avoid mechanical treatment and cooled afterward. In some cases, the packages are filled soon after ripening, followed by cooling.

(e) Breaking:

The set bulk cultured cream is agitated slowly to break it before it is pumped through a sieve to a packaging machine. The agitation also ensures that pockets of whey are mixed back into the cultured cream.

(f) Packaging:

After inoculation (and breaking), cultured/sour cream containing live organisms is packaged into suitable containers at incubation temperature of 20–24 °C. The packages are then carefully transferred to cold store, where they are kept at a temperature of about 4–6 °C overnight before distribution. This process allows the mass to reform and to achieve maximum body firmness.

11.9.3 Manufacture of Cheese

Cheese is a milk concentrate, which contains mainly casein protein and fat as the basic solids in which the whey protein/casein ratio does not exceed that of milk. Cheese is produced by coagulating (wholly or partly) milk, skimmed milk, partly skimmed milk, cream, whey cream, or buttermilk, through the action of rennet or other suitable coagulating agents, and by partially draining the whey resulting from such coagulation.

There are several types of cheese in the world: more than 400. Each variety is distinguished by several characteristics, such as structure (texture, body), flavor and appearance, which results from the ingredients (the type of milk, the choice of bacteria) and the processing technique employed. Cheese can be made using pasteurized or raw milk. Cheese made from raw milk gives different flavors and texture characteristics to the finished cheese. For some varieties of cheese, raw milk is heat-treated at a mild temperature (below pasteurization) before making the cheese to destroy some of the spoilage organisms and to provide the best conditions for the cheese cultures. In order to reduce the risk of exposure to disease causing microbes such as pathogens that might be present in the milk, cheese produced from raw milk must be aged for at least 60 days. Other cheese varieties required to be aged longer than 60 days.

Cheese is categorized into hard, semi-hard and soft cheeses (Table 11.5), which are based on the moisture content of the cheese. For example, hard cheese has low moisture content, etc. The casein and fat in the milk are concentrated about 10 times in the manufacture of hard and some semi-hard types of cheese.

The percentage of moisture in fat-free basis (FFB) and fat in dry solids basis (DSB) in cheese can be estimated by using the following equations:

$$\% \ \text{Moisture (FFB)} = \frac{\text{Weight of moisture in cheese}}{\text{Total weight of cheese} - \text{weight of fat in cheese}} \times 100$$

$$(11.3)$$

$$\% \ \text{fat (DSB)} = \frac{\text{fat content in cheese}}{\text{Total weight of cheese} - \text{weight of fat in cheese}} \times 100$$

$$(11.4)$$

Cheese can also be characterized as acid or rennet cheese, and natural or process cheeses. Acid cheeses are made by adding acid to the milk to cause the proteins to coagulate. For most types of cheese such as cheddar, coagulation of the milk requires the addition of an enzyme called *rennet* to the starter cultures. Natural cheese is a term used by dairy industries for cheese made directly from milk. Process cheese is made using natural cheese and additional ingredients that are cooked together to change the textural and/or melting properties and increase shelf-life.

Table 11.5 Categories of cheese

Category	% Moisture (fat-free basis)	% Fat (dry solids basis)
Extra hard	<41	>60
Hard	49–56	45–60
Semi-hard	54–63	25–45
Semi-soft	61–69	10–25
Soft	>69	<10

11.9.3.1 Raw Materials

Milk is the main ingredient in cheese. The variety of cheese required determines the type of coagulant to be used. For acid cheeses, an acid source such as acetic acid (the acid in vinegar) or gluconodelta-lactone (a mild food acid) is used. Calf rennet or rennet produced via microbial bioprocessing is used for rennet chesses. At times calcium chloride is added to the cheese to enhance coagulation properties of the milk. Depending on the type of cheese, flavorings may also be added.

11.9.3.2 Processing Steps

The type of cheese to be produced will result in a considerable variation of the cycle of processing steps, time, temperature, pH, salting/brining, block formation, and aging. The general processing steps for the production of hard and semi-hard cheese are illustrated in Fig. 11.10. Also, there are other treatment types that are specific to particular cheese varieties.

1. **Thermization**

Milk must be processed within about 12 h when it is delivered to the dairy to prevent liberation of lactic acid, especially the lower one, by lipase action which gives the milk a rancid flavor. To prevent the milk from going rancid, it is chilled to about +4 °C or, preferably, thermize. Thermization is moderate heat treatment of cheese milk at 63–65 °C for 15 s, followed by cooling to +4 °C, after which the milk is still phosphate positive. This helps to prevent the growth of psychrotrophic flora in the milk when it is stored for a further 12–48 h after arrival at the dairy. The "critical age" of raw milk kept at +4 °C normally falls between 48 and 72 h after milking.

2. **Standardization of milk**

Types of cheese are usually classified according to the fat content in dry solids basis (DSB). Hence, the fat content of the cheese milk should be adjusted accordingly to a fat: casein ratio of 1.47. The protein content of the cheese milk can also be standardized. The final fat, protein and dry solids ratios are important factors required to make a good quality cheese with a high yield.

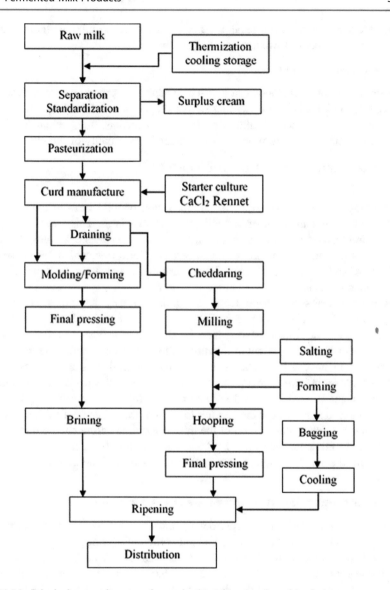

Fig. 11.10 Principal processing steps for production of hard and semi-hard cheese

Fat standardization

Standardization of fat can be performed in two ways: by in-line remixing after the separator, or by mixing whole milk and skim milk in tanks followed by pasteurization. Various automated milk standardization instruments are available. An example is infrared technology used to improve production efficiency as well as the quality and consistency of cheese. Fat content of raw milk can be reduced to a desirable level by using a separator. Another method is by mixing whole milk and skim milk in tanks followed by pasteurization.

Protein standardization

Membrane filtration methods can be used to adjust the protein content of the milk. Also, this can be achieved by adding skim milk powder. Total solids content in the milk is increased when ultrafiltration is used to increase the protein level. This affects the cheese making process, and ultimately the quality of the cheese.

3. Pasteurization or heat treatment of milk

Depending on the cheese required, pasteurization or mild heat-treatment may be required to reduce the number of spoilage microorganisms. This may also improve the environment required to grow the starter cultures. Typical examples of bacteria capable of affecting the quality of the cheese are Coliforms, which can cause early "blowing" and a disagreeable taste. Pasteurization may also kill most of the natural pathogenic bacteria.

Certain cheeses are made from raw milk, so they are not pasteurized or heat-treated. Such cheeses must be held at 1.67 °C or higher for at least 60 days before they can be consumed to reduce the chances of exposure to disease causing pathogens that may be present in the milk. Although cheese made from raw milk is considered to have a better flavor and aroma, for fresh (unripen) cheeses, most producers (except hard types) pasteurize the milk. Regular pasteurization requires heat treatment at 72–73 °C for 15–20 s or 63 °C for 30 min. Some producers use thermization treatment explained earlier to prevent the spoilage of milk stored over weekends, but full pasteurization of the milk is required before cheese making. The pasteurized milk is tempered to 31 °C and transferred to bring it to the temperature needed for the starter bacteria to grow. If raw milk is used the milk must be heated to 31 °C.

Mechanical separation as an alternative to pasteurization

Thermo-tolerant spore-forming microorganisms such as Clostridium tyrobutyricum survive pasteurization and can cause serious problems during the ripening process. Clostridium tyrobutyricum forms butyric acid and large volumes of hydrogen gas by fermenting lactic acid. The butyric acid has an unpleasant taste, and the gas destroys the entire texture of the cheese. More severe heat treatment would serve to reduce these microorganisms but would also denature whey proteins of the cheese making milk, and that is, impair its cheese making properties.

Conventionally, chemicals such as sodium nitrate ($NaNO_3$) have been added to cheese milk before production to prevent these problems caused by *Clostridium tyrobutyricum*. However, the use of chemicals has been generally criticized, so mechanical means of reducing the number of unwanted microorganisms has been adopted, particularly in countries where the use of chemical inhibitors is banned. A drawback of these chemical inhibitors on the process is that they can affect some of the added bacteria in the starter culture. Specially designed sealed tight separators for spore and bacteria removal have been used to separate bacteria, and especially the spores formed by specific bacteria strains, from milk, since their density is higher than that of milk. These separators normally separate the milk into a fraction that is free from bacteria, and a concentrate, which contains both spores and bacteria in general and amounts to up to 3% of the feed to the separator. The same temperature, typically 55–60 °C is often chosen for spore and bacteria removal as for separation.

There are two types of spore and bacteria removing separators: (i) **Two-phase type**—has two outlets at the top; one for continuous discharge of the heavy phase via a special top disc, and one for the spore and bacteria-reduced phase. (ii) **One-phase type**—has one outlet at the bowl, for bacteria-reduced milk. The sludge space of the bowl collects the concentrate, which is discharged through ports in the bowl body at pre-set intervals.

Microfiltration

A membrane filter can be used to filter bacteria from milk. However, the problem is that most of the fat globules and some of the proteins are as large as, or larger than, the bacteria may be removed with the bacteria. This causes the filter to foul quickly when membranes with a small pore size are chosen. Consequently, it is the skim milk phase that passes through the filter, while the cream needed for standardization of the fat content is sterilized, typically together with the bacteria concentrate obtained by simultaneous microfiltration. To lower the concentration of protein, membranes with a pore size of 1.4 μ are used. The protein also forms a dynamic membrane that contributes to the retention of microorganisms. Microfiltration has a high bacteria-reducing efficiency, which makes it possible to produce hard and semi-hard cheese without the need for chemicals to impede growth of Clostridia spores.

In practice, the microfiltration plant includes an indirect sterilization unit for combined sterilization of cream for fat standardization and of retentate from the filtration unit. The raw milk entering the microfiltration plant is pre-heated to a separation temperature, about 60–63 °C, upon which it is separated into skim milk and cream. The skim milk is transferred to a sterilization plant where the temperature is kept at the normal microfiltration temperature of 50 °C, before the milk enters the filtration unit. The milk is fractionated into a bacteria-rich concentrate (retentate), containing about 5% of the milk, and a bacteria-reduced phase (permeate). The retentate is then mixed with the cream to be standardized before sterilization at 120–130 °C for a few seconds. After sterilization, the mixture is cooled to approximately 70 °C before remixing with the permeate. After that, the mixed is pasteurized at 70–72 °C for about 15 s and cooled to renneting temperature of 30 °C.

4. **Inoculation with starter culture**

The starter cultures and any non-starter adjunct bacteria are added to the milk to accelerate ripening in hard and semi-hard cheeses. The repining stage is very essential. It allows the growth of the bacteria and start fermentation. This reduces the pH and develops the cheese flavor. The ripening process results in the breakdown of the protein into amino acids and peptides. This is caused by the combined proteolytic effect of the original enzymes of the milk and those of the bacteria in the culture in addition to the rennet enzymes.

The main task of the culture is to develop acid in the curd (i.e. acidification). Furthermore, starter cultures produce three principal characteristics in cheese making:

(i)   Produce lactic acid;
(ii) Break down the protein and, when applicable;
(iii) Produce carbon dioxide (CO_2).

Carbon dioxide creates holes in round-eyed cheeses and supports the openness of granular types of cheese. Formation of acid lowers the pH, which is important in assisting syneresis (contraction of the coagulum accompanied by expelling of whey). Additionally, salts of calcium and phosphorus are released, which impact the consistency of the cheese and help to increase the firmness of the curd. The acid-producing bacteria also suppresses bacteria that survives following pasteurization or recontamination bacteria, which either need lactose or cannot tolerate lactic acid. Production of lactic acid stops when all the lactose in the cheese (except in soft cheeses) has been fermented. Lactic acid fermentation is normally a relatively fast process. This process ought to be finished before the cheese is being pressed for some types of cheese such as Cheddar, whilst other types of cheese may be in a week.

Three main types of culture are used in cheese making:

(i)   *Mesophilic* cultures with an optimum temperature between 25 and 40 °C.
(ii) *Thermophilic* cultures, which develop at up to 50 °C.
(iii) *Adjunct* cultures, which are added for creating specific taste or texture. In many cases, the enzymes created by these bacteria gives these properties.

Mixed strain cultures are the most used cultures, as these comprises of more than two strains of bacteria, which may be either a mixture of mesophilic or thermophilic bacteria, or in some cheeses, a combination of both. The advantage is that mixed strain cultures can support each other in their functioning. *Single-strain cultures* are mainly used where the object is to develop acid and contribute to protein degradation, e.g. in Cheddar and related types of cheese.

5. Additives

Certain additives are added during cheese making. Anhydrous calcium chloride salt can be used in dosages of up to 0.02% level in milk (or 20 g/100 kg of milk) to speed up coagulation and improve cheese yield. Color can be added at the rate of 70 mL/1000 kg of milk to produce cheese of uniform color that can last for a year.

For white Cheddar cheese, carotenoids and norbixin (annatto seed extract) are added instead of color. In other cheeses, chlorophyll-based colorants and titanium dioxide are added to the milk to obtain a white color.

6. Coagulation

Coagulation of casein is an important process in cheese making. Coagulation takes place immediately rennet is added to the milk, when the culture growth is indicated by a pH drop of 0.05 pH unit. The rennet is an enzyme called *chymosin* which acts on milk proteins to form curd. After the rennet is added, the curd is not disturbed for about 30 min to form the coagulum. Also, other proteolytic enzymes can be used, as well as acidification of the casein to pH 4.6–4.7.

The renneting process works in several stages as follows:

 (i) Conversion of casein to paracasein through the influence of rennet;
(ii) Precipitation of paracasein in the presence of calcium ions.

The process is influenced by the temperature, acidity, and calcium content of the milk, as well as other factors. The optimum temperature for rennet is around 40 °C, but lower temperatures are usually used, mainly to control hardness of the coagulum.

7. Curd cutting

The curd is allowed to ferment until it reaches pH 6.4 to obtain a firm curd (usually 25 to 30 min). The curd is then cut using cheese knife into small cubes of curds, usually the size of 3 to 15 mm, depending on the type of cheese. Finer cuts result in lower moisture in the cheese produced. The cheese knife is made of two sets of stainless steel knives, made of wires arranged horizontally and vertically, separated by 9.5 cm (3/8-in). Before the curd is cut, a test for its whey-eliminating quality is usually carried out. This is normally done by sticking a knife into the surface of the clotted milk and then drawn slowly upwards until proper breaking occurs. If a glass-like splitting flaw is observed, then the curd is considered ready for cutting.

8. Cooking

During cheese making, heat treatment is required to control the size and acidification of the curd (which was explained earlier in step (c) above). The cut curds are heated to 38 °C for 30 min. The heating step helps to separate the whey from the curd.

Depending on the type of cheese, the heating can be carried out in the following ways:

(i)   Only saturated steam or hot water in the jacket.
(ii) Saturated steam in the jacket plus hot water added to the curd/whey mixture.
(iii) Only addition of hot water to the curd/whey mixture.

If pH of the whey at cutting is higher, cooking is extended to 60 min, or sometimes up to 75 min until the pH of whey drops to 6.2–6.3. The size of the curd should decrease to about half of the original size before cooking. The cooked curd should be firm with no soft or mushy center and should not stick together when manually pressed.

9. Draining of whey

About 50–70% of the whey is drained from the vat and the curd forms a mat. The curd is continuously stirred, and the remaining whey continues to drain to dry out the curd. The whey can be used to achieve the correct curd/whey ratio by using a continuous draining equipment. There are three forms of the whey in the curd/whey mixture:

(i)   Free whey—whey between the curd particles.
(ii) Whey incorporated in the curd grains.
(iii) Whey bound to protein.

Pressing to increase the compactness of the curd block easily drains the free whey. The whey inside the grains is more difficult to drain, but it can be released as a free whey by increasing the acidity and applying pressure on the grains. However, the whey bound to protein is not drainable under normal draining conditions.

When the whey is being drained, the static pressure in the column or the pre-press vat causes deformation of the curd grains and partly fuse together. Hence, the discharge of whey should be gentle and should not exert excessive force on the curd.

Four different types of cheese can be obtained depending on how the curd grains are tread after the whey is separated. If the grains are separated first from the whey, filled in molds and then turned and/or pressed, the result is a cheese with an open or granular structure. When the grains are collected in a layer for an acidification period, a cheese with a closed texture, such as Cheddar is produced. When the curd is kept under the surface of the whey during the combined draining and pre-pressed, the result will be a round-eyed type of cheese.

10. Cheddaring and milling

Cheddaring helps to remove more whey, allows the fermentation to continue until a pH of 5.1–5.5 is reached, and allows the curd mats to join and form a tighter matted structure. During cheddaring, the curd mats are cut into sections and piled on top of each other, which is occasionally flipped.

The curd mats are then milled, cut into smaller pieces, and piled 10.2–12.7 cm deep along the sides of the mat for matting/thickening. The whey discharges slowly through a trench. After all the whey has been drained, the curd is cut into thick sections, piled on top of each other and maintained at 30–35 °C for continued acidification and matting. During this period, the curd is formed into sections/slaps and are turned repeatedly at 5–15 min intervals for 2–2.5 h. This helps to expel more whey and allows the fermentation to continue until the pH of the cheddared curd has reached 5.1–5.5. This weaves the mats together to form a tighter matted structure. The curd slaps are then milled (cut) into smaller pieces, which are dry-salted before forming molds/hoops.

11. Salting

After 15–30 min, the smaller milled curd pieces are sprinkled with dry salt. The quantity of salt ranges from 2 to 3 kg/1000 kg of initial milk with a salt content of between 0.5 and 2.0% (w/w) in the cheese. The salt used in cheese making should not contain iodine in order to impact the true taste of salt. Addition of salt helps to drain more whey and moisture from the curd pieces through osmotic effect and salting effect on the proteins. The osmotic pressure is due to the formation of suction on the surface of the curd, causing moisture to be drawn out of the curd.

To form the cheese, the pieces of curd are put in cheese hoops after salting, and then pressed into blocks.

12. Final pressing

After the curds have been molded or hooped, the curd is lined up for final pressing at 75 kPa for 12–18 h. The initial pressure should be low to give a gradual pressing. This is because initial high pressure compresses the surface layer and can lock moisture into pockets in the cheese. After pressing, the blocks are removed from the hoops, vacuum packed in films and shrunk skin-tight by dipping in hot water and transferred to the ripening room.

The cheeses can be pressed per batch or per row in cases where a conveyor press is used. The pressure used depends on:

(i)   Cheese dimensions
(ii)   Type of mold
(iii)   Fat content
(iv)   Acidity level
(v)   Amount of residual whey in the cheese
(vi)   Curd temperature
(vii)   Available time for pressing.

The purpose of this final pressing is to:

(i)   Assist final whey expulsion;
(ii)   Reach desired acidification;
(iii) Provide texture;
(iv)   Shape the cheese; and
(v)   Provide a rind on cheeses with long ripening periods.

13. **Ripening**

After curdling, the cheese is allowed to ripe to generate the characteristics cheese flavor, which is caused by breakdown of lactose, lipid and protein through the action of bacterial enzymes. The ripening process starts with the decomposition of protein, which is caused by the enzyme system of microorganisms, rennet and plasmin. The ripening cycle differs between hard, medium-soft and soft cheeses. The ripening temperature for slow ripening is between 5 and 8 °C, whiles that for fast ripening varies from 10 to 16 °C. Ripening may take 3 to 9 months in order to achieve a sharper flavor. Slow ripening leads to better control of the cheese flavor.

14. **Storage, aging and packaging**

Storage helps to create the needed external conditions to control the ripening cycle of the cheese. The cheese must be stored in coolers to achieve the required age.

Different types of cheese require different temperatures, relative humidity, air circulation and air velocity in storage rooms during the various stages of ripening. For instance, cheddar cheese is ripened at low temperatures (4–8 °C) and relative humidity lower than 80%, because they are wrapped in a plastic film or bag and placed in cartons or wooden cases before transporting to the store. Wrapping the cheese prevents excessive water loss and protect the surface from infection and dirt.

11.10 Production of Ice Cream

Ice cream is a blend of air and a sweetened cream mixture with added flavorings, which is frozen. The basic ingredients of ice cream are milk fat, milk solids (protein, lactose and minerals), and air. The milk fat content must be at least 10% and no less than 20% total milk solids. Ice cream may contain egg yolks (emulsifier). Ice cream containing at least 1.4% egg yolks is labelled as custard. Reduced fat ice cream contains 25% fat less than the reference ice cream, whilst light ice cream contains 50% less fat than the reference. Low fat ice cream contains less than 3 g fat per serving and less than 0.5 g fat per serving for non-fat ice cream.

Ice cream is sold as soft or hard serve. Soft ice cream has only a small amount of water been frozen, and it is served directly from the freezer. Hard ice cream is package from the freezer after which it goes through a hardening process to freeze more of the water.

11.10.1   Ingredients

A wide variety of ingredients can be used to produce ice cream. Ice cream contains milk fat, milk solids non-fat (MSNF), emulsifiers, stabilizers, sugar and non-sugar sweeteners, flavors (natural and/or artificial) and colors.

Ice cream mix contains 37% (wt.) dry matter (i.e. 10% fat, 11.5% MSNF, 15% sugar and 0.5% emulsifiers/stabilizers) and 63% moisture. Frozen ice cream contains 5% fat, 5.7% MSNF, 7.5% sugar, 0.3% emulsifiers/stabilizers, 31.5% moisture and 50% air.

11.10.1.1   Fat
The minimum fat content is 10% for basic ice cream and about 16% for premium ice creams and may be milk or vegetable fat. Sources of milk fat include whole milk, cream, butter or anhydrous milk fat (AMF). Hydrogenated (hardened) palm kernel oil and coconut oil are the form of oils that are used as vegetable fat. Fat is essential in ice cream as it gives creaminess and richness to ice cream and improves its melting resistance by stabilizing the air cell structure of the ice cream.

11.10.1.2   Milk Solids-Non-Fat (MSNF)
The total milk solids comprise of milk fat and other milk solids. The (other) milk non-fat solids (MSNF) consist of proteins, lactose and mineral salts derived from whole milk, skim milk, condensed milk, milk powder and/or whey powder, and varies between 9 and 12% in ice cream. Sources of non-fat solids include milk, cream, dry milk, whey, evaporated milk and condensed milk. The MSNF has water-binding and emulsifying effect, which help to stabilize the structure of ice cream. The water-binding and emulsifying effect of MSNF stabilizes the air that is added during the freezing process, which helps to improve the body and texture or creaminess of ice cream.

11.10.1.3   Sugar/Sweeteners
Sweeteners used in ice cream are sugar (sucrose) and corn syrups. The sugar content of ice cream is usually between 12 and 20%. Sweeteners are added to give the level of sweetness consumers desire. Also, addition of sugar helps to increase the solids content of ice cream and allows a fraction of water in ice cream to remain unfrozen at serving temperatures as the freezing point is lowered. Although a lower freezing point makes ice cream easy to scoop and eat, too much sugar can make the ice cream too soft. Sweeteners such as aspartame, acesulfame K, and sucralose are sweeteners commonly used for producing sugar-free ice cream. The sweeteners are usually used in addition to bulking agent such as polydextrose, malto-dextrin, sorbitol, glycerol, lactitol or other sugar alcohols.

11.10.1.4   Flavors and Colors
A variety of flavorings are used in ice cream, which are classified as natural and artificial flavors. The most commonly used flavors are chocolate chunks, vanilla and

fruit such as strawberry. Flavors can be added at the mixing stage or after pasteurization. Natural or artificial colors are added to improve the appearance of the ice cream.

11.10.1.5 Emulsifiers

Emulsifiers are substances that enable immiscible liquids to become finely dispersed in each other by lowering the surface tension between the two phases. This results in a stable, smooth, and homogeneous emulsion. Emulsifiers commonly used in ice cream include egg yolks, mono- and di-glycerides of fatty acids often derived from a vegetable fat (triglyceride). The mono- and di-glycerides have two main functions during ice cream processing; firstly, by supporting displacement of the milk protein from the fat surface membrane to improve the churning during the freezing process. Secondly, the mono- and di-glycerides seed the crystallization of fat, which helps to evenly dispersed the milk fat in the ice cream during the freezing process and storage. A good dispersal of the fat helps stabilize the air added to the ice cream and provide a smooth product. Egg yolk is a common emulsifier, but it is very expensive and less effective than the mono- and di-glycerides types.

11.10.1.6 Stabilizers

Stabilizers are proteins or carbohydrates used in ice cream to add viscosity and control ice crystallization. Stabilizers help to isolate small ice crystals and prevent the growth of large crystals. Larger crystals make ice cream icy, crease and unpleasant to eat. Typical stabilizers used in ice cream include gelatins, gums (locust bean, guar), and alginates (carrageenan).

Stabilizers and emulsifiers are typically combined and added to the ice cream at dosages of 0.5%.

11.10.2   General Manufacturing Method of Ice Cream

The general steps in the manufacturing of ice cream are illustrated in Fig. 11.11.

1. Blending of ice cream mixture

The milk fat source, non-fat solids, stabilizers and emulsifiers are blended to ensure complete mixing of liquid and dry ingredients to form the "ice cream mix". High shear (speed) blenders are used to achieve rapid agitation of the ingredients. The dry ingredients, especially the milk powder is often added through a mixing unit through which water is circulated that creates an ejector effect that sucks the powder into the flow. The ingredients are selected based on the desired formulation.

2. Pasteurization of ice cream mix

The ice cream mix flows through a filter to a balance tank before it is pumped to a plate heat exchanger, where it is pasteurized at 68.3 °C for 30 min or 79.4 °C for 25 s. The conditions used to pasteurize the ice cream mix are stronger than those

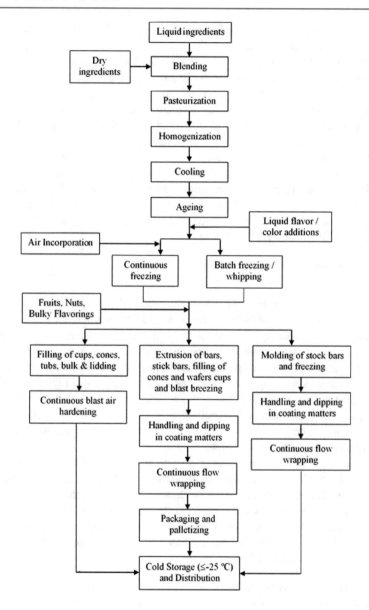

Fig. 11.11 General processing steps for production of ice cream

used for fluid milk because of increased viscosity from the higher fat, solids, and sweetener content, and the addition of egg yolks in custard products.

The purpose of pasteurization is to destroy pathogenic bacteria, reduces the number of spoilage organisms such as psychrotrophs, and helps to dissolve additives and ingredients. Pasteurization also helps to hydrate some of the components, especially proteins and stabilizers.

Batch pasteurizers and continuous methods are used in pasteurization of the ice cream mix. Batch pasteurization results in more whey protein denaturation, which improves the body of the ice cream. Batch tanks are usually operated in pairs so that as one is holding the mix the other is being prepared. In a batch pasteurizer, blending of the ingredient is performed in large jacketed vats equipped with some a heating system, usually steam or hot water. Automatic timers and valves are used to ensure the proper holding time is met.

Continuous pasteurization is typically performed in a high temperature short time (HTST) heat exchanger. Preheating to 30–40 °C is required to solubilize the components. The HTST system contains different sections for heating section, cooling, and regeneration. Cooling sections of ice cream mix HTST presses are usually larger than milk HTST presses.

3. **Homogenization of ice cream mix**

Following pasteurization, the mix is homogenized at the pasteurization temperature at high pressures (17–20 MPa) and then is passed across a plate or double or triple tube heat exchanger to cool the mix to refrigerated temperatures (4 °C). Homogenization breaks down or reduce fat globule in the milk to less than 1 μm to form a better emulsion, which improves the whipping property and texture of the ice cream mix to produce a smoother, creamier ice cream. Two-stage homogenization is usually ideal for ice cream mix. The high temperature used results in breaking up of the fat globules regardless of the applied pressure. This reduces fat clumping, the ability to thick, and heavy bodied mixes. Lower pressure is required for mixes containing higher fat and total solids.

If a two-stage homogenizer is used, a pressure of 14–17 MPa on the first stage and 3–7 MPa on the second stage would be required. A two-stage homogenization reduces clumping or clustering of the fat thereby producing a thinner, more rapidly whipped mix. Melt-down is also improved. Homogenization also ensures that the emulsifiers and stabilizers are well blended and evenly distributed in the ice cream mix before it is frozen. In general, the function of homogenization in ice cream manufacture are summarized as follows:

    (i)   Reduces size of fat globules
   (ii)   Increases surface area
  (iii)   Makes possible the use of butter, frozen cream, etc.
   (iv)   Forms membrane
    (v)   Makes a smoother ice cream
   (vi)   Increases resistance to melting
  (vii)   Better air stability
 (viii)   Gives a greater apparent richness and palatability.

4. **Ageing of the mix**

The ice cream mix is aged at a temperature of 2–5 °C (without freezing) for at least 4 h or overnight with a continuous gentle mixing in insulated or refrigerated storage tanks or silos. Ageing cools the mix down before freezing allows time for the milk fat to cool down and partially crystallize and gives the proteins and polysaccharides time to fully hydrate. This results in better air incorporation and improves melting resistance, body and texture of ice cream. Unaged mix is usually detected at the freezer.

The functions of ageing are achieved by allowing time for:

(i)   Fat crystallization, so the fat can partially coalesce.
(ii) Full protein and stabilizer hydration which results in a slight viscosity increase.
(iii) Rearrangement of the membrane and interaction of protein/emulsifier. Emulsifiers play essential role in the aging process by moving proteins from the fat globule surface, allowing a decrease in stabilization of the fat globule, which results in improved partial coalescence.

5. **Liquid flavors and colors**

Following ageing, the mix is transferred to a flavor tank where liquid colors, fruit purees, or flavors are added before freezing. Only liquid ingredients can be added at this stage to make sure the mix flows properly through the freezing equipment.

6. **Freezing/whipping of ice cream**

Ice cream contains a large amount of air, up to half of its volume, to give the product its characteristic lightness. The air content is termed its *"overrun"* and is normally between 80 and 100%, corresponding to 0.8–1.0 L of air per liter of mix, which can be calculated mathematically. Premium ice creams have less overrun (approximately 80%) and are more dense than regular ice cream.

Whipping controls the amount of air into the mix and freezing freezes significant part of the water content in the mix into many small ice crystals. The process involves freezing the ice cream mix and incorporation of air. Freezers can be either batch or continuous, with the continuous freezing process being faster than the batch process. The type of freezer used influences the applied conditions. Batch freezers consist of a rotating barrel that is usually filled with ice cream mix up to one-third to one-half full. A portion of the water is frozen and as the barrel turns, air is whipped into the frozen mix. The mix is pumped to the freezer and is drawn off the other end in 10–15 min. Ice cream freezers designed for home use are batch freezers.

Continuous freezers comprise a fixed barrel fitted with a blade fitted inside that scratches the surface of the barrel repeatedly. The ice cream mix is metered from a bulk tank into the freezing barrel by a gear pump and at the same time, a constant flow of air is added with another pump just before it enters the freezing barrel and the dashers inside the freezer help to whip the mix and incorporate air. The mix is drawn off the other end in about 30 s.

In general, the barrel freezer is a scraped-surface, tubular heat exchanger, which is jacketed with a boiling refrigerant such as ammonia or freon. In both batch and continuous freezers, the output temperature is -8 to -3 °C depending on the type of ice cream product, and about 30–55% of the water is frozen into ice crystals depending on the composition of the mix formulation. The rotating blades inside the barrel scratched off the ice from the surface of the freezer.

Calculation of Overrun

Overrun is defined as the percentage increase in volume of ice cream greater than the amount of mix used to produce the ice cream. That is, if 1 L of mix is used to make 1.5 L of ice cream, then the volume has increased by 50% (i.e. the overrun is 50%). Mathematically, overrun can be calculated as follows:

(i)   By volume, no particulates:

$$\%\text{Overrun} = \frac{\text{Vol. of ice cream} - \text{Vol. of mix used}}{\text{Vol. of mix used}} \times 100\% \qquad (11.5)$$

Example: If 500 L of mix produces 925 L of ice cream,

$$\%\text{Overrun} = (925 - 500)/500 \times 100\% = 85\% \text{ Overrun}$$

Note that any flavors added which become homogeneous with the mix can incorporate air and are therefore accounted for during Overrun calculations.

Example: If 90 L mix plus 10 L fruit syrup gives 185 L fruit syrup ice cream,

$$\%\text{Overrun} = (185 - (90 + 10))/(90 + 10) \text{ x } 100\%$$
$$= 85\% \text{ Overrun}$$

(ii)   By volume, with particulates (the particulates do not incorporate air):
Example: If 160 L mix plus 40 L dates produces 350 L dates ice cream,

$$\text{Actual ice cream} = 350 - 40 = 310 \, L$$

$$\%\text{Overrun} = (310 - 160)/160 \times 100 = 93.8\% \text{ Overrun}$$

(iii)   By weight, no particulates:

$$\%\text{Overrun} = \frac{\text{Wt. of mix} - \text{Wt. of same vol. of ice cream}}{\text{Wt. of same vol. of ice cream}} \times 100\% \quad (11.6)$$

In this case the density of the mix must be known (i.e. wt. of 1 L), usually 1.09 –1.1 kg/L.

Example: If   2   L   of   ice   cream   weighs   1150 g,   (i.e. wt. of mix $= 2\,L \times 1090\,kg/L = 2180\,g$)

$$\%\text{ Overrun} = (2180 - 1150)/1150 \times 100 = 89.6\% \text{ Overrun}$$

7. Addition of ingredients

After freezing and whipping, a pump transfers the semi-frozen slurry ice cream to an ingredient feeder or filling machine. A variety of ingredients can be added by the feeder: dry ingredients (e.g. nuts, cookies, chocolate); soft ingredients (e.g. pieces of fruits, cookie dough); liquid ingredients (e.g. marmalade, jam, caramel). The ingredient feeder adds ingredients continuously and accurately to the ice cream via a pump which feeds the ingredients gently into the ice cream flow from the freezer. The accuracy of the ingredients is controlled by using ingredient weighing cells under the ingredient feeder.

These bulky ingredients cannot be added before freezing as they would interfere with the smooth flow of the mix through the freezer. Mixing the ice cream with the bulky ingredients after freezing also prevents damage to the pieces and allow them to remain whole or in large chunks.

8. Filling

Filling machines:

A filling machine fills ice cream directly from the freezer into cups, cones and containers of different designs, shapes and sizes. A time-lapse filler, a volumetric filler or an extrusion filler is used for filling ice cream into containers. For extrusion filling, a cutting mechanism is provided. Decorations with the various ingredients can also be done after filling. Packs are covered with lids before leaving the filling machine, after which they passed through a hardening machine. The products can be manually or automatically packed in cartons or bundles before or after hardening. Plastic tubes or cardboard cartons can be filled manually by means of a can-filling unit equipped to supply single or twin flavors.

Molding stick:

The ice cream is supplied directly from the continuous freezer at a temperature of approximately −3 °C where it is molded in pockets. The filled molds are conveyed through a brine solution with a temperature of −40 °C, which freezes the ice cream through the mold wall. Sticks are inserted before the molds are completely frozen. The frozen products are removed from the molds by passing them through a warm brine solution which melts the surfaces of the products and enables them to be removed automatically by an extractor unit. After extraction, the products may be dipped in chocolate or vanilla and coated with nuts or other dry ingredients before being transferred to the wrapping machine. After wrapping, the products are place in cartons and are taken straight to the cold store.

Extrusion lines:

Extruded ice cream products are usually produced on a tray tunnel extruder. The ice cream can be extruded directly onto trays in different shapes and sizes or extruded into a cup or cone or onto a sandwich wafer. By using different filling and handling machines, a variety of products can be produced, such as ball-top cones, sandwich products, ice cream cakes, ice cream logs, filled wafer cups and bite-size products.

After decoration, the products are carried on the trays through a hardening tunnel. The hardened products are removed from the trays, after which they are wrapped and packaged in cartons. This can be done either manually or automatically. Depending on the capacity of the extruder and product type, between 5,000 and 43,000 units can be produced in an hour. Most extrusion lines allow great flexibility in switching between sticks and sandwiches, as well as handling up to 4 flavours and up to 12 different coatings.

9. Wrapping and packaging

The design of wrapping and packaging unit of an ice cream processing line depends on the capacity and type of product. Manual and automatic operation can be utilized. The packaging containers are packed in cartons. Hand-held products like sticks, cones and bars are wrapped in a single or multi-lane wrapping machine before being packed in cartons.

10. Hardening

The ice cream package is rapidly placed in a blast freezer at a static temperature of −25 °C. The type of storage freezer utilized determines the freezing temperatures and times. Fast cooling will cause rapid freezing of water, which will produce small ice crystals. Storage at −25 °C will help to stabilize the ice crystals without danger of ice crystal growth and maintain product quality. Ice cream can be stored for up to

9 months at −5 °C. At this temperature there is still a small portion of liquid water, which prevents the ice cream from becoming as hard as an ice cube. Above this temperature, the shelf-life of ice cream is limited as ice crystal growth can take place, and the rate of crystal growth depends on the storage temperature. In addition to maintaining a constant low temperature, the storage life of ice cream also depends on the type of product and the packaging.

Freezing must be rapid; hence, the freezing process involves low temperature (−40 °C) with either improved convection (freezing tunnels with forced air fans) or improved conduction (plate freezers). The faster the hardening, the better the texture.

For ice cream products made on an extrusion line or a stick freezer, the hardening operation is included in the process. However, products packed immediately after freezing must be transferred to a hardening tunnel. After hardening, the products are transferred to the cold store, where they are stored on shelves or pallet racks at the hardening temperature of −25 °C.

Factors affecting ice cream hardening

During the freezing process, the rate of heat transfer (Q) is affected by the exposed surface area of heat transfer (A), the temperature difference (dT) and the heat transfer coefficient (U). That is,

$$Q = UAdT \tag{11.7}$$

Factors that affect the rate of heat transfer also affect ice cream hardening, which are:

(i) Exposure of maximum surface area to cold air.
(ii) Rapid circulation of air—increases convective heat transfer.
(iii) Temperature of blast freezer—the colder the temperature, the faster the hardening, and the smoother the product.
(iv) Temperature of ice cream when placed in the hardening freezer—the colder the ice cream when drawn, the faster the hardening—the ice cream must go through packaging operations quickly.
(v) Composition of ice cream—this is related to freezing point depression and the temperature required to ensure a significantly high ice phase volume.
(vi) Method of stacking containers or packages must allow air circulation.
(vii) Package type should not impede heat transfer—e.g. corrugated cardboard or Styrofoam liner may protect against heat shock after hardening but will reduce heat transfer during freezing; hence not viable.
(viii) Evaporator must be free from frost—this acts as an insulator.

11.11   Production of Milk Powder

Whole (full cream) milk contains about 87% water and skim milk contains about 91% water. Milk powder is produced by removing this water via gentle boiling the milk under reduced pressure at low temperature in an evaporation process. The resulting concentrated milk is then spray dried to remove further moisture and to obtain a powder. The process must be under strict hygiene conditions while retaining all the desirable natural properties (color, flavor, solubility, nutritional value) of the milk. Milk powder that is mixed with water for direct consumption (recombination) should be easily soluble in water, and the nutritional value and taste must be appropriate. If the powder is to be used in chocolate production, some degree of caramelization of the lactose is useful by subjecting the powder to intensive heat treatment in a roller dryer.

One motive of drying milk is for preservation; milk power has a longer shelf life than liquid milk and because of its low moisture content, it does not need to be refrigerated. Another benefit of drying milk is to reduce its bulk for economy of transportation.

11.11.1   Milk Powder Manufacturing Process

The milk powder manufacturing process is shown in Fig. 11.12 and is described in detail below.

Fig. 11.12 General manufacturing steps for production of milk powder

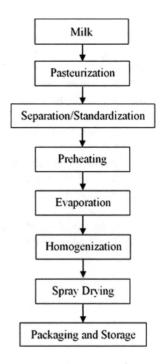

1. Pasteurization

Milk powder manufacturing starts with pasteurizing the raw milk received at the dairy factory by heating the milk at a temperature of 62.8 °C for 30 min (for a batch process) or 71.7 °C for 15 s (for a continuous process). This helps to inactivate most of lipolytic enzymes that would otherwise degrade the milk fat during storage. See Sect. 11.6.3 for details on pasteurization.

2. Separation/standardization

Following pasteurization, the milk is separated into skim milk and cream using a centrifugal cream separator. For production of whole milk powder, a portion of the cream is added back to the skim milk to produce a milk powder with a standardized fat content of between 26 and 30%.

3. Preheating

The next step in the process is preheating the standardized milk to temperatures between 75 and 120 °C for a few seconds or up to several minutes. Preheating can be performed via the following methods: indirect heating (using heat exchangers) or direct (via steam injection or infusion into the product), or a mixture of the two. Indirect heaters normally use heat recovered from other parts of the process as a way of saving energy. Preheating whole milk powder at higher temperatures improve storage quality but reduced solubility. Preheating before evaporation is very important as it causes a controlled denaturation of the whey proteins in the milk and it destroys bacteria, inactivates enzymes, generates natural antioxidants and imparts heat stability.

4. Evaporation

The preheated milk is transferred to the evaporator where it is boiled under a vacuum at temperatures below 72 °C in a falling film on the inside of vertical tubes, which removes the water as vapor. During the process, the milk is concentrated from 13% total solids content for whole milk and 9% for skim milk, up to 45–52% total solids. More than one stage or "effect" is used; and the vapor, which may be compressed mechanically or thermally, is then used to heat the milk in the next effect of the evaporator which is operated at a lower pressure and temperature than the previous effect. The evaporator may remove more than 85% of the water in the milk, depending on the number of effects used. For maximum energy efficiency, some evaporators have up to seven effects.

Evaporation of the milk before drying is more energy efficient as it is far cheaper to evaporate the water than to spray dry it. The energy used in multiple effect evaporators with steam vapor recompression is about 10 times less than spray drying.

5. Homogenization

Homogenization is usually performed between evaporation and spray drying to convert free fat into fat globules and regenerate membranes by protein adsorption on the globule surfaces. Whole milk powder must contain between 26 and 40% (wt.) milk fat on an "as is" basis and not more than 5% (wt.) moisture on a milk-solids-not-fat (MSNF) basis. Single stage homogenization can result in fat globules easily coalescing into larger globules. To prevent this, two-stage homogenization is usually use in milk powder manufacture, which is also considered important for infant formula manufacturing. Two-stage homogenization ensures minimal re-coalescing of fat globules. The solubility of milk powder is reduced when the protective membranes of the fat globules are damaged. This increases the possibility to oxidative rancidity.

6. Spray drying

Spray drying is the most widely used process for drying milk powders. Milk powders dried by this method have excellent solubility, flavor and color. It is a two-stage process that involves spray drying at the first stage with a static fluid bed integrated in the base of the drying chamber. The second stage is an external vibrating fluid bed.

During the drying process, the concentrated milk from the evaporator is atomized into fine droplets, which is carried out inside a large drying chamber in a flow of hot air (180–200 °C) using either a series of high-pressure nozzles or a centrifugal disc atomizer. Centrifugal atomizers produced powders with finer particle sizes than pressure nozzle atomizers. The milk droplets are cooled by evaporation and are never allowed to reach the temperature of the air. To evaporate all the moisture by using reasonably low temperatures, it is required to control the droplets size, the airflow and the air temperature. In order to increase the energy available for drying, the concentrate should be heated before atomization. Much of the remaining water is evaporated in the drying chamber, leaving a fine powder having a moisture content of about 6% with a mean particle size of < 100 μ diameter. Viscosity is generally maintained below 250 centipoise at the atomizer by increasing the feed temperature. This also increases the milk powder spray dryer capacity.

The final drying, which is the second drying stage takes place in a single or a series of fluidized beds, in which hot air is blown through a layer of fluidized powder removing most of the remaining water to give milk powder with a moisture content of 2–4%. The product is moved through the two-stage process as quickly as possible to prevent overheating of the powder. From the dryer, the milk powder enters a system of cyclones where it is cooled simultaneously. Safety measures must be put in place to prevent fires and to vent dust explosions should they occur in the drying chamber.

Heat treatment of milk powder depends on the specification of the finished product. For instance, heat treatment of infant formula milk powder is higher than most other powders. The aim is to destroy all pathogenic and other microorganisms present and to inactivate enzymes, particularly lipase, which can cause fat lipolysis during storage which results in off odors. Additionally, it minimizes the possibility of oxidative changes during storage.

7. **Packaging and Storage**

Milk powders must be protected from moisture, oxygen, light and heat to have a long shelf-life and maintain their quality. Milk powders are hygroscopic, that is they readily absorb moisture from air, which leads to lumping or caking and rapid loss of quality. Fat in whole milk powder can react with oxygen in the air to give off-flavors, especially at higher storage temperatures (>30 °C). A careful storage can be annulled by a few hours at elevated temperatures (>40 °C) during transport or shipment. Prolonged exposure to direct sunshine should be avoided during shipment or transport.

Milk powder is packed into either plastic-lined multi-wall bags, tin cans or bulk bins. Bags normally consist of several layers to provide strength and the necessary barrier properties. Packaging of whole milk powders are usually done under nitrogen gas to protect it from oxidation. This also helps to retain the quality and flavor of the milk powder.

11.11.2   Milk Powder Types

11.11.2.1   Whole Milk Powder

Whole milk powder can be produced in two ways: by removing water from pasteurized, homogenized whole milk; or by blending fluid, condensed or skim milk powder with liquid or dry cream or with fluid, condensed or dry milk. After standardization of the fat content, if the milk is thoroughly agitated without inclusion of air, homogenization will not be needed. Whole milk powders are used in bakery, confectionery, sauces, and dairy as:

  (i)   An easily transported and stored dairy ingredient.
 (ii)   An economical source of dairy solids, including milk fat.
(iii)   A convenient form of milk that can be stored safely without refrigeration and is easily reconstitution.

11.11.2.2   Non-Fat Dry Milk and Skim Milk Powder

Non-fat dry milk and skim milk powder are produced by removing water from pasteurized skim milk. Both milk powders contain 1.5% (wt.) or less milk fat (by weight) and 5% (wt.) or less moisture. The only difference is the protein content; skim milk powder has a minimum milk protein content of 34%, whereas non-fat dry milk has no standardized protein level. The typical composition/properties of whole and skim milk powders are shown in Table 11.6.

Table 11.6 Typical composition and properties of whole and skim milk powder

Component	Whole milk power	Skim milk powder
Protein (%)	24.5–27.0	34.0–37.0
Lactose (%)	36.0–38.5	49.5–52.0
Fat (%)	26.0–40.0	0.6–1.5
Moisture (%)	2.0–4.5	3.0–4.0
Ash (%)	5.5–6.5	8.2–8.6
Bulk density (g/ml)	0.55–0.6	0.65–0.75
Solubility (ml)	0.1–0.5	0.1–0.5

There are three main classifications of non-fat dry milk and skim milk powder, which are based on the heat treatment used in their production. The classifications are: high-heat (least soluble)—spray dried at 135 °C for 30 s; medium-heat—spray dried at 85 °C for 20 s; and low-heat (most soluble)—spray dried at 70 °C for 15 s. Both milk powders are available in two forms: ordinary or non-agglomerated (non-instant) and agglomerated (instant).

Agglomerated and instant milk powders

Spray dried milk powders, because of their fine dusty nature (i.e. particle size less that about 100 μ) do not reconstitute well in water; that is the powders tend to form lumps when mixed with water and require strong mechanical stirring in order to fully dissolve in the water. Agglomerated and instant powders are developed to solve this problem. Compared to standard milk powders, agglomerated and instant powders disperse in water more quickly. Also, agglomerated and instant powders are less dusty and easier to handle. To produce powders that are more readily dispersible, the specific surface of the powder must be reduced so that the water can penetrate more evenly around the particles.

Agglomerated milk powders

The standard process of evaporation and drying explained above is used in the initial stage of production of agglomerated milk powder. However, during spray drying small particles of powder leaving the drier (the "fines") are recovered in cyclones and returned to the drying chamber near the atomiser. The wet concentrate droplets collide with the fines and stick together, forming larger (0.1–0.3 mm), irregular shaped agglomerates.

Agglomeration is usually carried out in a vibrating fluidized bed where the re-wetting, collision and sticking of the particles as well as further gradual drying takes place. There must be enough agitation in the fluid bed to ensure distribution of the spray on the particle surfaces and to prevent lump formation. Water can be used as a re-wetting media for most dairy products and maltodextrine-based flavor formulations. The characteristics of the agglomerates can be varied by adjusting the drying temperatures, fluidizing velocity and spray rate.

Instant milk powder

Instant milk powder has extremely good wettability and solubility, and dissolves instantly, even in cold water. Thus, when added to water, instant powders should be able to wet, sink, and disperse quickly in the water, with minimal stirring. The resulting liquid should be like fresh milk and be free from undissolved particles. It can be produced from both whole milk and skim milk powder, where the powder is agglomerated into larger particles to increase the average grain size of the milk powder. An extra step is required after agglomeration, which consists of spraying tiny quantities of the natural surfactant or wetting agent, such as soy lecithin on the powder in the fluidized bed. This helps to overcome the hydrophobic (water-hating) nature of traces of free fat surface of the particles and give the product its instant dissolving property.

11.11.3   Applications of Milk Powder

Some powders are fortified with vitamins and minerals. Milk powders are used for various applications, such as:

- Production of ice cream; to increase the solids content.
- In baby foods (e.g. infant formulae).
- Direct consumer use (home reconstitution).
- Producing milk chocolate in the chocolate industry.
- Cheese milk extension (powder is added to local fresh milk to increase the yield of cheese).
- Added during bread making to increase the volume of bread and improve its water-binding capacity. Adding milk powder makes the bread remain fresh for a longer period.
- Recombination of milk and milk products.
- Animal feed, calf growth accelerator.

11.12   Manufacturing of Whey Products

Whey is the liquid by-product from the manufacture of cheese, casein and yoghurt; and comprises 80–90% of the total volume of milk entering the dairy process. Whey is one of the biggest sources of food protein and contains about 50% of the nutrients in the fresh milk, comprising serum proteins (whey protein), milk sugar (lactose), most of the water-soluble minor nutrients such as vitamins and minerals, and small amount of fat. It is found to be one of the most nutrient sources available today containing the following relevant properties for health: nutritional (e.g., high content of essential amino acids), biological (e.g., anticarcinogenic, antimicrobial, and immunomodulatory activities), and functional (e.g., gelation, foaming, and emulsifying agent). Whey is used as food ingredients in several food formulations such as confectionary, bakery, infant nutrition, health, and sports supplements; normally in dry form. Whey is natural sources of high-gelling β-lactoglobulin, human's milk equivalent protein α-lactalbumin, lactoferrin, and immunoglobulin and as a pre-cursor to the probiotic galactooligosaccharides (GOS).

About 92% of the whey is recovered as cheese whey, which is the liquid by-product from the production of cheese. Whey is classified in two different types: "sweet whey" produced from hard, semi-hard or soft cheese rennet casein with a pH of 5.6–6.6; and "acid whey", obtained from acid-coagulated cheese with a pH of below 5.1 (typically 4.3–4.6). Figure 11.13 illustrates the general processing steps for the manufacture of sweet whey, demineralized whey and lactose. The development of separation technologies relying on selective porous membranes are permitting a variety of higher valued products to be manufactured from whey, including whey protein concentrates, isolates and other fractions.

As indicated in Fig. 11.13, whey, upon separation from the cheese curd, is subjected to various processes to produce many whey products such as demineralized whey, "sweet" whey, and lactose.

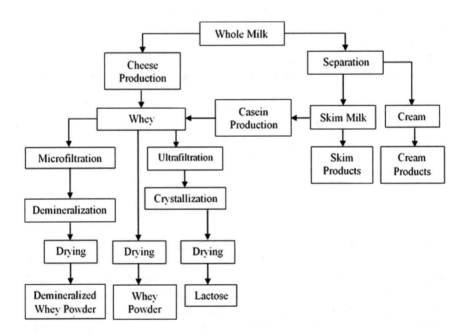

Fig. 11.13 General manufacturing steps for production of various whey powders

11.12.1 Composition of Whey

Whey is a turbid yellowish liquid, consisting of about 90–94% water with a high and varied organic content. Its composition and characteristics may differ depending on the origin and method of manufacture. The major organic component present in whey is lactose, followed by proteins, minerals, lactic acid, and lipids. Other minor components include vitamins of the B-complex, citric acid, and non-protein nitrogen compounds (urea and uric acid). The approximate compositions of whey from cheese and casein are summarized in Table 11.7.

11.12.2 Whey Products

The use of membrane filtration technologies has made it possible to achieve an efficient separation and concentration of liquid whey to produce many whey products, such as whey protein concentrates (WPC), demineralized whey, lactose, reduced lactose whey, whey protein isolates (WPI), milk mineral concentrates and ethanol. The process uses membranes with specific pore sizes, which allows the permeate to flow through while the retentate is blocked based on the size or molecular weight. The liquid whey processing techniques for producing the whey products are illustrated in Fig. 11.14.

By using different membranes or a combination of sequential steps, different concentration of proteins can be produced, which after spray drying gives whey protein concentrate (WPC). The whey protein concentrates contain different protein concentrates, usually ranging from 25–80% as well as a significant amount of minerals and lactose. Further purification using other techniques such as ion exchange (IE) and/or microfiltration/ultrafiltration followed by spray drying results in the production of whey protein isolates (WPI), which contains at least 90%

Table 11.7 Approximate composition of whey	Component	Cheese whey (%)	Acid casein whey (%)
	Total solids	6.00–7.74	5.00–6.40
	True protein	0.60–1.00	0.40–0.80
	Fat	0.05–0.5	0.05–0.09
	Ash (minerals)	0.20–0.59	0.50–0.80
	Non-protein nitrogen	0.20	0.20
	Lactose	4.50	4.60
	Calcium	0.035	0.12
	Phosphorous	0.040	0.065
	Lactic acid	0.05	0.05
	Chloride	0.09	0.11
	Sodium	0.045	0.050
	Potassium	0.14	0.16

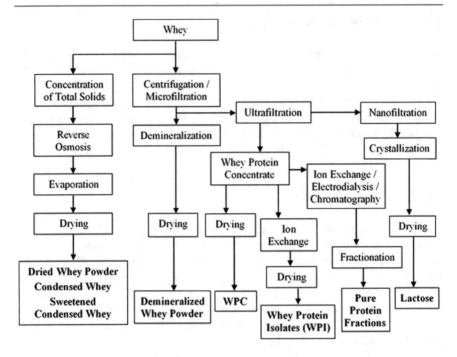

Fig. 11.14 Processing technologies for the production of protein-based ingredients from liquid whey

protein with no lactose fraction and higher purity. Application of more advanced techniques such as chromatography and selective precipitation plus centrifugation and electrodialysis permit the production of pure protein fractions.

11.12.3   Processing of Whey Fractions

The whey obtained from the cheese curd must be processed immediately as its composition and temperature promote the growth of bacteria that causes degradation of protein and formation of lactic acid. Primary treatments before further processing include clarification, separation, pasteurization and cooling (<5 °C). The whey should be concentrated by membrane filtration if it will be transported, to reduce transportation costs.

11.12.3.1   Fat Separation and Casein Fines Recovery

The first step in the treatment of whey to produce the various fractions involve filtering the curd particles remaining in the whey followed by separation of fat and casein fines, using either centrifugal separators or vibrating/rotating screens or cyclones. These constituents interfere with successive treatments and should be removed first. This also increase the economic yield of the whey products.

11.12.3.2   Concentration and Drying of Whey

Whey is concentrated and dried for several motives; for example, to reduce storage and transportation costs, and to induce crystallization of lactose.

I. Concentration

The first step in concentration of whey is using reverse osmosis (RO) or a combination of reverse osmosis and nanofiltration (RO-NF) to increase the initial solids content from about 6.5 to 18–25%. The whey is then transferred to an evaporator where the remaining water is boiled off to increase the solids content to 45–65%.

A falling film evaporator is typically used to concentrate whey, which consists of a bundle of tubes where the whey flows as a thin film inside the tubes, which are surrounded by steam heating jacket maintained under vacuum. To minimize energy costs and that is reduce the amount of steam needed, multiple-effect evaporation is used. A temperature difference of 15 °C is maintained between effects and almost the same amount of water is removed in each effect. The partially concentrated whey from the first effect is separated from the vapor using a cyclone and are pumped to the second effect, and so on. Vapor from the first effect which has a temperature of about 70 °C is used to heat the second effect, etc.

Mechanical vapor recompression (MVR) or thermo-compressor is often used to increase the thermal efficiency of the evaporator. The compressor acts as a heat pump to increase the vapour pressure and hence the temperature of the first effect from 70–85 °C. This compressed vapor is then used to heat the first effect again, and that is increases the thermal efficiency.

The concentrated whey super saturated lactose solution is hygroscopic, so becomes lumpy as it absorbs moisture. To give a non-hygroscopic product when spray-dried, after evaporation, the concentrated whey is flash cooled rapidly to 30–40 °C to start nucleation of lactose crystals and further cooled and stirred in specially-designed crystallization tanks for 4–8 h to obtain a uniform distribution of small lactose crystals.

II. Drying

Similarly, as in milk powder manufacture (see Sect. 11.11.1), spray drying is the most widely used process of drying whey concentrates. The concentrated whey containing lactose crystals is fed to the top of the dryer through an atomizer, which produces a range of droplets in the drying chamber. The droplets are dried in the drying chamber via drying air supplied at temperatures of up to 225 °C, obtaining powder particles of different sizes with a moisture content of about 6%. The air leaving the drying chamber contains fines, which are separated from the air in a cyclone and returned to the drying chamber to agglomerate. The uniformly distributed powder particles are further dried in the fluidized bed dryer of the drying chamber and again, transferred to an external fluidized bed dryer consisting of a casing with perforated bottom that allows hot air to circulate through the powder layer to reduce the moisture content of the powder to 3–4%. After drying, the powder is cooled and packed in airtight sacks.

11.12.3.3   Demineralization (Desalination) of Whey

Whey contains high salt content, about 8–12% (calculated on dry matter), which affect its taste and limits the application of whey in food products. This problem can be solved by demineralizing the whey, which involves desalting the whey through removal of inorganic salts, together with reduction of organic ions such as lactates and citrates.

Partial demineralization (25–30%) is achieved by utilization of crossflow membranes that permit particles having radii in the nanometer range (10–9 μm) through them, called nanofiltration (NF) or ultra-osmosis. Small particles such as certain monovalent ions (e.g. sodium, potassium, chloride and small organic molecules like urea and lactic acid) permeate through the membrane, plus the aqueous permeates. During nanofiltration, leakage of lactose must be maintained at a minimum of <0.1% to avoid high levels of biological oxygen demand (BOD) in the wastewater (permeates), which is a problem.

High degree demineralization (90–95%) is realized by two techniques: **ion exchange**, and **electrodialysis**.

I. Electrodialysis

Electrodialysis is the transport of ions through non-selective permeate membranes under the driving force of a direct current (DC) and an applied potential. The membranes have both cation and anion exchange functions, hence the electrodialysis process is able to reduce the mineral content of the whey. The whey feed is circulated through dilution cells (stack), and a 5% acidified brine solution passes through the concentrated cells. The extent of demineralization is determined by the ash content of the whey, current density, residence time in the stack and flow viscosity.

The electrodialysis plant can be run as a batch or continuous system. The batch system is normally used for demineralization rates above 70%, which can consist of one membrane stack over which the whey is circulated until a certain ash level is reached. The holding time in a batch system can be in the range 2–3 h for 90% demineralization at 10–15 °C. The whey should be pre-concentrated to 20–30% dry matter before it enters the electrodialysis unit to minimize electricity consumption and capacity. The continuous system consists of five stacks in series with a holding time as short as 10–40 min. The maximum demineralization rate is typically about 60–70%. A continuous plant has a much larger membrane area capacity than a batch plant. A cooling stage is required to cool down the process liquid, which heats up during the process, and that is, maintain the process temperature. Electrodialysis is the best technique for demineralization levels below 70%, compared with iron exchange.

The main disadvantage of using electrodialysis in demineralization of whey is the cost of replacing fouled membranes, spacers and electrodes, which constitutes about 35–40% of the plants total running costs. The membrane fouling is caused by deposition of protein on the surfaces of the anion exchange membrane and precipitation of calcium phosphate on the surfaces of the cation exchange membrane. The problem of calcium phosphate deposit can be solved by proper flow design over the membrane surface and regular acid cleaning. Protein deposit problem can be resolved by polarity reversal to dislodge the deposits from the membrane, and disassembly of the stack for manual cleaning at 2–4 weeks intervals.

Other costs of electrodialysis is the processing cost, which depends on the demineralization rate. Operating costs of the demineralization plant include:

- Electric power, which account of 10–15% of the operating cost.
- Wastewater treatment—phosphates removed from the whey accumulates in the waste stream, at 90% demineralization lactose leaks through the membranes at a rate of 7–10%.
- Steam used for pre-heating the product and cooling costs for control of process temperature accounts for 10–15%, depending on the level of demineralization.
- Chemicals used in the process, mostly hydrochloric acid amounts to less than 5%.

II. Ion Exchange

The ion exchange process utilizes resin beads to adsorb ionizable minerals (solids) from whey in an exchange of other ionic species. Ion exchange resins are macromolecular porous plastic materials, which are formed into beads with diameters in the range 0.3–1.2 mm. The resins are usually used in fixed columns. The resins have a fixed capacity, so the adsorbed minerals must be removed when the resins are completely saturated for regeneration before reuse.

Whey from sweat whey has a pH of 5.9–6.6, while the pH of an acid casein/cheese is between 4.3 and 4.6. Apart from the differences in acidity in the two wheys, acid whey contains high level of calcium phosphate.

During demineralization, the whey first enters the strong cation exchanger, loaded in H^+ form, and continuous to the anion exchange in a weak base anion-exchanger in its free base form. The ion exchange columns are rinsed and regenerated separately with dilute hydrochloric acid and sodium hydroxide (or ammonia), respectively. The exchange columns must be disinfected once a day with a small amount of active chlorine solution.

Ion exchange columns are designed to have a conical shape to allow for swelling of the bed during transition from the free base form to the salt form. Counter-current flow is normally used for regeneration of the cation exchanger to reduce consumption of regeneration chemicals by 30–40%, but this requires a complicated design. Typically, two or three parallel ion exchange systems are needed for a continuous flow of whey. A normal cycle time is six hours, but four hours are used for regeneration. The composition of concentrated and demineralized cheese whey powders is shown in Table 11.8.

Table 11.8 Composition of concentrated and demineralized cheese whey powders

Constituents (wt. %)	Concentrated whey powder	Demineralized whey powder	
		Ion exchange	Electrodialysis
Total proteins	13.0	13.0	14.5
Lipids	1.0	1.0	1.0
Lactose	73.0	80.0	77.0
Mineral (ash)	8.0	1.0	4.5
Moisture	3.0	3.0	3.0
Phosphorous	0.6	0.15	0.5
Magnesium	0.2	0.02	0.1
Calcium	0.6	0.05	0.5

Ion exchange processes have the capacity to demineralize the whey by up to 90%. However, the following disadvantages limit the use of ion exchange:

(i)   Disposal of the mineral waste is expensive. The liquid whey contains high salt load, which results in short cycles during ion exchange. The costs of regeneration chemicals are high if they are not recovered. This accounts for 60–70% of the operating costs of the process.

(ii) Losses of whey proteins occur on the columns due to denaturation/absorption, which is caused by great pH variation in the whey during the ion exchange process.

(iii) Rinse water consumption is high, especially for washing out excess sodium hydroxide from the weak anion resin.

11.12.3.4 Protein Recovery from Whey Using Membrane Processes

Whey proteins are isolated by using various membrane separation or fractionation techniques, which include microfiltration (MF), ultrafiltration (UF), nanofiltration (NF), and reverse osmosis (RO), or chromatographic process such as ionic exchange (IE). Whey proteins obtained by these processes have good functional properties, namely foaming, solubility, emulsion and gelling, and can be highly nutritional. Membrane separation takes place via semipermeable membranes of controlled pore size (as shown in Fig. 11.15) using a hydrostatic pressure gradient as the driving force. Figure 11.16 is a conceptual illustration of the membrane separation/filtration process of whey.

The solids sizes in the whey are characterized on different particle sizes according to their molecular weight (shown in Fig. 11.15). When membranes are charged, they possess additional characteristics which are essential for the separation of ions of a charge.

As illustrated in Fig. 11.16, membrane filters are sealed on a porous support. Different configurations are applied in filtration units, such as tubular, plate and

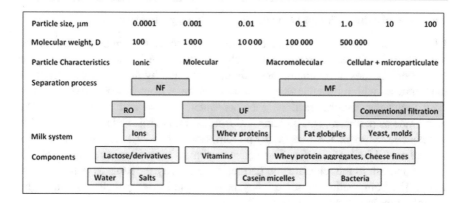

Fig. 11.15 Approximate particle sizes of membrane processes for separation of whey fractions

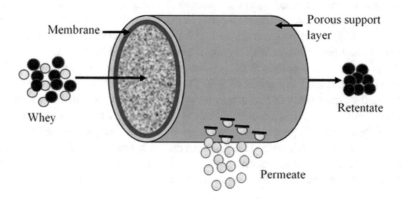

Fig. 11.16 Conceptual illustration of membrane filtration process of whey

frame, hollow fiber and spiral wound. The tubular systems consist of membranes arranged to units of many tubes in series. Plate and frame design consist of membranes packed together between membrane support plates which are arranged in stacks, identical to a plate and frame heat exchanger.

Microfiltration uses membranes with wide pore sizes (>0.1 μm) to remove bacteria and milk fat globules from whey. Ultrafiltration is used to separate whey proteins based on their molecular weight. Nanofiltration is applied to fractionate smaller molecules. A typical application is partial demineralization of whey products. Nanofiltration can be used as an alternative desalination process instead of electrodialysis. Reverse osmosis is used for the removal of water against osmotic pressure (or osmosis), but it is not a filtration process. Reverse osmosis requires higher pressures than the other membrane processes. In reverse osmosis, the membranes act more as a layer of material that allows only water to dissolve and pass through by diffusion, retaining other whey components.

Recovery of whey protein concentrates (WPC) by ultrafiltration

Ultrafiltration is used for the separation of whey proteins. Protein concentrates have a very good amino acid profile, with high percentages of lysine and cysteine. The separation abilities of the membranes are characterized by the molecular weight of the fractions, with molecular weight limit in the range from 10,000–50,000 Dalton. Low molecular weight fractions such as water, lactose and minerals pass the membrane as permeate, and proteins and residual fat are rejected and remain inside as retentate. Most of the true (total) protein, normally >99%, is retained together with almost 100% of the fat. The concentration of non-protein nitrogen (NPN), lactose and ash in the retentate and permeate are usually the same as in the original whey. The composition of some whey protein concentrate powders produced by ultrafiltration are shown in Table 11.9.

Whey protein concentrates (WPC) are powders produced by drying the retentate from ultrafiltration of whey. All types of WPC can be produced using ultrafiltration, which are described according to their protein contents (percentage of dry matter or total solids), which ranges from 35–80%. WPC products may be classified as low protein WPC (protein between 25 and 45%), medium WPC (protein between 45 and 60%) and high protein WPC (protein between 60 and 80%).

To produce a 35% protein WPC product, the liquid whey is concentrated about six-fold to about total dry solids of 9%. To obtain a more than 80% protein concentrate, the liquid whey is first concentrated 20- to 30-fold by direct ultrafiltration to obtain solids content of about 25%; this is considered as the as maximum solids content for economic operation. Higher degrees of whey fractionation result in very high viscosity of the retentate. This problem can be solved by the addition of water during ultrafiltration in a process called "diafiltration", to remove low molecular weight components (basically salts and lactose), which pass through the membranes.

Table 11.9 Compositions of permeate and some WPC powders form ultrafiltration

Component (%)	Permeate	WPC-35	WPC-50	WPC-65	WPC-80
Total protein in dry matter	13.7	35.0	50.0	65.0	80.0
Crude protein	3.3	36.2	52.1	63.0	81.0
True protein	9.7	29.7	40.9	59.4	75.0
Non-protein nitrogen	3.3	3.4	–	3.6	4.8
Moisture	4.0	3.8	3.8	3.8	3.8
Lactose	81.3	46.5	30.9	21.1	5.3
Fat	1.0	2.1	3.7	5.6	7.2
Mineral (ash)	8.2	7.8	6.4	3.9	3.1
Lactic acid	3.2	2.8	2.6	2.2	1.2

The composition of WPC depends on the following:

(i)   The type of membrane and/or membrane properties;
(ii)   The characteristics of the whey feed (such as pre-concentrated after demineralization, etc.);
(iii)   The water added to the feed (diafiltration);
(iv)   The flux; and
(v)   The duration of the filtration process.

Chromatographic separation of whey proteins

Chromatographic techniques can be used to separate proteins from whey in line with specific properties on the proteins. Typical methods are gel filtration chromatography, which is based on differences in molecular sizes, affinity chromatography which uses specificities in molecular interactions, and ion exchange chromatography which is based on charge differences.

Gel filtration is mostly used as the final purification step of defatted WPC. Gel filtration uses the diffusion rate of molecules of difference sizes through a column filled with porous resin beads to separate whey proteins from other components in whey. The porous resin beads allow only small molecules in the whey to pass through the column, such as lactose and part of the non-protein nitrogen (NPN) to enter the pores of the resins. The larger whey proteins pass through the column at a higher rate as they cannot enter the resin beads.

In affinity chromatography, the separation process is based on specific reversible interactions between two biologically active substances such as antigen affinity for an antibody. The system involves a (chemical) covalent bond between ligand and matrix (usually sepharose beads). The ligand binds the protein which is to be isolated from the liquid whey that is loaded on the column. Buffer can desorb the retained molecules, which modifies the recognition site.

Ion exchange chromatography (IEC) is a useful method for separating specific whey proteins. There is an opposite charge between the whey protein molecules of interest and the resin particles from the ion exchanger. The binding proteins and the beads of the IEC column are affected by parameters such as pH of the whey, the net charge of the protein, and the binding capacity of the beads; hence these should be controlled. Variation of these parameters allows the selective elution of proteins from the beads, which is achieved by using a salt of increasing concentration. A typical example is the separation of lactoperoxidase and lactoferrin from whey following microfiltration for the removal of fat and other particles. Lactoperoxidase and lactoferrin are positively charged proteins at the normal pH of 6.5; hence, their molecules bind to negatively charged beads of the ion exchange column. Other proteins of whey are negatively charged proteins at this pH, so will not bind to these beads. The lactoperoxidase and lactoferrin are eluded from the beads using salt solutions of different concentrations, which may then be recovered separately in different eluates. Further processing using ultrafiltration and diafiltration yields products of about 95% purity, which are then sterilized by microfiltration (0.1–0.2 μm pores), and finally spray dried.

Production of whey protein isolates (WPI)

Whey protein isolate is the purest form of whey protein compared with other whey products, and contains more than 92% protein, and thus, the most expensive. WPI has minimal lactose content (<1%) and almost no fat. They are processed to remove the lactose and fat, but generally contain lower bioactive compounds. WPI are used in food products such as body-building supplements in which fat and other non-protein components are not required. WPI supplements absorbed rapidly, and thus very fast-acting. WPI is also used in acidic fruit beverages and as a replacement for egg white in whipped products such as meringues.

The production of WPI is carried out by one of two processes:

(i) Ion exchange (IE), and subsequent concentration, followed by spray drying; or
(ii) Microfiltration followed by ultrafiltration (MF/UF) and subsequent spray drying.

The ion exchange process for producing WPI involves adjusting the pH of the pre-treated fluid whey to 3–3.5 with phosphoric acid and then transferring into a stirred resin tank where some of the proteins are adsorbed. After removing the remaining partially deproteinized whey from the tank, the adsorbed proteins are removed from the resin (i.e. eluded), the pH is then raised using a sodium hydroxide solution and the resin is regenerated for reuse. The pH of the resulting solution is adjusted as required, and then concentrated by ultrafiltration, reverse osmosis and/or vacuum evaporation followed by spray drying.

The microfiltration/ultrafiltration process for producing of WPI is a pressure driven membrane filtration process which consists of two molecular sieves. In each step, different components are removed and/or concentrated to achieve the desired product. In the process, the pre-treated fluid whey is first pumped to an ultrafiltration step where it is concentrated to about 35% protein in dry matter. The retentate is then pumped to a microfiltration plant, while the permeate is transferred to a collecting tank following concentration with reverse osmosis and cooling. All the fat (about 0.05%) in the pre-treated liquid whey is removed during the microfiltration step as they are retained by the membrane. The retentate from microfiltration step, which contains most of the fat and bacteria, is collected separately, and the defatted permeate (whey) is transferred for further ultrafiltration with diafiltration, where the proteins can be concentrated to a level of greater than 90%. The resulting WPI proteins is then further concentrated using high concentration nanofiltration to 35–38% (dry matter) followed by spraying drying to reduce the moisture content to a maximum of 4% before bagging.

11.12.3.5 Lactose Recovery from Whey

Lactose is the main component of whey and it is recovered by crystallization. There are two basic methods of recovery of lactose, depending on the raw material and its composition:

(i) Crystallization of the lactose in untreated but concentrated whey;
(ii) Crystallization of lactose in whey with the protein removed using UF, or another method, before it is concentrated.

Both methods produce a mother-lye, called *molasses*, which can be dried and used as feed. The feed value can be increased by desalination of the molasses and addition of high-quality proteins.

The crystallization process is determined by factors such as the crystal surface available for growth, degree of saturation, viscosity, purity of the solution, agitation of the crystals in the solution, and temperature. Most of these factors, for example, viscosity and degree of saturation, are mutually related. The flow diagram for the production of lactose from whey is shown in Fig. 11.17. The whey is first pasteurized and concentrated by evaporation to 60–65% total solids, after the removal of whey cream and casein fines and then transferred to the crystallization tanks at a temperature of about 50 °C, where crystals are added. Crystallization of the concentrated whey takes place slowly, according to a predetermined time/temperature program for cooling down to 10 °C. The concentration of lactose in the concentrated whey is about 40–45%, whiles its solubility decreases from 17.5% at 50 °C to approximately 6% at 10 °C. The crystallization tanks have cooling jackets and a system for control of the cooling temperature.

After crystallization, the slurry proceeds to a decanter centrifuge and a sieve centrifuge for separation of the crystals mass and delactozed liquid. The delactozed whey is rich in protein, so may be used as animal feed, or after demineralization, applied as a replacer of skim milk in food products. The lactose crystals are washed in a more efficient second decanter separator, which increases the lactose content from 80–99% by weight total solids, producing what is called "*edible lactose*", which contains less than 9% moisture. For efficient and simple separation of lactose crystals from the mother liquor, the crystals should exceed 0.2 mm in size, as larger crystals are better for separation. The residual moisture content is then reduced to about 0.1–0.5% after drying to a powder in a fluidized bed dryer. The moisture content depends on the future use of the product. The drying temperature should not exceed 93 °C, as β-lactose is formed at higher temperatures. Immediately after drying, the lactose is usually ground to powder (typically in a hammer mill) and sieved to the desired particle size and then packed.

Refining of lactose

Edible lactose is used in the food industry, such as ingredient in substitute for human milk. A higher degree of purity or a very white lactose is required for some applications, such as pharmaceutical uses. This requires further refining steps. To produce pharmaceutical lactose, the edible quality lactose from the decanter is re-dissolved in hot water (100 °C) at a pH 4 to a concentration of 50%. The solution is then mixed with active carbon, phosphate, and filtration agents and pumped to a filter press. After filtration, the solution is recrystallized, and the refined lactose is then separated by a decanter centrifuge before drying, milling and packaging as described above for edible lactose. The purity of the pharma-lactose produced is no less than 99.8% and may be modified for other specific applications.

Fig. 11.17 Process flow diagram for the recovery manufacture of lactose from whey

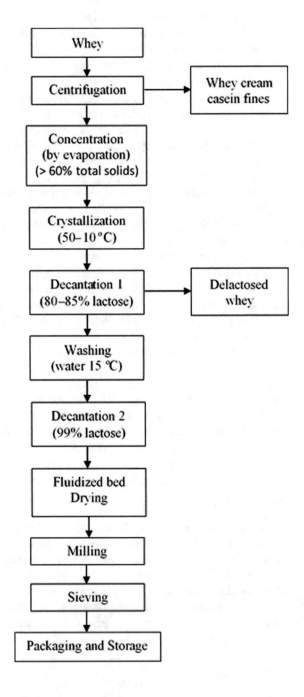

11.13   Environmental Impact from Dairy Products Manufacturing

Significant amount of water is used in the processing of milk and dairy products, mainly to clean equipment and for heating and cooling, pasteurization, and homogenization of the milk. About 2–5 L of water is used per liter of processed milk; hence, large amount of wastewater and other by-products are produced per day. Processing of different types of milk products (e.g. milk, butter, cheese and whey) generate different types and amount of waste. Production of milk, cream and yoghurt create approximately 3 L of wastewater per kg of milk processed, whilst cheese generate 4 L of wastewater per kg of processed milk. Dried milk powder and whey concentrates generate about 5 L of wastewater per kg of milk processed.

11.13.1   Sources of Dairy Waste and Environmental Components

The main wastes generated from dairy processing are wastewater and particulate emissions. In addition, noise and odors are issues of environmental concern. Table 11.10 summarizes the types of waste generated during the processing of dairy products as well as other environmental components.

About 65% of dairy factory losses end up in wastewater discharge streams. The amount of water can be reduced using improved technology such as reverse osmosis.

Odors and particulates are the main emissions from dairy processing factories. The odors are caused by biological decomposition of the organic matter in milk, mostly in wastewater and are due to improper management of wastewater treatment facilities as well as long storage of strong wastes such as whey. The particulate emissions are usually produced during spray drying of milk and whey, which may deposit and accumulate on drainage systems, rainwater tanks, roofing and vehicles. Rain will wash the settled particles affecting the quality of stormwater discharge from the dairy processing site and nearby residents.

11.13.1.1   Cleaning-In-Place (CIP)

In the dairy processing industry, wastewater is generated from holding tanks at start-up and shut down, batch pasteurization units and washing lines. The fat content and protein of milk if not removed from holding tanks and milk processing equipment can cause fouling deposits (containing protein, milk fats, minerals) which must be cleaned. However, these deposits when removed during cleaning increases the organic load of the wastewater. During dairy processing, CIP comprises of the following steps: (i) recovery of any residual products as practicably possible; (ii) a pre-rinsing step with water, which is usually hot water; (iii) cleaning using detergents to remove deposits; (iv) washing with clean water to remove any residual detergent(s); and (v) sanitization/disinfection.

Table 11.10 Composition of wastewater from cheese manufacturing

Process and product	Waste generated	Waste minimization strategy
Milk preparation stages	Spills and leakages from storage tanks, hoses and pipes Wastewater from cleaning operations	– Prevent leakages and spillages—make sure storage tanks are filled to the maximum allowable capacity. Holes and pipes should be properly tightened – Recovery of downgraded products
Liquid milk production	Wastewater, odor, noise, solid waste, cleaning-in-place (CIP) chemicals Deposits on surfaces of pasteurization and heating equipment	– Purging of raw materials and product lines – Recovery and reuse of raw materials and product – Prevent spillage from overflows, malfunctions, and poor handling procedures – Maintain aerobic conditions for wastewater processing to prevent decomposition of organics – Automated CIP system
Butter production	Butter milk, butter wash water	– Avoid spillages—use facilities large enough to hold all buttermilk – Use correct preparation before filling – Dry and reuse of butter milk as animal feed – Use solids recovered form butter wash water as stock feed
Cheese production	Wastewater, whey, curd	– Reduce spillages—avoid overfilling cheese vats to stop curd loss – Remove whey and curds completely from vats before rinsing, and separate all whey drained from cheese – Collect all whey and cheese including fines from all liquid streams
Powdered milk production	Wastewater, milk, energy, spills of powder, particulates	– Use effluent entrainment separators during condensation to prevent carry-over of milk droplets – Reprocess washings with > 7% solids and use < 7% solids for stock feed – During evaporation, maintain a liquid level low enough to prevent product boil-over during evaporation – Recirculate low concentration milk until the desired concentration is achieved – Run to specified run length—overly long runs than required leads to blocked tubes, which are difficult to clean, and results in high pollution – Minimize emission of particulates by using wet scrubbers or fabric (bag) filters

Once the residual products are recovered, the pre-rinsing with hot water is immediately performed to remove residues of milk fat. The initial rinsing step helps to remove all loosely bound materials, and it is performed until the return water becomes clear. This is an important step of the CIP process, as effectual pre-rinsing can reduce the quantity of detergent that is used in the successive cleaning steps, and thus, improves the overall efficiency of the CIP process.

The fouling deposits are insoluble in water. Alkaline solutions are used to clean the fat and protein fractions of the deposits, whiles the mineral content washed with either water or acidic solutions. The protein content is also slightly soluble in acidic solutions. The cleaning solutions are usually heated to a temperature of 70–80 °C to improve the effectiveness of the chemicals and reduce the amount of time needed to achieve proper cleaning. The alkaline solution mostly used in CIP is sodium hydroxide (NaOH), also known as 'caustic soda'. It is typically used in a concentration range of 0.5–2.0%, although high concentrations such as 4.0% may be used for highly fouled equipment. Using caustic soda in CIP wash cycles breakdown and softens the fats, and hydrolyze the proteins, making them easier to remove. It also partially eliminates pathogenic organisms. Other alkaline solutions used for detergent applications in diary CIP are potassium hydroxide (KOH, also known as 'caustic potash'), alkaline silicates and phosphates, and sodium carbonate (Na_2CO_3). Nitric acid is the most commonly used wash for the removal of milk scales in dairy CIP and requires less heating. At a concentration of about 0.5%, it can be used effectively at lower temperatures than caustic solutions. Nitric acid when used in dairy CIP strengthens metals and maintains their appearance via a chemical process. It brightens up discolored stainless steel by removing calcified mineral stains. However, great care must be taken when using nitric acid because it can attack some elastomers such as gaskets and valve seats resulting in failure or premature degradation. Phosphoric acid is sometimes used, but it is more expensive than nitric acid, and rather less common. Sulfuric and hydrochloric acids can also be used in dairy CIP, but they can corrode pipes.

Care must be taken when applying alkaline and acid washes to remove milk deposits. For example, acid wash should precede a caustic wash as the acid could precipitate the protein and make the deposits more difficult to remove. Other considerations include the concentration of the detergent solution, the temperature of the solution (which has been explained above), and the duration of cleaning. The cleaning duration for the circulation of the detergent solution depends on the level of fouling and the temperature of the detergent solution. At the end of the detergent cleaning, the system is washed with clean water to remove any residual chemicals and suspended solids.

Sanitization also referred to as disinfection is the final CIP step, which is carried out to reduce microorganisms within the system to acceptable levels where they do not pose risk to health. Disinfection is carried out by either thermally or using chemical means. Thermal disinfection is performed by the circulation of either hot water (at 80–96 °C for 5 min) or low-pressure steam (at a condensate temperature of >80 °C for 15 min) through the system. However, both thermal processes result in high energy costs. Chemical-based disinfection is carried out using solutions of chlorine or hypochlorite, quaternary ammonium, and iodophors (iodine-containing compounds). The disadvantage of these chemicals is that they can be harmful to the environment and water bodies. They can also be harmful to the system; for example, chlorine can be damaging to stainless steel by causing staining, corrosion, and pitting. In recent years, more eco-friendly alternatives are used to minimize the impact of CIP on the environment. Examples include peracetic acid (PAA) and enzymes. PAA is a combination of hydrogen peroxide and acetic acid, and it is a strong disinfectant even at low temperatures. It is effective against all microorganisms including spoilage organisms, pathogens and bacterial spores. One disadvantage of PAA is that it has a strong pungent smell, so it should be used in well-ventilated areas. Protease and lipase enzymes have been identified as a replacement for the conventional high temperature caustic-based or chemical-based CIP methods. Enzyme-based CIP can be carried out effectively at low temperatures (<40 °C), and thus, economical than caustic based cleaning. Additional benefits include safety improvement, a reduction in energy and water consumption, decreased waste generation, greater compatibility with ensuing wastewater treatment processes, and a reduction in the environmental impact of the CIP process as enzymes are biodegradable.

11.13.2   Dairy Wastewater Composition

The wastewater generated by the processing of milk and dairy products contains milk solids that contain milk fat, salt, proteins, lactose, lactic acid and residual chemicals resulting from cleaning processes, which potentially affect water quality. These are characterized by high organic load such as fatty acids and lactose, increased in suspended solids between 24.0 and 5,700 g/L, and disparities in pH in the range 4.2–12.0. Chemical Oxygen Demand (COD) or Biological Oxygen Demand (BOD) is generally used as the indicator of water quality, which is measured by the level of organic compounds (mostly milk) present in the water. Dairy wastewater before treatment contains high BOD_5 concentration of up to 5,312.0 mg/L and dissolved solids levels of about 1,800 mg/L. The composition of water generated from cheese making is presented in Table 11.11. Discharging wastewater containing high concentrations of nitrogen and phosphorous lead to eutrophication of receiving water bodies.

Production of milk powder generates a large amount of condensate, which can be contaminated with milk residues and hence, should be treated. The production of WPC creates whey permeate as a residue, which has almost a high BOD level as a whole raw whey fluid; because it contains nearly all the lactose present in whey, and hence, will pose a problem to dispose to the environment. A solution to this is to produce alcohol from the whey permeate, which is use as fuel.

Table 11.11 Composition of dairy industry wastewater

Component	Composition (mg/L)
BOD_5	450.0–5,312.0
COD	5,312.0–20,559.0
Fats	96.0–463.0
Suspended solids	24.0–5,700.0
Nitrogen	15.0–180.0
Phosphorous	11.0–160.0
Sodium (mg/L)	60.0–807.0
Calcium	57.0–112.0
Chloride	48.0–469.0
Potassium	11.0–160.0
Magnesium	25.0–49.0
pH	4.2–12.0

11.13.3 Dairy Processing Wastewater Treatment

As explained above, diary wastewater contains a heavy organic load due to the presence of proteins, fats and solids. The wastewater also contains nutrients such as phosphorous and nitrogen, which need to be removed to avoid eutrophication effects to receiving water bodies. Wastewater generated in dairy processing plants can be recovered for reuse. This can reduce the volume of freshwater requirements and decrease the costs of fresh water and wastewater treatment. Whey can be recovered from waste steams to produce whey powder or stock feed. Fat recovery or removal from the wastewater is very crucial in the treatment process. The organic load comprises of about 50% fat, which needs to be separated or removed. The treatment of dairy wastewater typically involves primary (to remove solids and fats), secondary (to remove the organic loads), and tertiary methods (for complete removal of phosphorous).

11.13.3.1 Primary Treatment of Dairy Wastewater

Primary treatment consists of several steps such as screening/sedimentation for removing coarse milk solids and debris; balancing to produce a stable flow and concentration in the wastewater; dissolved air flotation (DAF) method for the removal of the fats, oils and greases; and equalization and pH control.

The wastewater from the dairy plants is separated to remove useful components. For example, whey can be reused for the production stock feed or converted into whey powder. Wastewater containing high level of saline can be discharge to an evaporation pond to recover the salts for recycling. CIP wastewater should be separated from the other wastewaters for treatment to recover the cleaning chemicals. To control the pH, spent acid and alkali cleaners are used to neutralize each other. Alternatively, a balance tank or pond is used to balance the differences in pH and temperature.

11.13.3.2   Secondary Treatment of Dairy Wastewater

The organic matter in diary wastewater may be colloidal or dissolved; hence, the primary treatment processes are mostly ineffective in removing it. The organic matter provides a high demand for oxygen which should be reduced to render the effluent safe for disposal into receiving water bodies. The secondary treatment involves removing, stabilizing, and making safe the remaining components in the wastewater that could not be removed by the primary treatments. These are done to ensure that oxygen supplied to the bacterial is consumed under controlled conditions to remove most of the BOD in the treatment plant instead of in the water bodies.

The most commonly used secondary treatments for diary wastewater include aerobic processes (carried out using dissolved oxygen) and anaerobic processes (carried out in the absence of oxygen), which can be used separately or in combination with each other. Aerobic treatment methods include activated sludge systems, aerobic/aerated lagoons, sequencing batch reactors (SBRs), and trickling filters/biological film systems or membrane bioreactors (MBRs).

Aerobic Secondary Treatment Techniques

(i) **Activated Sludge Process**:

In this system, the wastewater is made to contact microorganisms suspended in an aerated tank in the form of a flocculant. This type of process has a long retention time and a low net sludge yield, and it aids in biological nitrogen removal from the wastewater. The following processes take place in the aerated tank: an aeration stage, settling stage—solids-liquid separation after aeration, and effluent withdrawal or sludge recycle stage. The stages in the activated sludge processes are illustrated in Fig. 11.18. During the process, the wastewater following primary treatment is sent to an aeration tank where the organic matter is brought into close contact with the sludge from the secondary clarifier which is loaded with micro-organisms having an active state of growth. Transfer of air into the tank is done in the form of bubbles either by surface aerators or through diffusers. The oxygen in the air is utilized by the micro-organisms while converting the organic matter into stabilized, low-energy compounds namely SO_4, CO_2, and NO_3, and make new bacterial cells.

A settling tank, also called secondary settler or a clarifier, is used to separate the discharge from the aeration tank containing the sludge. The separated sludge in the settling tank has no contact with the organic matter and becomes activated. Part of the activated sludge is recycled to the aeration tank as a seed, and the remaining part is processed and disposed of. The reason to recycle only a portion of the activated sludge is that, recycling all the activated sludge will increase the bacterial mass and clog the system with solids.

Oxidation ditch, an activated sludge system is a highly effective extended aeration process that uses the processes of adsorption, oxidation, and decomposition by using a large circular basin equipped with aerators. It has minimal operating costs.

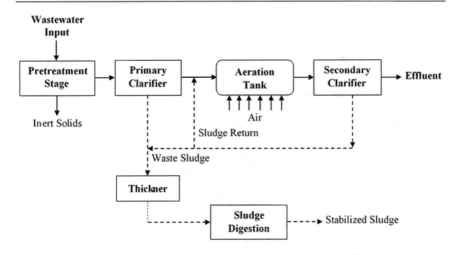

Fig. 11.18   An illustration of an activated sludge treatment plant

The disadvantages of aerobic activated sludge treatment are that it is very energy intensive depending on the aeration technology used. Also, a stable performance cannot be achieved continuously because of several factors such as variations in seasonal flow, bulking of the sludge, and overloading.

(ii)   **Trickling Filters**:

Trickling filters, also known as percolating filters or biological film system, is the second commonly used treatment process for biological loads in dairy wastewater. In the trickling filter process, the wastewater is contacted with a mix microbial population, which is by way of a film of mucus attached to the surface of a solid support system. Trickling filters are made up of a rock bed with a depth of about 1 to 3 m with adequate openings between the rocks to permit a free passage of air to prevent the generation of odors. However, trickling filters are pH sensitive, and thus, low or high pH may result in killing the biomass.

During treatment, the wastewater is sprinkled over the bed packing, which is coated with the biological slime. As the wastewater trickles over the bed packing, oxygen and the dissolved organic matter diffuse into the film, which is then metabolized by the micro-organisms in the slime layer. Products such as CO_2, NO_3, etc. diffuse from the film into the filter effluent. The thickness of the slime film increases as the micro-organism use the organic matter, and this reaches a stage where it can no more be supported on the solid media and becomes detached from the surface. The detached bacterial film and suspended matter are removed using a settling tank after the trickling filter. To ensure safe disposal, the sludge effluent containing about 2% solids should be taken through the following operations: thickening or concentration, digestion, conditioning, dewatering, and oxidation.

(iii)   **Aerated Lagoons**:

Aerated lagoons, which resemble the activated sludge system use floating aerators which force oxygen input to the wastewater. This helps to achieve low odor generation and the highest quality wastewater.

(iv) **Sequencing Batch Reactor (SBR)**:

A sequential batch reactor is an activated sludge type system for wastewater treatment in which all the activated sludge process takes place in a single mix fill-and-draw reactor. The reactor is a tank into which air is pumped to supply sufficient oxygen to aid aerobic biochemical process. During the treatment, carbonaceous biochemical oxygen demand (cBOD), nitrification, denitrification, and settling are all achieved in the same reactor. Typical operation of SBR involves five or more stages: feeding, reaction, settling, decanting or effluent withdrawal, and idling. This is illustrated in Fig. 11.19.

During stage 1, the influent wastewater is added to the reactor basin. In stage 2, intermittent mixing is applied to provide uniform distribution of substrates and solids. In this stage, air/oxygen is added to allow microorganisms to consume or oxidize the dissolved organic matter (or cBOD) in the wastewater and nitrify total nitrogen. Anoxic mixing under anoxic conditions usually precedes the aerated mixing to denitrify the wastewater which remains in the reactor from the previous cycle. In stage 3, both mixing and aeration are stopped to allow the sludge to settle to the bottom of the reactor after a specified period of time. The settled sludge at the bottom of the tank consists of the microorganisms that have fed on the organics in the wastewater. During stage 4, a fraction of the treated content is discharged from the reactor above the level of the settled sludge. The sludge remaining in the reactor is rich in microorganisms, which helps to instantly recreate microorganisms in the reactor as the next batch of wastewater is transferred into it. This reduces the time required to treat each batch of wastewater. In general, a longer retention period

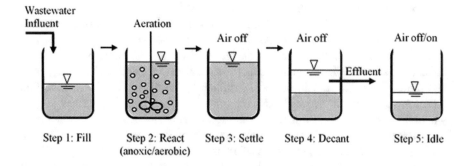

Fig. 11.19 Key steps in SBR process cycle

produces less sludge and cleaner effluent. The final stage is a flexible duration step that allows changes in the flow.

The advantages of SBR are that:

1. It is flexible and can be operated continuously using multiple reactors so that while one batch of wastewater is being treated, the next one is ready to start filling.
2. It produces effluent low in organic compounds, and thus can be used to meet strict effluent standards.
3. SBR use small space, allowing for a smaller footprint, and easy to expand by adding additional reactors.
4. The treatment cycle can be altered to perform aerobic, anoxic, and anaerobic treatments for removal of nutrients nitrogen and phosphorous.

The operation of SBR is, however, more complex than other aerobic treatment systems. Also, it is more costly to construct and operate than most others, but it requires fewer maintenance over its lifespan.

(v) Membrane Bioreactors:

Membrane Bioreactor (MBR) technology is increasingly used for dairy wastewater treatment. MBR systems provide effluents of high-quality because they are effective for the removal of various pollutants in the wastewater, including micropollutants, such as dissolved organics with low molecular weights. MBR is a combination of activated sludge system with membrane filtration system, usually low-pressure microfiltration (MF) or ultrafiltration (UF) membranes. In the activated sludge system, solid–liquid separation is accomplished using secondary and tertiary clarifiers plus tertiary filtration, but the membranes are used in MBRs. There are two types of MBR systems, which are vacuum or gravity-driven, and pressure-driven systems. The vacuum or gravity systems utilize flat sheet or hollow fiber membranes fitted and submerged in either the bioreactor or a subsequent membrane tank. Pressure-driven systems use in-pipe cartridge systems that are situated externally to the bioreactor.

Anaerobic Secondary Treatment Techniques

Anaerobic digestion is a biological process for treatment of dairy industry wastewaters where microorganisms degrade organic contaminants in the absence of oxygen into biogas and digestate. In a treatment cycle, the bioreactor which is a Continuous-Stirred Tank Reactors (CSTRs) contains sludge consisting of anaerobic bacteria and other microorganisms. The continuous or intermittent agitation provided by the CSTR increases the contact surface between components of the wastewater and the anaerobic microbes, which promotes high biodegradation. The biodegradable matter present in the wastewater is digested by the anaerobic microorganisms, which causes the effluent to have lower levels of COD, BOD, total suspended solids (TSS), and by-products in the produced biogas.

Advantages of anaerobic treatment of dairy wastewater are:

1. Low land requirement.
2. Low energy consumption compared with aerobic processes because aerated equipment is not needed.
3. Lower amount of excess sludge.
4. Energy generation through the production of methane or biogas.

The anaerobic process involves two stages; firstly, an acidification phase which is followed by a methane production phase. The acid forming phase involves hydrolysis of the complex organic matter by the anaerobes into simple soluble organic compounds (sugars, fatty acids, amino acids). The methane production phase consists of two stages: (i) acetogenesis in which the anaerobic microorganisms synthesize the soluble organic compounds to form volatile fatty acids, which further produces acetate (or acetic acid), formate (or formic acid), hydrogen gas, and carbon dioxide; (ii) methanogenesis, where anaerobes synthesize the generated acidogenic compounds to produce methane and carbon dioxide.

Different types of anaerobic wastewater treatment systems exist. Technologies that are used for dairy wastewater treatment include anaerobic membrane bioreactor (An-MBR), and anaerobic blanket reactors (ABR).

(i) Anaerobic Membrane Bioreactor

The anaerobic membrane bioreactor (An-MBR) is similar to the MBR process explained earlier; the only difference is that An-MBR takes place under anaerobic conditions, whilst MBR is an aerobic process. An-MBR system uses a microfiltration grade membrane of > 0.1 μm pore size. The membrane component allows the An-MBR system to operate at lower temperatures (such as 15 °C under optimal conditions) than the conventional anaerobic process which requires temperatures higher than 25 °C. An-MBR has a low growth rate, but the membrane allows it to operate at high solid concentrations with high solid retention times and can achieve COD removal of more than 90%. Other benefits of An-MBR are that the technique requires less energy and produces biogas, which can be used to power the dairy processing plant. It generates less sludge than the aerated MBR system and does not also require much space.

(ii) Anaerobic Migrating Blanket Reactor

An anaerobic migrating blanket reactor (AMBR) uses granular biomass, allowing very long retention times and produces biogas. Like the An-MBR technique, AMBR process can be operated at lower temperatures (20 °C) and can achieve COD removal of up to 95%.

11.13.3.3 Tertiary Dairy Wastewater Treatment

Tertiary treatment of dairy wastewater includes phosphorus removal through the utilization of chemical precipitation with chemicals such as ferric sulfate or aluminium chloride before discharging the treated wastewater to water bodies or municipal wastewater treatment systems. Another process for phosphorous removal is through the formation of struvite crystals. Unlike the chemical precipitation method, this process allows crystallization of phosphorous to obtain a functional product. The struvite process involves transferring the sludge to an upflow fluidized bed reaction together with chemical additives typically magnesium to produce a magnesium ammonium phosphate ($MgNH_4PO_4$) precipitate, properly known as MAP or struvite. The struvite crystals formed are suspended in the flow until they grow to the required size after which they settle to the bottom of the reactor and are collected. A controlled addition of magnesium and the upflow rate are needed to achieve efficient formation of the crystals. This process can achieve up to 85% recovery of the initial phosphorous in the wastewater as well as removal of nitrogen.

11.13.3.4 Advanced Treatments for Resuse

The organic components in the wastewater can be treated by using the physical techniques explained in the previous sections such as reverse osmosis, microfiltration, and flotation. The components removed by these methods can be recycled or reused.

Membrane Filtration

This process can be used for the recovery and recycling of milk fat, acid and alkali by pumping the effluent through membranes of pore sizes less than 0.1 μm, resulting in quality wastewater. The membranes are regularly cleaned using high pressure backwash with water and/or air.

Reverse Osmosis

This method is used for desalination of the wastewater using osmotic pressure on one side of the membrane. This is used for the removal of dissolved solids by pumping the wastewater through a semi-permeable membrane that restricts the movement of salts.

Further Readings

Arvanitoyannis, I. S., & Giakoundis, A. (2006). Current strategies for dairy waste management: A review. *Critical Reviews in Food Science and Nutrition, 46*(5), 379–390.

Burton, H. (1988). *Ultra-high-temperature processing of milk and milk products*. London: Elsevier Applied Science.

Bylund, G. (1995). Dairy processing handbook. Tetra Pak Processing Systems AB S-221 86, Lund.

Chandan, R. C. (2014). Dairy–fermented products. In S. Clark, S. Jung, & B. Lamsal (Eds.), *Food processing: Principles and applications* (2nd ed., pp. 405–431). Chichester: John Wiley & Sons Ltd.

Datta, N., Elliott, A. J., Perkins, M. L., & Deeth, H. C. (2002). Ultra-high temperature (UHT) treatment of milk: Comparison of direct and indirect modes of heating. *Aust J Dairy Technol, 57*, 211–227.

Deeth H (2012) UHT processing of milk. New food magazine. Issue 3.

Ergas, S. J., & Aponte-Morales, V. (2014). Wastewater treatment and reuse. In S. Ahuja (Ed.), *Comprehensive water quality and purification* (1st ed., pp. 123–149). Amsterdam: Elsevier.

EPA. (1997). *Environmental guidelines for the dairy processing industry* (p. 570). State Government of Victoria, Publication.

Goff, H. D. (2018). *The dairy science and technology*. University of Guelph, Canada.

International Dairy Federation (IDF) (2013) Protein determination. IDF Factsheet.

IDF Standard 185: 2002(E), ISO 14891:2002(E). Milk and milk products. Determination of nitrogen content routine method by combustion according to the Dumas principle.

ISO 8968–1/2 IDF 020–1/2- Milk – Determination of nitrogen content – Part 1/2: Determination of nitrogen content using the Kjeldahl method.

ISO 8968–4 IDF 020–4:2001- Milk – Determination of nitrogen content – Part 4: Determination of non-protein nitrogen content.

Ramos, O. L., Pereira, R. N., Rodrigues, R. M., Teixeira, J. A., Vicente, A. A., & Malcata, F. X. (2016). Whey and whey powders: Production and uses. In B. Caballero, P. Finglas, & F. Toldrá (Eds.), *The encyclopedia of food and health* (Vol. 5, pp. 498–505). Oxford: Academic Press.

Robertson, G. L. (2011). Ultra-high temperature treatment (UHT): Aseptic packaging. In: JW Fuquay, PX Fox, PLH McSweeney (Eds) Encyclopedia of dairy sciences (Vol. 2, 2nd edn, pp 708–713). Academic Press, San Diego

Ryan, M. P., Walsh, G. (2016). The characterization of dairy waste and the potential of whey for industrial fermentation. EPA Research Programme 2014–2020, EPA Research Report (2012-WRM-MS-9). Environmental Protection Agency, Ireland.

de Wit, J. N. (2001). *Lecturer's handbook on whey and whey products*. Brussels: European Whey Products Association.

Printed in the United States
by Baker & Taylor Publisher Services